I0034226

Emerging Two Dimensional Materials and Applications

This book details 2D nanomaterials, and their important applications—including recent developments and related scalable technologies crucial to addressing strong societal demands of energy, environmental protection, and worldwide health concerns—are systematically documented. It covers syntheses and structures of various 2D materials, electrical transport in graphene, and different properties in detail. Applications in important areas of energy harvesting, energy storage, environmental monitoring, and biosensing and health care are elaborated.

Features:

- Facilitates good understanding of concepts of emerging 2D materials and its applications.
- Covers details of highly sensitive sensors using 2D materials for environmental monitoring.
- Outlines the role of 2D materials in improvement of energy harvesting and storage.
- Details application in biosensing and health care for the realization of next-generation biotechnologies for personalized health monitoring and so forth.
- Provides exclusive coverage of inorganic 2D MXenes compounds.

This book is aimed at graduate students and researchers in materials science and engineering, nanoscience and nanotechnology, and electrical engineering.

Emerging Materials and Technologies

Series Editor:
Boris I. Kharissov

For more information about this series, please visit: www.routledge.com/Emerging-Materials-and-Technologies/book-series/CRCEMT

Emerging Two Dimensional Materials and Applications

Edited by
Arun Kumar Singh,
Ram Sevak Singh, and Anar Singh

CRC Press
Taylor & Francis Group
Boca Raton London New York

CRC Press is an imprint of the
Taylor & Francis Group, an **informa** business

First edition published 2023
by CRC Press
6000 Broken Sound Parkway NW, Suite 300, Boca Raton, FL 33487–2742

and by CRC Press
4 Park Square, Milton Park, Abingdon, Oxon, OX14 4RN

CRC Press is an imprint of Taylor & Francis Group, LLC

© 2023 selection and editorial matter, Arun Kumar Singh, Ram Sevak Singh and Anar Singh; individual chapters, the contributors

Reasonable efforts have been made to publish reliable data and information, but the author and publisher cannot assume responsibility for the validity of all materials or the consequences of their use. The authors and publishers have attempted to trace the copyright holders of all material reproduced in this publication and apologize to copyright holders if permission to publish in this form has not been obtained. If any copyright material has not been acknowledged please write and let us know so we may rectify in any future reprint.

Except as permitted under U.S. Copyright Law, no part of this book may be reprinted, reproduced, transmitted, or utilized in any form by any electronic, mechanical, or other means, now known or hereafter invented, including photocopying, microfilming, and recording, or in any information storage or retrieval system, without written permission from the publishers.

For permission to photocopy or use material electronically from this work, access www.copyright.com or contact the Copyright Clearance Center, Inc. (CCC), 222 Rosewood Drive, Danvers, MA 01923, 978–750–8400. For works that are not available on CCC please contact mpkbookspermissions@tandf.co.uk

Trademark notice: Product or corporate names may be trademarks or registered trademarks and are used only for identification and explanation without intent to infringe.

Library of Congress Cataloging-in-Publication Data
A catalog record for this book has been requested

ISBN: 978-1-032-16287-4 (hbk)
ISBN: 978-1-032-16288-1 (pbk)
ISBN: 978-1-003-24789-0 (ebk)

DOI: 10.1201/9781003247890

Typeset in Times
by Apex CoVantage, LLC

Contents

Editors' Biographies

Dr. Arun Kumar Singh is working as an associate professor at Department of Pure and Applied Physics, Guru Ghasidas Vishwavidyalaya, Bilaspur (CG), India. He received his MSc degree in physics from Banaras Hindu University (BHU), Varanasi, India, and obtained his PhD degree from School of Materials Science and Technology, IIT (BHU), India. After his PhD, he joined post-doctoral research work at Graphene Research Institute, Sejong University, South Korea. He got India's most prestigious research award, Inspire Faculty Award, from the Department of Science and Technology, India. He also visited many countries for research/collaborative work. He got many fellowships/awards from different scientific societies of India and abroad. He has published many research papers in renowned international journals—including *Advanced Functional Materials, ACS Applied Materials & Interfaces, Journal of Materials Chemistry C, Applied Materials Today, Sensors and Actuators-B*, and *Journal of Applied Physics*—and also written a book/book chapters. His research work basically includes materials in science and engineering; organic semiconductors; nanomaterials, including carbon nanotubes, graphene, and other two-dimensional materials; electronic/optoelectronic devices; and materials for energy conversion and storage. He is a member of many scientific societies, like American Chemical Society, Materials Research Society of India, Indian Science Congress Association, Electron Microscope Society of India, Indian Physics Association, etc. He is also a reviewer of many scientific journals.

Dr. Ram Sevak Singh is currently working as an associate professor in the Department of Physics, OP Jindal University, Raigarh, Chhattisgarth, India. He received his PhD in physics from National University of Singapore, MTech in materials science and engineering from IIT Kharagpur, and MSc in physics from Banaras Hindu University. He also served as an assistant professor in Physics Department, NIT Kurukshetra, India, and as a post-doctoral research fellow in Centre for Nano and Soft Matter Sciences, Bangalore, India, Nanyang Technological University, Singapore, and National University of Singapore. He has also received prestigious NUS Research Scholarship, Singapore, and IETE-CEOT (94) Award (Biennial)-2014, India. He is a member of American Chemical Society, Materials Research Society of India, and the Graphene Council. Dr. Singh has several years of research and teaching experiences in the areas of physics and nanotechnology and has published many research articles in journals of international repute, including *ACS Nano, Nano Letters, Carbon, Renewable Energy*, and *Applied Physics Letters*, and book chapters with

Wiley and Elsevier. He is also reviewer of several reputed international journals. His areas of interest include materials physics; nanomaterials, including nanotubes, graphene, and other two-dimensional materials; optoelectronic devices; and materials for sensors, corrosion protection, energy conversion, and storage.

Dr. Anar Singh earned his MSc degree in physics with specialization in solid state physics from Banaras Hindu University (BHU), Varanasi, in 2004. He received the Prof. B. Dayal Gold Medal and BHU Medal for securing the highest percentage of marks in the MSc examination. In July 2006, he joined the PhD program in the School of Materials Science and Technology (SMST), Indian Institute of Technology (BHU), Varanasi. After receiving his PhD in 2013, he joined the Paul Scherrer Institute, Switzerland, as a post-doctoral fellow. He also worked as a research fellow at National University of Singapore, Singapore, from July 2014 to June 2015. He received Inspire Faculty Award from DST, India, in 2015 and joined Institute of Chemical Technology, Mumbai, as Inspire Faculty. In June 2016, he joined the University of Lucknow as assistant professor. He has visited large-scale facility centers FRM II, Germany, and ANSTO, Australia. He has also been part of the Indo-Japanese Science Collaboration Program with Hiroshima University and Tohoku University, Japan. In 2019, he has been offered a visiting research position at University of Johannesburg, South Africa, for three years. His research interests are phase transitions in bulk and thin films, crystallography (nuclear and magnetic structures), multiferroic materials, ferroelectrics, and piezoelectrics.

Contributors

Sharmila Kumari Arodhiya
Department of Physics
Rajiv Gandhi Government College for
    Women
Bhiwani, Haryana, India

Department of Physics
National Institute of Technology
Kurukshetra, Haryana, India

Kamalakanta Behera
Department of Applied Chemistry
    (CBFS-ASAS)
Amity University Gurugram
Manesar, Haryana, India

Department of Chemistry, Faculty
    of Science
University of Allahabad
Prayagraj, India

Samir K. Beura
Department of Zoology, School of
    Biological Sciences
Central University of Punjab
Bathinda, Punjab, India

Abhinaba Chatterjee
Department of Zoology, School of
    Biological Sciences
Central University of Punjab
Bathinda, Punjab, India

Bijoy Kumar Das
Centre for Automotive Energy
    Materials
International Advanced Research Centre
    for Powder Metallurgy and New
    Materials
Taramani, Chennai, India

Sachindranath Das
Department of Instrumentation Science
Jadavpur University
Kolkata, India

Ghulam Dastgeer
Department of Physics and Astronomy
Sejong University
Seoul, Korea

Renu Dhahiya
Department of Applied Sciences
National Institute of Technical Teachers
    Training and Research
Chandigarh, India

Jonghwa Eom
Department of Physics and Astronomy
Sejong University
Seoul, Korea

R. Gopalan
Centre for Automotive Energy Materials
International Advanced Research
    Centre for Powder Metallurgy and
    New Materials
Taramani, Chennai, India

Shilpee Jain
Society for Innovation and
    Development, Innovation Centre
Indian Institute of Science
Bengaluru, Karnataka, India

Neetu Jha
Department of Physics Nathalal
    Parekh Marg
Institute of Chemical Technology
    Mumbai
Matunga, Mumbai

Rajesh Katoch
Institut national de la recherche
    scientifique
Centre Énergie Matériaux et
    Télécommunications
Varennes, Québec, Canada

H. M. Waseem Khalil
Department of Electrical Engineering,
    College of Engineering and
    Technology
University of Sargodha
Sargodha, Punjab, Pakistan

Muhammad Asghar Khan
Department of Physics and
    Astronomy
Graphene Research Institute-Texas
    Photonics Center International
    Research Center (GRI–TPC IRC),
    Sejong University
Seoul, Korea

Muhammad Farooq Khan
Department of Electrical Engineering
Sejong University
Neungdong-ro, Gwangjin-gu, Korea

Ashok Kumar
Department of Applied Sciences
National Institute of Technical Teachers
    Training and Research
Chandigarh, India

Astakala Anil Kumar
Nanomaterials for Photovoltaics and
    Biomaterials Laboratory
Godavari Institute of Engineering and
    Technology
Rajahmundry, India

Deeksha Nagpal
Department of Physics
Chandigarh University
Gharuan, Mohali (Punjab), India

Sobia Nisar
Department of Industrial Engineering
    and Management
University of Engineering and Technology
Taxila, Pakistan

Abhishek R. Panigrahi
Department of Zoology, School of
    Biological Sciences
Central University of Punjab
Bathinda, Punjab, India

Shyam Sundar Pattnaik
Media Engineering
National Institute of Technical Teachers
    Training and Research
Chandigarh, India

Sayli Pradhan
Department of Physics
Nathalal Parekh Marg, Institute of
    Chemical Technology Mumbai
Matunga, Mumbai

Shashank Priya
Materials Research Institute
Penn State University
State College, Pennsylvania

Varun Rai
Department of Chemistry
National University of Singapore
3 Science Drive 3, Singapore

Department of Chemistry, Faculty
    of Science
University of Allahabad
Prayagraj, India

Shania Rehman
Department of Electrical Engineering,
    College of Engineering and
    Technology
University of Sargodha
Pakistan

Moumita Saha
Department of Chemistry
Institute of Science, Banaras Hindu
    University
Varanasi, India

Samik Saha
Department of Physics
General Degree College Dantan-II
West-Midnapore, West Bengal

Sourav Sarkar
Department of Instrumentation
    Science
Jadavpur University
Kolkata, India

Pankaj Sharma
Department of Applied Sciences
National Institute of Technical Teachers
    Training and Research
Chandigarh, India

Anar Singh
Department of Physics
University of Lucknow
Lucknow, India

Arun Kumar Singh
Department of Pure and Applied
    Physics
Guru Ghasidas Vishwavidyalaya
Bilaspur, India

Jashandeep Singh
Department of Physics
Gulzar Group of Institutes
    Khanna
Khanna, Punjab, India

Ram Sevak Singh
Department of Physics
OP Jindal University
Raigarh, Chhattisgarh, India

Sunil K. Singh
Department of Zoology, School of
    Biological Sciences
Central University of Punjab
Bathinda, Punjab, India

Ajay Vasishth
Department of Physics
Chandigarh University
Gharuan, Mohali (Punjab), India

Jyoti Yadav
Department of Zoology, School of
    Biological Sciences
Central University of Punjab
Bathinda, Punjab, India

Pooja Yadav
Department of Zoology, School of
    Biological Sciences
Central University of Punjab
Bathinda, Punjab, India

Preface

Nanomaterials are the key components of nanotechnology. Compared to their bulk counterpart, physicochemical properties of the nanomaterials dramatically changes due to the quantum confinement effect and increase in surface-to-volume ratio. Two-dimensional (2D) materials are a new class of emerging nanomaterials with thickness down to atomic scale. The discovery of graphene in 2004 has stimulated tremendous research interest in finding out other new 2D materials and their potential applications. Owing to fascinating properties, 2D materials have potential applications in various sectors, including energy conversion and storage, electronics and optoelectronics, corrosion, polymer electrolyte membrane fuel cells, memory storage, and biomedical fields. In comparison with 0D and 1D nanomaterials, the 2D nanomaterials possess a large surface area, superior electron mobility, more surface active sites, and serve as excellent photocatalytic supports and good electron transfer platforms.

The book covers almost all aspects of 2D nanomaterials and their important applications. The chapters are arranged systematically to make the topic more understandable. Chapter 1 gives a brief introduction to 2D materials, covering their history of development and current practical applications. Chapter 2 comprises syntheses and structures of various 2D materials. Chapter 3 focuses on electrical transport in graphene, which has stimulated potential interests in the discovery and study of other 2D materials. Chapters 4–7 cover different properties in detail, focusing more on recent developments and applications. Chapters 8–11 shed light on more detailed applications in important areas of energy harvesting, energy storage, environmental monitoring, biosensing, and health care.

The book aims to attract a large number of readers throughout the world as topics covered in the book are the most recent, emerging, and of great interest to the current scientific community. The book also encourages solutions to many problems faced by industries. This book differs from other published books on 2D materials due to the following grounds. First, it deals with many aspects of 2D materials and their potential applications systematically to help the readers understand the topic in a better way. Second, it includes the most up-to-date published results as research in this area is being progressed day by day. Third, the chapters of the book have been written by authors having prolonged research and academic experiences from all over the world. Simplified and illustrative figures, tables, and language are other attractive features of this book. The book will be very helpful to scientists, engineers, and students of science and engineering backgrounds to explore the world of nanotechnology and its innovative roles for a better world in the future.

Arun Kumar Singh
Ram Sevak Singh
Anar Singh

1 Overview of 2D Materials

Arun Kumar Singh, Ram Sevak Singh,
and Anar Singh

CONTENTS

INTRODUCTION

Two-dimensional (2D) materials are the materials consisting of a single layer of atoms. Whenever we think about two-dimensional materials, the first material that comes to mind is graphene. Graphene is single sheet of sp^2-bonded carbon atoms arranged in a honeycomb lattice.[1] It is a basic building block for all other dimensionalities of carbon nanostructures. We can wrap it to form 0D fullerenes, rolled into 1D nanotubes, or stacked into 3D graphite.[1–3] Before graphene, scientists argued that strictly 2D crystals were thermodynamically unstable and could not exist.[4–5] In 2004, Geim and Novoselov were successfully isolated single-layer graphene and received the 2010 Nobel Prize in physics for their outstanding work. This breakthrough led to the intensive research not only by physicists and chemists but also inspired renewed interest in carbon-based electronics from device engineers and other applications.[6–9] Graphene was the first 2D material to represent a major advancement in science and technology due to its unique properties, such as electronic, mechanical, optical, and thermal property. Graphene possesses unique linear band structure, remarkable high charge carrier mobility at room temperature, high optical transmittance, flexibility in very low manufacturing cost, which makes graphene a suitable candidate for many exciting applications.[10–13] Many interesting ideas have been demonstrated in the laboratory, and some of the promising applications include flexible and transparent conducting electrodes, sensors, supercapacitors, low power switches, solar cells, battery, spin devices, nanocomposites, and tunable plasmonic devices for THz and mid-infrared applications.[14–17] Some more attractive products may materialize at the industrial scale in the coming years.

Graphene has received widespread attention since it was first isolated from graphite and has been demonstrated in many technological applications in nanoelectronics, optoelectronics, energy harvesting, and sensors. Graphene has a conical Dirac spectrum and linear dispersion characteristic. However, the absence of an energy gap in the electronic band structure limits its practical applications in nanoelectronics and optoelectronics. Recently the field has rapidly expanded beyond graphene and many other 2D materials—such as the transition-metal dichalcogenides (TMDCs), transition-metal oxides and boron nitride—have been investigated.[18–20] MXene, a new 2D

DOI: 10.1201/9781003247890-1

material, has been discovered more recently, which consists of a-few-atoms-thick layers of transition metal carbides, nitrides, or carbonitrides.[21] The formula for TMDCs is MX_2, where M is a transition metal element of group IV, V, VI, and X is a chalcogen (S, Se, or Te). These materials form layered structures of the form X-M-X, with the chalcogen atoms in two hexagonal planes separated by a plane of metal atoms.[18] Many of TMDCs have finite bandgaps, such as MoS_2, WS_2, $MoSe_2$, WSe_2, and many more. Bulk MoS_2 is an n-type semiconductor with an indirect bandgap of 1.2 eV (where single-layer MoS_2 has a direct bandgap of 1.8 eV) and has attracted increasing interest for its novel electronic and optoelectronic properties.[22–24] Fascinatingly, the library of 2D layered materials grows very rapidly every day; currently, it consists of more than 150 interesting families. Research regarding 2D materials is in a growing phase, and many new materials are introduced and added to their list every year. TMDCs have many attractive properties, such as layer-dependent properties, electronic, and optical. Some of the TMDCs are metallic, some are semiconducting, and some are even superconducting.[18] Another class of materials is hexagonal boron nitride (h-BN), which is an insulating 2D material.[25] The layer structure h-BN is similar to graphite. It has excellent thermal and chemical stability and perfect dielectric materials for many applications.

The combination of these 2D materials, we can fabricate many fascinating devices. For example, among many applications, the field-effect transistor (FETs) is the most important and is a basic element of any electronic circuit.[26] To fabricate the FETs, we require three types of materials, metal, semiconductor, and insulator. By using 2D materials, we can fabricate very thin FETs. Since graphene can be used as thinnest (thickness of single-layer graphene is 0.34 nm) and transparent electrode material, 2D TMDCs semiconductors can be used as channel materials, and h-BN can be utilized as gate dielectric, which is also very thin. Overall thickness of device would be within 2–3 nm. As such, miniaturization of electronic devices (reducing the size of devices) can be achieved more effectively using 2D materials, which dramatically improves the performance and reduces the cost of a device. Therefore, with the combination of many 2D materials, we can fabricate the high performance for flexible as well as very thin devices. It is easy to do integration and miniaturization of devices using 2D materials as compared to one-dimensional or zero-dimensional nanomaterials. Even various homo and hetero junctions in both planner and vertical configuration has already been demonstrated by researchers all around the world.[27]

The 2D materials have unconventional and multifunctional features that will trigger further research and will hopefully overcome the constraints appeared along with the various applications shown in Figure 1.1. The ability to isolate, mix, and match highly distinct atomic layers at mono or few-atom layer thickness can facilitate thrilling opportunities to manipulate the confinement and transport of electrons, holes, photons, excitons, and phonons at the limit of single-atom thickness and to enable novel devices with extraordinary performance or entirely new functions. Although 2D materials have many exciting properties, however, it is limited with low yield synthesis and reproducibility of fabricated devices. These problems can be solved using large and wafer scale synthesis as well as device fabrication using existing semiconductor fabrication techniques.

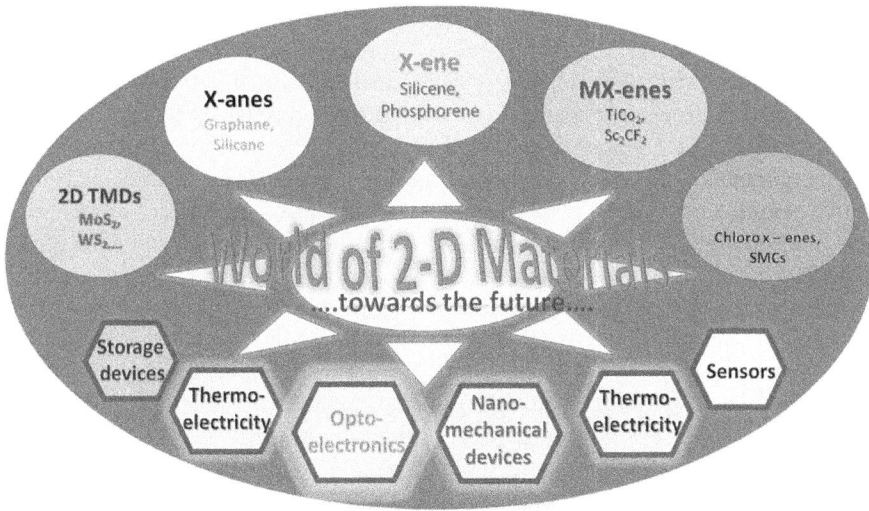

FIGURE 1.1   Schematic diagram of various 2D materials and their applications.

ORGANIZATION OF BOOK

This book originated from the need to bridge the gap of knowledge for researchers of any discipline of science or engineering when they try to start the research in the area of 2D materials. The book covers synthesis, properties, and various applications of 2D nanomaterials. The chapters are arranged systematically and divided into self-sufficient eleven chapters. Chapter 1 presents an overview of 2D materials with glimpses of various chapters in the book. Chapter 2 focuses on the different synthesis methods of 2D materials. This chapter also illustrates the structure of various 2D materials. Chapter 3 deals with the electrical property of different forms of graphene and its dependence on external parameters like temperature, pressure, field, doping, etc. The electrical conductivity of graphene changes dramatically as it is grown from monolayer to multilayer. A discussion on effect of twisting angle between two layers, chemical potential, and impurity on the electrical properties of bilayer and multilayer graphene is also included in this chapter. Chapter 4 comprises the electrical transport and challenges of the electronic devices based on post-graphene 2D semiconductor materials. The contact engineering of metal-2D semiconductor materials is discussed in detail, which plays an essential role in charge transportation. This chapter also addresses electric properties of lateral and vertically stacked van der Waals heterostructure based on 2D semiconductor materials and highlights the role of gate-assisted electric transport through these devices.

Chapter 5 deals with 2D materials and their role in the manufacturing of photonic and optoelectronic devices, such as broadband photodetectors, photovoltaic devices, optoelectronic memory, bio-photonic, light-emitting diode (LED), and MXenes-based photodetectors. Also, a great podium will be provided here to investigate the TMDCs and MXenes for their advantages, challenges, and capabilities to play an important role in future technology. Chapter 6 focuses on magnetic properties of 2D layered materials, including half-metals like graphene, transition metal dichalcogenides (TMDs), transition metal thiophosphates (TMPX3), transition metal halides (TMHs), and topological insulators (TIs). Also discussed is the various characterization techniques for probing magnetic properties in 2D materials and novel applications. Chapter 7 deals with the mechanical properties of 2D materials such as graphene, transition metal dichalcogenides, hexagonal boron nitride, phosphorus, and MXene, and the influence of defects on the mechanical characteristics of 2D materials has also been discussed. Chapter 8 focuses on the characteristic attainments regarding the application of the 2D materials in the various energy-harvesting devices, such as perovskite and dye sensitized solar cells, and thermoelectric and piezoelectric devices.

Chapter 9 deals with 2D materials as electrode materials in the energy storage devices. The 2D material morphology minimizes the structural and volumetric transformations associated with the intercalation/deintercalation of mobile charges associated. This chapter also discusses the different parameters of 2D materials based energy storage devices. Chapter 10 deals with the role of 2D materials in environmental monitoring, which is essential for healthy and sustainable life. In Chapter 11, the authors critically evaluate the recent advances of promising 2D nanomaterials and their characteristic features for both their use in biosensors as well as tissue engineering. We have also discussed emerging approaches for the interaction of 2D nanomaterials and their application in cardiac tissue engineering considering their associated limitations. The aim of the book is to attract and help readers from all over the world as it is of great interest to the current scientific field and to encourage solutions to real-world problems. The present book is significantly different from other published books on 2D materials as it covers almost all aspects of 2D materials and their potential applications with the inclusion of the most up-to-date published results. Importantly, the chapters of the book have been written by the best authors from all over the globe. Simplified language with illustrative figures and tables makes the book attractive. The book will be very helpful to scientists, engineers, and students of science and engineering backgrounds to explore the world of nanotechnology and its innovative roles for a better world in the future.

REFERENCES

1. Geim A. K. and Novoselov K. S. The rise of graphene. *Nature Materials* 6, 183–191 (2007).
2. Kumar P., Singh A. K., Eom J. and Singh J. Graphene: Synthesis, properties and application in Transparent Electronic Devices. *Reviews in Advanced Sciences and Engineering* 2, 238–258 (2013).
3. Novoselov, K. S., *et al.* Electric field effect in atomically thin carbon films. *Science* 306, 666–669 (2004).
4. Peierls, R. E. Quelques proprietes typiques des corpses solides. *Ann. I. H. Poincare* 5, 177–222 (1935).
5. Landau, L. D. Zur Theorie der phasenumwandlungen II. *Physikalische Zeitschrift der Sowjetunion* 11, 26–35 (1937).

6. Singh R. S., Wang X., Ariando, Chen W. and Wee A. T. S. Large room- temperature quantum linear magnetoresistance in multilayered epitaxial graphene: Evidence for two-dimensional magnetotransport. *Applied Physics Letters* 101, 183105 (2012).

7. Singh A. K. and Eom J. Negative magnetoresistance in vertical single layer graphene spin valve at room temperature. *ACS Applied Materials & Interfaces* 6, 2493–2496 (2014).

8. Singh A. K., Iqbal M. W., Singh V. K., Iqbal M. Z., Lee J. H., Chun S. H., Shin K. and Eom J. Molecular n-doping of chemical vapor deposition grown graphene. *Journal of Materials Chemistry* 22, 15168–15174 (2012).

9. Singh A. K., Ahmad M., Singh V. K., Shin K., Seo Y. and Eom J. Tailoring of electronic properties of exfoliated graphene layer by molecular doping. *ACS Applied Materials & Interfaces* 5, 5276–5281 (2013).

10. Singh R. S., Rasheed A., Gautam A., Singh A. K. and Rai V. Enhanced optical and electrical properties of graphene oxide-silver nanoparticles nanocomposite film by thermal annealing in the air. *Russian Journal of Applied Chemistry* 94, 402–409 (2021).

11. Singh R. S., Gautam A. and Rai V Graphene-based bipolar plates for polymer electrolyte membrane fuel cells. *Frontiers of Materials Science* 13, 217 (2019).

12. Pandey R. K., Singh A. K. and Prakash R. Enhancement in performance of polycarbazole-graphene nanocomposite Schottky diode. *AIP Advances* 3, 122120 (2013).

13. Singh R. S., Li D., Xiong Q., Santoso I., Chen W., Rusydi A. and Wee A. T. S. Anomalous photoresponse in the deep-ultraviolet due to resonant excitonic effects in oxygen plasma treated few-layer graphene. *Carbon* 106, 330 (2016).

14. Singh R. S., Jansen M., Ganguly D., Kulkarni G. U., Ramaprabhu S., Choudhary S. K. and Pramanik C. Shellac derived graphene films on solid, flexible, and porous substrates for high performance bipolar plates and supercapacitor electrodes. *Renewable Energy* 181, 1008 (2022).

15. Singh R. S., Nalla V., Chen W., Ji W. and Wee A. T. S. Photoresponse inepitaxial graphene with asymmetric metal contacts. *Applied Physics Letters* 100, 093116 (2012).

16. Singh A. K., Hwang C. and Eom J. Low-voltage and high performance multilayer MoS_2 field-effect transistors with graphene electrodes. *ACS Applied Materials & Interfaces* 8, 34699–34705 (2016).

17. Andleeb S., Eom J., Naz N. R. and Singh A. K. MoS_2 field effect transistor with graphene contacts. *Journal of Materials Chemistry C* 5, 8308 (2017).

18. Wang Q. H., Zadeh K. K., Kis A., Coleman J. N. and Strano M. S. Electronics and optoelectronics of two-dimensional transition metal dichalcogenides. *Nature Nanotechnology* 7, 699 (2012).

19. Singh A. K., Kumar P., Late D. J., Kumar A., Patel, S. and Singh J. 2D layered transition metal dichalcogenides (MoS_2): Synthesis, applications & theoretical aspects. *Applied Materials Today* 13, 242–270 (2018).

20. Khan K., Tareen A. K., Aslam M., Wang R., *et al.* Recent developments in emerging two dimensional materials and their applications. *Journal of Materials Chemistry C* 8, 387 (2020).

21. Naguib M., Kurtoglu M., Presser V., Lu J., Niu J., Heon M., Hultman L., Gogotsi Y. and Barsoum M. W. Two-dimensional nanocrystals produced by exfoliation of Ti_3AlC_2. *Advanced Materials* 23, 4248–2453 (2011).

22. Singh A. K., Andleeb S., Singh J., Dung H. T., Seo Y. and Eom J. Ultra violet light induced reversible and stable carrier modulation in MoS_2 field effect transistors. *Advanced Functional Materials* 24, 7125–7132 (2014).

23. Singh A. K., Pandey R. K., Prakash R. and Eom J. Tailoring the charge carrier in few layers MoS_2 field-effect transistors by Au metal adsorbate. *Applied Surface Science* 437, 70–74 (2018).

24. Chaudhary V., Pandey R. K., Prakash R., Kumar N. and Singh A. K. Unfolding photo-physical properties of poly(3-hexylthiophene)-MoS$_2$ organic-inorganic hybrid materials: An application to self-powered photodetectors. *Nanotechnology* 32, 385201 (2021).

25. Singh R. S., Tay R. Y., Chow W. L., Tsang S. H., Mallick G. and Teo E. H. T. Band gap effects of hexagonal boron nitride using oxygen plasma. *Applied Physics Letters* 104, 163101 (2014).

26. Chaudhary V., Pandey R. K., Shahu P. K., Prakash R., Kumar N. and Singh A. K. MoS$_2$ assisted self-assembled poly(3-hexylthiophene) thin films at an air/liquid interface for high-performance field-effect transistors under ambient conditions. *The Journal of Physical Chemistry C* 124, 8101–8109 (2020).

27. Jin C., Ma E. Y., Karni O., Regan E. C., Wang F. and Heinz T. F. Ultrafast dynamics in van der Waals heterostructures. *Nature Nanotechnology* 13, 994–1003 (2018).

2 Synthesis and Structure of 2D Materials

Bijoy Kumar Das and R. Gopalan

CONTENTS

2.1 GENERAL INTRODUCTION TO 2D MATERIALS

Currently, immense research and development (R&D) activities are focused on developing advanced materials with improved functional properties, which can be successfully integrated into devices [1–3]. The functional properties of these materials are mainly dependent on their composition, microstructure, and sizes. Developing materials at nanoscale can change the functional properties compared to their bulk [4, 5]. The reduction of size to nanoscale leads to confinement of electrons, known as quantum confinement, thereby leading to better functional properties [4, 5]. Based on the size reduction of materials to nanoscale, they can be broadly classified into zero-dimension (0D), such as quantum dot; one-dimension (1D), such as nanowire; two-dimension (2D), such as nanosheet; and three-dimension (3D), such as bulk material (TABLE 2.1) [4, 5]. The discovery of graphene by Geim and Novoselov in 2004, which led them to receive a Nobel Prize in 2010, has raised the possibility of obtaining stable 2D materials of single-layer or multi-layer through exfoliation of van der Waals solids [6]. Such 2D materials can show improved functional properties such as quantum hall effect, transport properties, and excellent optical transparency [4–7]. The successful development and implementation of graphene into device application has created plenty of interest to further develop 2D materials beyond graphene. Such interest has resulted in the

DOI: 10.1201/9781003247890-2

TABLE 2.1
Classifications and Examples of Materials

Sl. No.	Dimension	Classifications	Examples
1	0D	Nanoparticles	Quantum dot
2	1D	Nanowire/nanotube	Carbon nanotube, Si nanowire
3	2D	Nanosheet	Graphene, MoS_2
4	3D	Bulk material	Any micron-size object

discovery of 2D materials like MoS_2, MXene, silicene, germanene, etc. with interesting properties and applications.

A 2D material consists of covalently bonded atoms arranged to form a sheet-like structure in XY direction, and the growth of such structure is not limited to nanometer scale; whereas the thickness of the sheet is confined to one or few atoms in Z-direction via weak van der Waals interaction [8–10]. The atomic arrangement shows in-plane strong bonding; whereas the interlayers are connected by the out-plane weak bonding. Such structural arrangement of atoms in 2D materials makes them unique due to their unprecedented chemical and physical properties that are unparalleled compared to their bulk counterpart [8–10]. For example, graphene consists of the arrangement of carbon atoms bonded covalently in-plane, forming a sheetlike structure. The layer is limited mostly to one atom and can be extended to a few atoms. Graphene shows better functional properties, such as higher electronic mobility and mechanical strength, compared to the bulk counterpart, graphite [11]. The 2D materials are unique compared to their bulk counterpart for the following reasons:

1. **High surface-area-to-volume ratio:** The surface area, which is defined as the open area for chemical interactions, of a 2D material is higher compared to its bulk counterpart, leading to higher surface-area-to-volume ratio. High surface area 2D materials have more atoms of unsaturated bonds, which is beneficial in terms of faster chemical reaction, higher catalytic effect, and better sensitivity to external species [12].
2. **Exclusion of weak van der Waals interaction:** A normal bulk material having layered crystal structure shows arrangement of atoms with in-plane covalently bonding, forming sheetlike structure, where these sheets (or layers) are held together by weak van der Waals interaction [8–10]. The application of external forces can easily outplay the weak van der Waals interaction, leading to breaking of these materials. In case of a monolayer or few-layer 2D materials, the in-plane atoms are connected by strong covalent bonds and seem to have better mechanical strength due to the removal of weak van der Waals interaction or the negligible presence of weak van der Waals interaction. For example, graphene, which is a single or few-layered 2D material exfoliated from graphite, shows higher mechanical strength, i.e., tensile strength 1,000 times greater than graphite. It can be noted here that the graphene even appears 100 times stronger than the steel [13].

3. **Localization of electron, known as quantum confinement:** As mentioned earlier, the functional properties of a material are the function of the geometrical dimensions. Restriction in size/dimension of a material to nanometer scale leads to electron confinement in the direction where the dimension has been reduced. Such electron confinement can modify the band structure of a material, thereby changing the functional properties. Band structure of a material has great impact on the functional properties of a material, mainly electrical and optical properties. It has been seen that graphene shows extremely high electrical conductivity compared to graphite due to modified band structure, where the electrons are confined in the perpendicular direction of graphene surface [14]. The fluorescence and sensing properties of MoS_2 monolayer has been improved compared to bulk MoS_2[15].

With the change in properties due to limiting in dimension, 2D materials are often well accepted for industrial application, where their counterpart bulk materials fail to perform. They find a wide range of applications, such as flexible transistor, sensor, photodetector, and energy storage (electrode for batteries and supercapacitors) [8–10]. The next section shows some interesting applications, where 2D materials outperform their bulk counterpart.

1. **Flexible transistor:** 2D materials having ultra-low thickness are considered as potential class of materials for flexible electronics. Their versatile electronic properties and high breaking strain make them suitable for flexible electronics applications. The 2D metal chalcogenides, such as MoS_2 and WSe_2, show semiconducting properties, which make them suitable as channel materials for flexible electronics. Graphene showing high electrical conductivity is suitable for electrode materials. The 2D materials that have a high bandgap, such as 2D hexagonal boron nitride (h-BN), are suitable for dielectric gate applications [16].
2. **Pressure or strain sensor:** 2D materials are subjected to change in bandgap under applied pressure or strain. The existing distortions in 2D material may contribute to the additional scatterings, leading to lower electrons' mobility. The combined factors modulate the resistivity of the 2D material, known as the piezoresistive effect. The pressure or strain sensor is based on the principle of piezoresistive effect [17].
3. **Photodetectors:** For photodetectors, materials having suitable bandgap energy close to optical or near infrared region and excellent transport properties are highly essential [18]. In 2D materials, bandgap energy can be easily tuned while controlling the growth of the number of layers. Various metal chalcogenides, such as MoS_2, $MoSe_2$, and WSe_2, show bandgap energy close to optical or near infrared region with excellent transport properties, which is required for ideal photodetector materials [19]. The device based on monolayer MoS_2 and combined graphene/MoS_2 heterostructure has shown 103–108 A/W sensitivity in 400–680 nm spectral range [19].
4. **Electrode materials batteries and supercapacitors electrodes:** Materials having high electronic conductivity and high specific capacity are required

for both batteries and supercapacitors [20, 21]. 2D materials are known to have high electronic conductivity along with high surface-to-volume ratio, which helps to store more ions, leading to high specific capacity. In addition, they will provide a faster charge transport along their surface, which will enhance the rate performance. MoS_2 and graphene have received huge attention as electrode materials for batteries and supercapacitors [20, 21].

As discussed, the 2D materials find wide range of applications due to their enriched functional properties, which fully depend on the synthesis routes. The synthesis remains the key to develop the high-quality 2D materials in large scale for industrial applications.

2.2 SYNTHESIS OF 2D MATERIALS

Synthesis of materials is the key to design advanced materials of different shapes and sizes to achieve improved functional properties. In literature, various approaches were adopted to prepare different classes of materials from bulk to nanometer scale, limiting the geometrical dimensions; it may be in one dimension or in all three dimensions [8–10]. Basically, 2D materials are synthesized either from its bulk counterpart by thinning or nucleating to one or few atomic layers. To create 2D materials, mainly layered materials are preferred due to the presence of weak van der Waals interaction between two layers. Hence, such materials can easily be exfoliated or thinned to one or few-layered 2D materials. The synthesis routes used for the preparation of 2D materials are broadly classified into two different categories, such as **top-down** (which explains synthesis of 2D materials by thinning or exfoliating the bulk materials) and **bottom-up** (which explains synthesis of 2D materials by the nucleating or assembling of atoms) [22]. These broad categories of synthesis routes include various specific approaches adopted for the synthesis of 2D materials, and their advantages and disadvantages are explained.

2.2.1 TOP-DOWN APPROACH

2.2.1.1 Mechanical Exfoliation

Mechanical exfoliation—the term refers to exfoliation or disintegrating the smaller unit from the bulk by means of mechanical force applied to the parent samples. It can also be termed as thinning of the sample to small units. Basically, this approach has been used for the preparation of 2D materials in a laboratory. Mainly layered materials having weak interlayer van der Waals interaction are chosen for this synthesis routes [23]. A Scotch tape is used to peel the material from the bulk sample by means of mechanical force. The sticky tape is pressed onto the surface of the bulk material, which peels the smaller units due to the adhesive force. This method was first used to prepare the 2D graphene from graphite. Novoselov et al. reported the synthesis of single- or few-layered graphene from graphite using the Scotch tape approach [23]. Jayasena et al. reported the use of diamond wedge for mechanical exfoliation of graphene from graphite, known as highly oriented pyrolytic graphite (HOPG) [24]. They used ultra-sonication in addition to mechanical exfoliation to get high-quality graphene, similar to the approach adopted by Geim et al. [6]. The control in ultra-sonication oscillations led to the consistency in the properties of graphene. The synthesis of defect-free graphene has been reported by Keith et al., where they replaced the ultra-sonication by the shear exfoliation in stabilizing N-methyl-2-pyrrolidone (NMP) [25]. In this process of exfoliation, the graphite is immersed in NMP and mixed at shear rate higher than $10,000 \text{ s}^{-1}$. However, mechanical exfoliation of graphene has various shortcomings, such as uncontrolled synthesis of graphene with formation of many layers and defects in graphene created during exfoliation. For these reasons, mechanical exfoliation remains popular for lab-based studies; however, it is not scalable for integration into new technologies.

2.2.1.2 Exfoliation via Ion Intercalation

Unlike mechanical exfoliation, a better route, known as ion intercalation exfoliation, was extensively used for the synthesis of 2D materials like graphene and MoS_2 [26]. The exfoliation through ion intercalation has been adopted as a potential route to synthesize 2D materials, where a smaller size alkali ion, like lithium, is inserted into the interlayer gap of 2D material, which helps in widening the gap and weakening the van der Waals interaction [26]. The ion intercalated materials are subjected to reaction with various aqueous/nonaqueous solutions, starting from water to organic solvent materials, in order to release the gas, which helps to achieve 2D materials suspended in solution. In a few cases, heating of the solution is carried out to achieve better results. However, the prolonged reaction time (two to three days) and uncontrolled ion interaction into materials are the two big challenges in the synthesis of 2D material though this route. Single- or few-layers MoS_2 have been synthesized through this route [26].

2.2.2 Bottom-Up

2.2.2.1 Soft Chemical Synthesis Route

Solution-based chemical route is widely used for synthesis of nanomaterials. The atomic scale mixing of the precursor materials can provide high-quality nanomaterials. Such unique synthesis route has been extended to synthesize the 2D materials. Mainly, graphene and various transition metal dichalcogenides (TMDCs) are synthesized using solution-based chemical routes [27]. These include interface mediated growth, solvothermal, hydrothermal, and microwave-assisted chemical routes. Solution-based chemical routes are beneficial in terms of large-scale synthesis of 2D materials, low-cost, and their versatility. However, residual solvent remains a major issue [27].

2.2.2.2 Chemical Vapor Deposition (CVD) Route

The CVD technique is an effective route to synthesize the nanomaterials of high quality and controlled morphology. Through controllable process parameters offered by CVD techniques, 2D materials of high quality and improved functional properties can be achieved. The 2D materials like graphene and various TMDCs have been synthesized through this approach [28]. Not only pure 2D materials, their composites can be synthesized through CVD by using various precursor materials, even at large scale. CVD involves growth of 2D materials on the surface of a heated substrate placed within a furnace. The temperature of the substrate is fixed based on the type of materials to be grown. The gaseous precursors along with the career gases are passed through the heated furnace, where the precursors react with the heated substrate to form the film [28]. The synthesis process has successfully been used to synthesize graphene, TMDCs, etc. To achieve the high-quality 2D materials of required thickness and composition, the process parameters can be easily tuned, which is a great advantage in CVD.

Based on the previously discussed synthesis routes, different 2D materials were synthesized to integrate them into devices. However, the discussion on the synthesis of few 2D materials likes TMDCs, MXenes, and silicene are in the scope of this chapter.

2.3 SYNTHESIS OF TRANSITION METAL DICHALCOGENIDES (TMDC)

Transition metal dichalcogenides (TMDC) are the potential class of materials due to their wide variety of applications, such as sensor and optical materials [29–31]. The properties of these TMDCs can be improved through designing their nanostructures, while the 2D TMDCs have shown excellent properties [29–31]. Synthesis of 2D TMDCs remains crucial to achieve high-quality materials with improved functional properties. Various synthesis routes, such as mechanochemcial exfoliation [32], solution-based exfoliation [33], ion intercalation exfoliation [34], and chemical vapor deposition (CVD) [35], were extensively used for the synthesis of 2D TMDCs. The mechanical exfoliation is considered as one of the best approaches to synthesize 2D TMDCs. The 2D MoS_2 synthesized through this approach have shown excellent photoluminescence properties to date [32]. In addition, solvent-based exfoliation has been used to synthesize the 2D TMDCs. Coleman et al. [33] reported the preparation of few-layer MoS_2 and WS2 using a surfactant-free liquid-exfoliation route. Low boiling point solvent mixture was also used for the exfoliation process for the synthesis of 2D TMDCs. Another effective route for the synthesis of 2D TMDCs is the ion intercalation exfoliation. Zeng et al. [34] reported the synthesis of single-layer MoS_2, WS2, TiS2, and ZrS2 though lithium ion interaction exfoliation process. However, the process involves prolonged reaction time with environmental impact. In addition to these synthesis routes, chemical vapor deposition (CVD) [35] and pulse laser deposition (PVD) [36] were extensively used for the synthesis of single or few-layer 2D. The detailed synthesis of 2D TMDCs using various chemical and physical routes has been explained elsewhere [22].

2.4 SYNTHESIS OF MXENES

Recently, MXenes have been considered as potential class of materials, which are basically two-dimensional inorganic compounds consisting of a-few-atoms-thick layers of transition metal carbides, nitrides, or carbonitrides [37, 38]. MXenes are prepared from the MAX having hexagonal crystal structure with P63/mmc space group, where the M is transition metal, A is mainly Al (or elements from group 13 or 14 of the periodic table) and X is C, N, or CN [37, 38]. MXenes are typically prepared from MAX by selective etching of A layer using an etchant like HF [37]. During this process, the MAX powder is mixed with HF solution and stirred for a few hours at room temperature to expedite the process of etching, where selective etching of A layers takes place. While converting MAX to MXene, the metallic bonding is replaced by the weak interlayer bonding. Non-MAX phase, such as Mo_2Ga_2C, is also used for the synthesis of Mo_2C MXene, where the Ga layer is being etched [39]. Apart from selective etching, MXenes are also prepared by intercalating the guest molecules or ions into the layered MAX phases that have weak interlayer bonding. The insertion of guest molecules or ions causes lattice expansion in MAX and, hence, increases the c-lattice parameter, further weakening the interlayer bonding and finally resulting in exfoliation of MXene layers [38]. For instance, hydrazine (N_2H_4) has been intercalated into $Ti_3C_2(OH)_2$, forming Ti_3C_3MXene [38].

2.5 SYNTHESIS OF SILICENE

The silicene, which is the graphene version of silicon, was first investigated theoretically by Shiraishi and Takeda in 1994 [40, 41]. The silicene is different from graphene in crystal structure point of view, where the two sublattice A and B does not lie in the same plane and is shifted in the perpendicular direction to the atomic plane [40, 41]. In 2007, Guzman-Verri and Lew Yan Voon named such structure as silicene [40]. The synthesis of silicene still remains a challenge, unlike graphene, due to various difficulties, such as lack of availability of layered silicon and also the highly insulating nature of silicon oxide; hence, mechanical and chemical exfoliation techniques cannot be employed. Mainly, CVD technique has been employed as potential synthesis route to prepare silicene [40]. Si nanomaterials with spherical shape, consisting of freestanding 2D Si nanosheets, were prepared by CVD technique. The thickness of these Si nanosheets was maintained below 2 nm [40]. Solution-based chemical routes were also adopted to prepare 2D silicene [41]. Nakano et al. [42] reported the synthesis of silicene through the use of magnesium doped hexagonal $CaSi_2$. The exfoliation of $CaSi_{0.85}Mg_{0.15}$ in the presence propylamine hydrochloride (PA·HCl) solution resulted in silicene with composition of Si: Mg: O = 7.0: 1.3: 7.5 as confirmed using XPS. Okamoto et al. [43] reported the synthesis of oxygen-free Si nanosheets of thickness below 2 nm via exfoliation of layered Si_6H_6, while reacting with n-decylamine. The obtained Si nanosheet exhibited excellent photoluminescence properties. The synthesis of silicene (or silicon nanosheets) via solution-based routes appears more difficult compared to other 2D materials.

2.6 SUMMARY

In summary, 2D materials are considered as potential class of materials due to their rich functional properties achieved through the presence of high surface-to-volume ratios and excellent electronic transport. They find wide applications as a catalyst, flexible transistor, pressure sensor, and electrode material for supercapacitors and batteries. The localization of electrons in the direction perpendicular to the surface further adds to their excellent optical properties due to the modulation of bandgap. Hence, it is highly essential to opt for advanced synthesis routes to prepare high-quality 2D materials in large scale. Two broad synthesis approaches, such as top-down and bottom-up, have been used for the synthesis of 2D materials. The top-down approach includes micromechanical, ultra-sonication, and ion intercalation exfoliations. The bulk materials having layered structures are preferred for the synthesis of their 2D materials through top-down approach. In the case of bottom-up approach, the synthesis of 2D materials occurs by the nucleation or the assembling of atoms. The bottom-up approach includes chemical vapor deposition (CVD), soft chemical routes, such as microwave-assisted chemical, solvothermal, and hydrothermal routes. In future, efforts should be made to develop advanced synthesis routes to prepare high-quality and defect-free 2D materials in large scale for industrial applications.

ACKNOWLEDGMENT

The authors would like to acknowledge the financial support from the Department of Science and Technology through Technical Research Centre (TRC project: AI/1/65/ARCI/ 2014) and DST project: DST/TMD/MES/2K17/46, Government of India, for the completion of the work. The authors would like to thank Dr. Tata Narasinga Rao, director (additional charge), ARCI, for his support to complete the work.

REFERENCES

1. R. Dong, T. Zhang, X. Feng, Interface-Assisted Synthesis of 2D Materials: Trend and Challenges. Chem. Rev. 118 (2018) 6189–6235.
2. L. Zhang, J. Dong, F. Ding, Strategies, Status, and Challenges in Wafer Scale Single Crystalline Two-Dimensional Materials Synthesis. Chem. Rev. 121 (2021) 6321–6372.
3. N. C. Frey, J. Wang, G. I. V. Bellido, B. Anasori, Y. Gogotsi, V. B. Shenoy, Prediction of Synthesis of 2D Metal Carbides and Nitrides (MXenes) and Their Precursors with Positive and Unlabeled Machine Learning. ACS Nano 13 (2019) 3031–3041.
4. M. Garg, A. Gupta, A. L. Sharma, S. Singh, Advancements in 2D Materials Based Biosensors for Oxidative Stress Biomarkers. ACS Appl. Bio Mater. 4 (2021) 5944–5960.
5. M. G. Stanford, P. D. Rack, D. Jariwala, Emerging nanofabrication and quantum confinement techniques for 2D materials beyond graphene. NPJ 2D Mater. App. 2 (2018) 20.
6. A. K. Geim, Graphene: Status and Prospects. Science 324 (2009) 1530.
7. H. Zhang, Introduction: 2D Materials Chemistry. Chem. Rev. 118 (2018) 6089–6090.
8. K. Khan, A. K. Tareen, M. Aslam, R. Wang, Y. Zhang, A. Mahmood, Z. Ouyang, H. Zhang, Z. Guo, Recent developments in emerging two-dimensional materials and their applications. J. Mater. Chem. C 8 (2020) 387–440.
9. S. Haol, X. Zhao, Q. Cheng, Y. Xing, W. Ma, X. Wang, G. Zhao, X. Xu, A Mini Review of the Preparation and Photocatalytic Properties of Two-Dimensional Materials. Front. Chem. 8 (2020) 582146.
10. A. Zavabeti, A. Jannat, L. Zhong, A. A. Haidry, Z. Yao, J. Zhen Ou, Two-Dimensional Materials in Large-Areas: Synthesis, Properties and Applications. Nano-Micro Lett. 12 (2020) 66.
11. X. Xu, Z. Zhang, J. Dong, D. Yi, J. Niu et al., Ultrafast epitaxial growth of metre-sized single-crystal graphene on industrial Cu foil. Sci. Bull. 62 (2017) 1074–1080.

12. Z.-Fei Li, H. Zhang, Q. Liu, L. Sun, L. Stanciu, J. Xie, Fabrication of High-Surface-Area Graphene/Polyaniline Nanocomposites and Their Application in Supercapacitors. ACS Appl. Mater. Interfaces 5 (2013) 2685–2691.

13. Y. Huang, E. Sutter, N. N. Shi, J. Zheng, T. Yang, D. Englund, H.-J. Gao, P. Sutter, Reliable Exfoliation of Large-Area High-Quality Flakes of Graphene and Other Two-Dimensional Materials. ACS Nano 9 (2015) 10612–10620.

14. C. D. Nunez, P. A. Orellana, L. Rosales, Electron localization due to side-attached molecules on graphene nanoribbons. J. Appl. Phys. 120 (2016) 164310.

15. H. Deng, X. Yanga, Z. Gao, MoS_2 nanosheets as an effective fluorescence quencher for DNA methyltransferase activity detection. Analyst 140 (2015) 3210–3215.

16. G.-H. Lee, X. Cui, Y. D. Kim, G. Arefe, X. Zhang, C.-H. Lee, F. Ye, K. Watanabe, T. Taniguchi, P. Kim, J. Hone, Highly Stable, Dual-Gated MoS_2 Transistors Encapsulated by Hexagonal Boron Nitride with Gate-Controllable Contact, Resistance, and Threshold Voltage. ACS Nano 9 (2015) 7019–7026.

17. S.-H. Bae, Y. Lee, B. K. Sharma, H.-J. Lee, J.-H. Kim, J.-H. Ahn, Graphene-based transparent strain sensor. Carbon 51 (2013) 236–242.

18. W. Zhang, C.-P. Chuu, J.-K. Huang, C.-H. Chen, M.-L. Tsai, Y.-H. Chang, C.-T. Liang, Y.-Z. Chen, Y.-L. Chueh, J.-H. He, M.-Y. Chou, L.-J. Li, Ultrahigh-Gain Photodetectors Based on Atomically Thin Graphene-MoS_2 Heterostructures. Sci. Rep. 4 (2014) 3826.

19. O. Lopez-Sanchez, D. Lembke, M. Kayci, A. Radenovic, A. Kis, Ultrasensitive photodetectors based on monolayer MoS_2. Nat. Nanotech. 8 (2013) 497–501.

20. J. Hassoun, F. Bonaccorso, M. Agostini, M. Angelucci, M. Grazia. Betti, R. Cingolani, M. Gemmi, C. Mariani, S. Panero, V. Pellegrini, B. Scrosati, An Advanced Lithium-Ion Battery Based on a Graphene Anode and a Lithium Iron Phosphate Cathode. An Advanced Lithium-Ion Battery Based on a Graphene Anode and a Lithium Iron Phosphate Cathode. Nano Lett. 14 (2014) 4901–4906.

21. M. Acerce, D. Voiry, M. Chhowalla, Metallic 1T phase MoS_2 nanosheets as supercapacitor electrode materials. Nat. Nanotech. 10 (2015) 313–318.

22. R. Dong, T. Zhang, X. Feng, Interface-Assisted Synthesis of 2D Materials: Trend and Challenges. Chem. Rev. 118 (2018) 6189–6235.

23. K. Novoselov, D. Jiang, F. Schedin, T. Booth, V. Khotkevich, S. Morozov, A. K. Geim, Two-dimensional atomic crystals. P. Natl, Acad. Sci. USA 102 (2005) 10451.

24. B. Jayasena, S. Subbiah, A novel mechanical cleavage method for synthesizing few-layer graphenes. Nanoscale Res. Lett. 6 (2011) 95.

25. K. E. Whitener Jr., P. E. Sheehanb, Graphene synthesis. Diam. Relat. Mater. 46 (2014) 25–34.

26. Z. Zeng, Z. Yin, X. Huang, H. Li, Q. He, G. Lu, F. Boey, H. S. Zhang, Single-Layer Semiconducting Nanosheets: High-Yield Preparation and Device Fabrication. Angew. Chem. 50 (2011) 11093–11097.

27. W. Cheng, J. He, T. Yao, Z. Sun, Y. Jiang, Q. Liu, S. Jiang, F. Hu, Z. Xie, B. He, W. Yan, S. Wei, Half-Unit-Cell α-Fe_2O_3 Semiconductor Nanosheets with Intrinsic and Robust Ferromagnetism. J. Am. Chem. Soc. 136 (2014) 10393–10398.

28. K. K. Kim, A. Hsu, X. Jia, S. M. Kim, Y. Shi, M. Hofmann, D. Nezich, J. F. Rodriguez-Nieva, M. Dresselhaus, T. Palacios, J. Kong, Synthesis of Monolayer Hexagonal Boron Nitride on Cu Foil Using Chemical Vapor Deposition. Nano Lett. 12 (2012) 161–166.

29. A. B. Laursen, S. Kegnas, S. Dahl, I. Chorkendorff, Molybdenum sulfides—efficient and viable materials for electro - and photoelectrocatalytic hydrogen evolution. Energy Environ. Sci. 5 (2012) 5577.

30. J. Puthussery, S. Seefeld, N. Berry, M. Gibbs, M. Law, Colloidal Iron Pyrite (FeS_2) Nanocrystal Inks for Thin-Film Photovoltaics. J. Am. Chem. Soc. 133 (2011) 716.

31. J. Feng, X. Sun, C. Z. Wu, L. L. Peng, C. W. Lin, S. L. Hu, J. L. Yang, Y. Xie, Metallic Few-Layered VS2 Ultrathin Nanosheets: High Two-Dimensional Conductivity for In-Plane Supercapacitors. J. Am. Chem. Soc. 133 (2011) 17832.

32. K. F. Mak, C. Lee, J. Hone, J. Shan, T. F. Heinz, Atomically Thin MoS_2: A New Direct-Gap Semiconductor. Phys. Rev. Lett. 105 (2010) 136805.

33. J. N. Coleman, M. Lotya, A. O'Neill, S. D. Bergin, P. J. King, U. Khan, K. Young, A. Gaucher, S. De, R. J. Smith, I. V. Shvets, S. K. Arora, G. Stanton, H. Y. Kim, K. Lee, G. T. Kim, G. S. Duesberg, T. Hallam, J. J. Boland, J. J. Wang, J. F. Donegan, J. C. Grunlan, G. Moriarty, A. Shmeliov, R. J. Nicholls, J. M. Perkins, E. M. Grieveson, K. Theuwissen, D. W. McComb, P. D. Nellist, V. Nicolosi, Two-Dimensional Nanosheets Produced by Liquid Exfoliation of Layered Materials. Science 331 (2011) 568.

34. R. J. Smith, P. J. King, M. Lotya, C. Wirtz, U. Khan, S. De, A. O'Neill, G. S. Duesberg, J. C. Grunlan, G. Moriarty, J. Chen, J. Z. Wang, A. I. Minett, V. Nicolosi, J. N. Coleman, Large-Scale Exfoliation of Inorganic Layered Compounds in Aqueous Surfactant Solutions. Adv. Mater. 23 (2011) 3944.

35. W. K. Hofmann, Thin films of molybdenum and tungsten disulphides by metal organic chemical vapour deposition. J. Mater. Sci. 23 (1988) 3981.

36. J. S. Zabinski, M. S. Donley, S. V. Prasad, N. T. McDevitt, Synthesis and characterization of tungsten disulphide films grown by pulsed-laser deposition. J. Mater. Sci. 29 (1994) 4834.

37. M. Naguib, M. Kurtoglu, V. Presser, J. Lu, J. Niu, M. Heon, L. Hultman, Y. Gogotsi, M. W. Barsoum, Two-Dimensional Nanocrystals Produced by Exfoliation of Ti_3AlC_2. Adv. Mater. 23 (2011) 4248–4253.

38. M. Naguib, V. N. Mochalin, M. W. Barsoum, Y. Gogotsi, MXenes: A New Family of Two-Dimensional Materials. Adv. Mater. 26 (2014) 992–1005.

39. J. Halim, M. R. Lukatskaya, K. M. Cook, J. Lu, C. R. Smith, L.-Å. Näslund, S. J. May, L. Hultman, Y. Gogotsi, P. Eklund, M. W. Barsoum, Transparent Conductive Two-Dimensional Titanium Carbide Epitaxial Thin Films. Chem. Mater. 26 (2014) 2374–2381.

40. G. G. Guzmán-Verri, L. C. Lew, Y. Voon, Electronic structure of silicon-based nano-structures. Phys. Rev. B 76 (2007) 075131.

41. H. Okamoto, Y. Sugiyama, H. Nakano, Synthesis and Modification of Silicon Nanosheets and Other Silicon Nanomaterials. Chem. Eur. J. 17 (2011) 9864.

42. H. Nakano, T. Mitsuoka, M. Harada, K. Horibuchi, H. Nozaki, N. Takahashi, T. Nonaka, Y. Seno, H. Nakamura, Soft Synthesis of Single-Crystal Silicon Monolayer Sheets. Angew. Chem. 45 (2006) 6303.

43. H. Okamoto, Y. Kumai, Y. Sugiyama, T. Mitsuoka, K. Nakanishi, T. Ohta, H. Nozaki, S. Yamaguchi, S. Shirai, H. Nakano, Silicon Nanosheets and Their Self-Assembled Regular Stacking Structure. J. Am. Chem. Soc. 132 (2010) 2710.

3 Electrical Properties of Graphene
Comprehensive Study

Sourav Sarkar, Samik Saha, and Sachindranath Das

CONTENTS

3.1 INTRODUCTION

Graphene is a common name in the research field due to its exceptional and interesting mechanical, thermal, optical, and electrical properties. It is a stable form of carbon atoms, wherein atoms are arranged in a two-dimensional hexagonal lattice in a picturesque manner. Single-layer graphene was theoretically defined by P. R. Wallace in 1947 [1]. It has been studied since its inception; however, it was first unambiguously produced and identified in 2004 [2] by Andre Geim and Konstantin Novrselove. They received the Nobel Prize in physics in 2010 for their groundbreaking experiments.

DOI: 10.1201/9781003247890-3

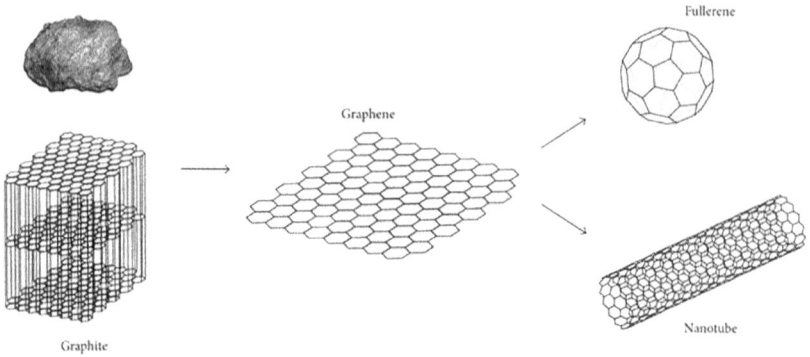

FIGURE 3.1   Different forms of carbon.

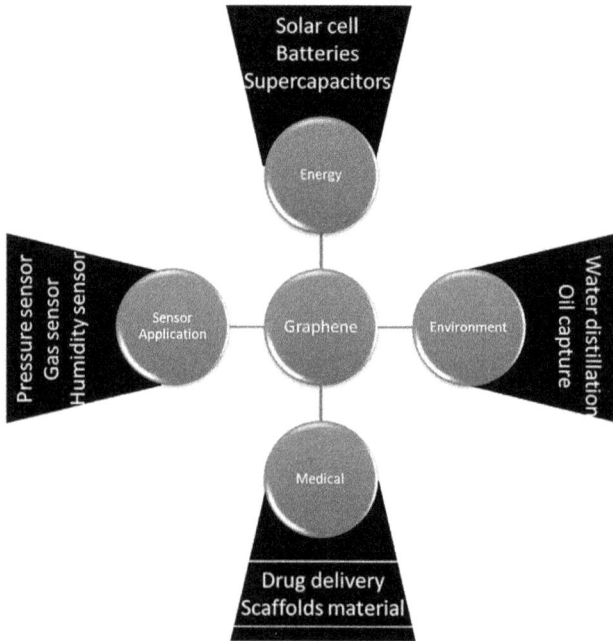

FIGURE 3.2   Use of graphene in different fields.

Graphene belongs to the family of different graphitic materials like carbon nanotube (CNT) and fullerene, as shown in Figure 3.1. Graphene has a large specific surface area, high thermal conductivity, high electrical conductivity, and good flexibility. Graphene first attracted the curiosity of physicists owing to its particular electronic behavior under magnetic fields and at low temperatures. Graphene is stronger than steel but is a much lighter material. Because of its attractive properties, it has many potential uses in different fields. In the field of energy research, graphene has been used as an electrode material for different electrochemical energy storage devices. Graphene can be used to make supercapacitors. Graphene has also found applications in the field of biomedical and environmental research. Figure 3.2 shows the different fields where graphene can find potential applications. In ideal graphene, there is no gap between the valence band and conduction band [3]. Hence, unmodified graphene is not useful in the field of semiconduction. To modify the band structure of graphene, nitrogen (N), boron (B), etc. have been used as dopant. It is an almost transparent material with high electrical conductivity. It is one of the promising materials for optoelectronic and nanophotonic devices. Graphene can also be used to fabricate sensors, like biosensors, pressure sensors, and magnetic sensors. Graphene can be categorized as very-few-layer graphene (vFLG), having one to three layers of carbon; few-layer graphene (FLG), with two to five layers of carbon; and multilayer graphene (MLG), with more than five layers of carbon. In this chapter, we are going to focus on graphene's electrical properties and the impact of different factors on it.

3.2 ORIGIN OF THE EXCEPTIONAL ELECTRICAL PROPERTY OF GRAPHENE

Experimental studies on graphene have revealed exotic transport properties. As we have mentioned earlier, graphene is a two-dimensional carbon allotrope. Carbon atoms form a honeycomb-like arrangement. The hexagonal lattice of graphene can also be regarded as two interleaving triangular lattices. Each carbon atom is situated at a distance 1.42 Å from its three nearest neighbor carbon atoms. The carbon atom has four outer cell electrons. Three out of these four electrons occupy sp^2 hybrid orbital in graphene. Three electrons reside in a 2D plane, and the carbon atom wants to share these three electrons with their neighboring three carbon atoms and thus form σ bonds, as shown in Figure 3.3 (a). These bonds make graphene a strong and flexible material. The fourth electron makes a π-bond, which is oriented in the out-of-plane direction. Each carbon atom has one of this out-of-plane bond. Geometrically, the π bond is like a dumbbell-shaped symmetric lobe having the carbon nucleus at the center and oriented along the z-axis. The fourth

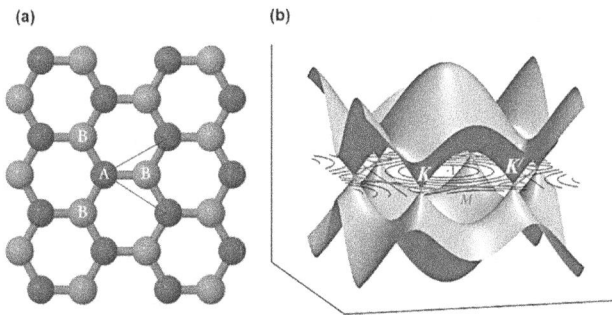

FIGURE 3.3 (a) Sublattice structure of graphene. (b) Band structure of graphene in the first Brillouin zone.

Source: Reproduced from ref. [5], Cooper et al. is an open access distributed under the creative attribution license, copyright 2012.

electron can reside in the lobe oriented along either the positive z-axis or the negative z-axis. This electron is free to move along the plane and is responsible for the
exotic electronic properties. In 1947, Wallace theoretically determined the band
structure of single-layer graphene [4]. In ideal graphene, there is no gap between
the valence band and conduction band as these bands touch at the Dirac points. The
Dirac points are located at the edges of the Brillouin zone. In Figure 3.3 (b), the
band structure of graphene in the first Brillouin zone is shown. There are six Dirac
points, and these points can be divided into two sets, each having three points. The
two sets are completely different from each other and can be pointed out by K and
K'. The two sets of Dirac points give graphene a valley degeneracy of g = 2 [5].
Transport in graphene displays a novel chirality. Each graphene sublattice forms a
branch of dispersion. These dispersion branches make very weak interactions with
one another. The chiral effect determines the presence of a pseudospin quantum
number for the charge carrier. This quantum number has similarity with spin, but
it is not dependent on the real spin. Using pseudospin, the contributions from each
of the sublattices can be recognized [6].

According to Dirac Hamiltonian, the charge carriers cannot be constricted using
electrostatic potentials. In semiconducting material, if an electron encounters an electrostatic potential barrier with a barrier height higher than electron's kinetic energy,
then the electron's wave function would decay exponentially within the potential barrier and may reach the other side depending on the height and width of the potential
barrier. This can be shown for wider and taller barriers, the tunneling probability for
the electron reduces. However, if the particle follows the Dirac points, then the transmission probability will increase with increasing barrier height. In the case of graphene, this situation would be different from quantum tunneling. When electrons run
upon a barrier, it would change into a hole that moves through the barrier [5]. When
the hole reaches another side of the barrier, then it would change into an electron.
This phenomenon is defined as Klein tunneling. In this case, the mode matching
between the electron outside the barrier and hole within the barrier increases with the
height of the barrier. It is exactly the opposite of quantum tunneling probability. In
the case of perfect mode matching, perfect transmission can be seen. In graphene, the
transmission probability can be controlled with the angle of the incident to the barrier [7]. In brief, ideal graphene has zero bandgap, which can be tuned by doping. It
shows Klein tunneling and follows linear dispersion relation. These facts are behind
the attractive physical and electrical properties of graphene.

3.3 ELECTRICAL PROPERTIES OF SINGLE-
LAYER GRAPHENE ON A SUBSTRATE

The suspended graphene offers higher charge carrier mobility compared to graphene
deposited on a substrate. However, from the device point of view, it is essential to
study the electrical properties of graphene deposited on a substrate. Different substrates like silicon dioxide (SiO_2), silicon carbide (SiC), hexagonal boron nitride
(h-BN), etc. have been used as a substrate to hold graphene as a flat layer limiting the
self-corrugation. The electrical properties thus measured depend on several parameters and are discussed in the next section.

3.3.1 IMPACT OF TEMPERATURE

Schiefele et al. theoretically studied the effect of temperature on the electrical conductivity of graphene deposited on hexagonal boron nitride (h-BN) and compared with the same deposited on SiO_2 and SiC. Different mechanisms influence the conductivity and carrier mobility of deposited layer of graphene. The nature of substrate determines such mechanisms. It is found that optical surface phonon plays a major role in determining the charge carrier mobility of graphene deposited on SiO_2 substrate. In such case, the temperature dependence of the conductivity consists of two contributions: First one is the scattering by acoustic phonons in graphene, and the second one is the scattering by surface phonons of the substrate. In the case of h-BN, remote-phonon scattering plays a major role at room temperature. Resistivity of graphene sheet on h-BN increases with the increase in temperature because of the dependence of scattering of electrons with the surface phonons. With the increase of temperature, the surface phonons' population increases, which increases the rate of electron scattering. Figure 3.4 shows the variation of resistivity of graphene layer on different substrates [8].

3.3.2 EFFECT OF ELECTRIC FIELD

To study the electrical properties of graphene in the presence of an external electrostatic field, generally, a four-probe method based on field-effect transistor (FET) principle is used. The electrical properties can be tuned using an electrostatic field. FET consists of a source terminal and a gate terminal, as shown in Figure 3.5 (a). A conducting channel exists between these two terminals. Usually, the channel contains a thin layer of graphene. There is a third electrode known as gate, which is kept separated from the channel by a dielectric material. The gate electrode and the channel form a geometry resembling a parallel plate capacitor. When a gate voltage (V_g) is applied to the gate terminal, the capacitor, like geometry, induces charge into the channel. The presence of this electric field changes the position of the Fermi level and the carrier density, and thus, the electrical properties of the channel could be changed. Reportedly, a change of carrier density (two-dimensional) of the order

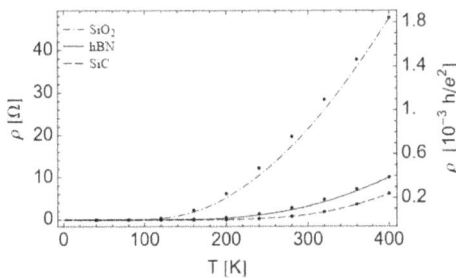

FIGURE 3.4 Variation in graphene resistivity with temperature (for three different substrates).

Source: Reproduced from ref.[8] with permission from American physical society, copyright 2012

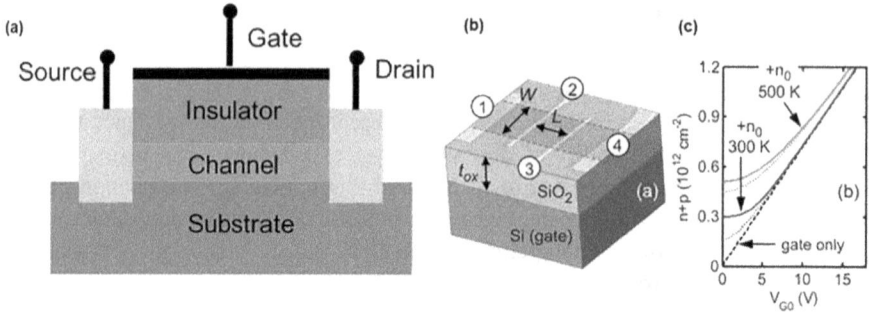

FIGURE 3.5 (a) Geometry of FET. (b) Four-probe method. (c) Variation of carrier density vs. gate voltage.

Source: Reproduced from ref.[10] with the permission American Institute of Physics, 2010

of 10^{12}–10^{13}/ cm^2 can be incorporated in the channel using a dielectric separator of SiO$_2$ [9]. This process of modulation of the electric field is known as the electrostatic doping. As the process is physical and involves no direct charge transfer, it is highly reversible and induces no defect in the channel lattice.

Dorgan et al. examined the mobility and saturation velocity of charge carriers in graphene on SiO$_2$ substrate [10]. For this purpose, a four-probe method was used on a modified FET-like structure, where s highly doped Si substrate acted as a back gate. The schematic of the setup is shown in Figure 3.5 (b). From Figure 3.5 (c), it can be seen that the carrier density can efficiently be controlled by the application of the gate voltage.

3.3.3 Effect of Pressure

Graphene can sustain ultrahigh pressure (~ 130 GPa) because of the strong in-plane bondings [11]. However, application of high pressure can modify the local electrical properties of graphene layer deposited on a substrate. Ares et al. studied the effect of pressure on monolayer graphene deposited on SiO$_2$ substrate [12]. Ultrahigh pressure was obtained using an atomic force microscopy (AFM) having a diamond tip canti-lever. This simple setup can be used to reach a pressure level of ~ 40 GPa. Initially, graphene flakes were deposited on SiO$_2$/Si substrate. Optical microscopy along with the use of Raman spectroscopy enabled the authors to locate the different areas of the substrate where a single layer of graphene has been deposited. Using the diamond tip of the AFM cantilever, ultrahigh pressure was applied to these different zones having a single layer of graphene. Due to the application of pressure, these regions were modified. An AFM image revealed a formation of valleys in these regions. The depths of the valleys were different for different applied pressures. It was seen that for pressure less than 13 GPa, no valley could be formed. For pressure within 16 to 25 GPa, a valley with a depth of ~ 1 nm was observed. Upon applying even more pres-sure (29–37 GPa), the depth of the valley increased to 1.3 nm. However, application of pressure greater than 40 GPa broke the graphene sheet. The modified regions were later analyzed using Kelvin probe force microscopy (KPFM). KPFM is a noncontact type variant of AFM, which can be used to measure the work function of the surface at an atomic scale. It was seen that as the pressure was increased, the contact potential difference reduced. This signifies that the Fermi level moves down as the pressure is increased. This is equivalent to hole doping, and thus the whole process can be called pressure-dependent doping in graphene. It was asserted that the application of pressure resulted in the formation of strong covalent bonds between graphene and the SiO_2 substrate. These bonds shifted the position of Fermi energy, and thus electrical properties of graphene can be tuned. However, when the same experiment was repeated on the few-layer graphene, no such shift in Fermi energy was observed. This is mainly because of the lack of bonds between the different carbon layers.

3.3.4 Effect of Doping and Absorption of Nanoparticle

We know that the mobility of ideal graphene is much higher than that of semiconducting material. Consequently, graphene is considered as a promising material for applications in electronics. However, as already mentioned, pristine graphene has no bandgap, and most electronic applications are hindered for this reason [3]. Doping of atoms is an efficient way to control the graphene bandgap. There are several methods, using which, we can tune the same.

Heteroatom doping: In this method, boron (B) and nitrogen (N) atoms are used [13–18]. The size of B atom and N atom are similar to the size of the carbon atom, and these atoms have hole-acceptor and electron-donor characters. These properties make these atoms suitable as doping elements. After doping with B and N atoms, graphene behaves like *p*- and *n*-type semiconducting material, respectively. First principle density functional theorem shows that the electronic bands of these doped graphene are almost similar to pristine graphene within 1 eV of the Fermi energy and displays linear dispersion relation. It is relatively difficult to introduce foreign atoms in the perfect 2D structure of graphene; however, there are a few methods of B- and N-doping in graphene, like the arc discharge method, chemical vapor deposition method (CVD), etc. The resistance of the doped graphene decreases with the increase in temperature, indicating its semiconducting property and the carrier mobility of doped graphene is lower than that of the pristine graphene [3]. Temperature-dependent resistivity measurement of N-doped graphene shows that the resistance reduces by a factor of 80 when the temperature is increased from liquid N_2 temperature to 300 K.

Chemical modification: In this case, graphene electrical properties are modified by NO_2 and NH_3 treatment [3]. Using NO_2 and NH_3, bandgap can be created in graphene. After doping, graphene behaves like a *p*-type and *n*-type semiconductor, respectively. Water vapor absorption also significantly influences the electrical properties of graphene. Graphene film, when exposed to humidity, develops a maximum bandgap of 0.206 eV. Ability to induce a different bandgap depending on different chemicals makes graphene a strong contender for sensor application.

Impact of absorption of nanoparticle: Nanoparticle (both organic and inorganic) absorption on single-layer graphene can tune the electrical properties. Tanaka et al. synthesized a small stripe of single-layer graphene, also known as graphene nano ribbon (GNR), from double-walled carbon nanotubes (DWNTs). As the name suggests, DWNTs have two concentric cylinders, which can be made by rolling two stacked graphene layers in the form of a cylinder. While synthesizing GNR, the authors used chemical treatment followed by sonication to unzip the two cylinders. Thus, a stripe of bilayer graphene was produced. Further sonication separates the bilayer graphene into two separate single-layer graphene. They further deposited nanoparticles of an organic molecule (C15-NDI) on the single layer of graphene. The nanoparticles were attached to the graphene layer via π-π stacking. As the p-electron is the major contributor to the electronic conductivity of graphene, hence by varying the number of nanoparticles, the electrical properties can be tuned [19]. If the absorbed molecular number on GNR is higher than that of conduction electrons, the electrons would be trapped by the nanoparticles. The conduction path of the electron will become narrower as the size of the absorbed nanoparticle increases. The I-V characteristic curve of GNR reveals its semimetallic nature, whereas after the adsorption of nanoparticles, it behaves like a p-type semiconductor. The bandgap of GNR is similar to the semiconductor [20].

3.3.5 EFFECT OF DEFECT

Ideal graphene has one atomic layer where carbon atoms are situated at the vertices of two-dimensional hexagonal lattice structures. However, it is not possible to experimentally synthesize the symmetric hexagonal lattice of carbon atoms. Experimentally obtained graphene contains different types of defects. These defects break the structural symmetry. The defects in graphene can be classified as intrinsic and extrinsic. When the crystal structure is perturbed without the presence of foreign atoms, the defect is known as intrinsic defect. On the other hand, when external foreign atoms induce defects in the lattice structure, then the defect is known as extrinsic defect. The defects are not always stationary and can migrate over the crystals. This propagation of defect can influence the different properties of the crystal [21].

Stone-Wales (SW) defect is a kind of point defect in the graphene structure, which is intrinsic in nature. Sometimes within the graphene lattice, there could be a few areas that can contain nonhexagonal rings. This is known as SW defect. Single or multiple vacancies is also an intrinsic-type defect. If in the lattice of graphene there are missing atoms in certain areas, then it is known as a vacancy defect. Single vacancy (SV) is formed when only one carbon atom is missing from a lattice. Multiple vacancies arise with two or more SV coalescence. Theoretically the formation energy of DV is of the same order as SV, and it is around 8 eV.

There are different types of extrinsic defects. When a foreign atom is incorporated in the graphene lattice, it is known as the defect due to foreign adatoms. The foreign atom may form covalent bonding with the nearest carbon atom if the interaction between that atom and a carbon atom is strong. This is known as chemisorption. However, if the interaction is weak, then only physisorption of the atom occurs through van der Waals interaction. How the property of the graphene will be modified depends entirely on the type of bond the atom forms with the carbon atoms. Foreign atoms can also be incorporated into the lattice, replacing one or more carbon atoms from the lattice. Boron or nitrogen is usually used in this type of substitution.

Topological defect arises when the graphene deviates from a planar structure and forms hills or valleys at some places, imposing curvature in the lattice structure. Due to curvature, the bond angle is changed, which changes the properties of the graphene.

The electronic properties of graphene are strongly influenced by the presence of defects. In theory, to obtain the quantum mechanical solution of the electronic wave function, we have to use Dirac equation in place of Schrodinger equation. However, in the presence of a defect, the Dirac equation has to be modified, which changes the electronic wave function of the electrons in the lattice. All these defects act as scattering centers for the electron wave and changes the trajectories of the electrons. Thus, decrease in conductivity is observed with the increasing defect density. Point defects create local states near the Fermi energy level in the sp^2 orbital. SW defect can create local bandgap up to 0.3 eV in width. However, the effect on conductance due to more complicated defects (which comprises of nonhexagonal rings) has not been studied till now, and more research is necessary to gain more insight into this field.

3.4 ELECTRICAL PROPERTIES OF BILAYER AND MULTILAYER GRAPHENE

3.4.1 BILAYER GRAPHENE

Bilayer graphene (BLG) is a pile of two monolayer graphene, where two layers are stacked vertically. Fabrication of monolayer graphene is difficult. On the other hand, bilayer graphene can be fabricated relatively easily, and thus it is used instead of monolayer graphene in different electronics devices. Electrical property of bilayer graphene varies with different parameters, like sample length, twisting angle between two layers, temperature, frequency, chemical potential, etc.

3.4.1.1 Impact of Sample Length

The ensemble average conductance of the opened bilayer graphene depends on the length of the sample and d, depth from starting point. Some attractive properties originate from the edge state of graphene, like valley-dependent transport and electric field tunable magnetism. It is difficult to fabricate perfect edge; some disorder must be there in the edge of graphene. Removing the tight binding randomly between two layers of graphene creates structural edge disorder in the bilayers, till depth d from starting point. Transport property of low energy states of BLG has been discussed. The conductance of gapped bilayer graphene can be tuned. The ensemble average conductance of bilayer graphene can be altered with the length of the sample (L) for different values of edge disorder (d). The conductance of different types of edges, like zigzag edge, armchair, and other edges, depends on the length of the sample (L) for different values of edge disorder (d). Zigzag, armchair, and other edges can be created by cutting bilayer graphene along arbitrary crystallographic direction. In Figure 3.6, zigzag edge and armchair edge have been shown [22]. Carbon atom of zigzag edge is not chemically stable. It is prone to combine with other reactants because it has an unpaired electron. But the carbon atom of an

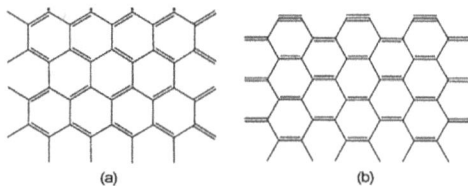

FIGURE 3.6   (a) Zigzag edge of graphene. (b) Armchair edge of graphene.

Source: Reproduced from ref. [22] with permission from American Chemical Society, copyright 2010

armchair edge is chemically stable owing to its triple covalent bond. Edge disorder has influence over transport properties of graphene nanoribbon with different types of edge character like zigzag and armchair. Zigzag edge and armchair edge show different types of standing wave pattern, and it can be probed by using scanning tunneling spectroscopy [23]. For long-range impurities scattering zigzag graphene nanoribbon shows perfect conductance. In single-mode regime, we get different transport outlines for armchair graphene nanoribbon and zigzag graphene nanoribbon [24]. The transport properties in multimode regime have been discussed [24]. Practically, it has been found that disorder has less influence over zigzag edge of BLG when it is close to an edge. This behavior is completely different when chiral edge is at zigzag boundary [25]. It is related to the effective boundary condition at the zigzag edge. In this case, zigzag edge does not mix valleys. That is why edge disorder has no influence on conductance over zigzag edge. However, edge disorder has influence on conductance over armchair edge because an armchair edge mixes valleys. In a many-mode regime, there is a small difference between the conductance of armchair and zigzag edge. Chemical disorder has randomly distributed energy +1 eV for chemical species of outermost binding of carbon atoms. The presence of chemical disorder or defect has influence on the variation in conductance with length. The conductance of bilayer graphene decreases with the sample length exponentially, and it does not depend on the width of the graphene strip. Nevertheless, there is some exception; if the strip's length is less than 5 nm, then the bulk contribution to conductance would be significant [26]. Moreover, the strong disorder has an impact on the variation in conductance of bilayer graphene with the length of the strip. In every case, the conductance varies similarly with sample length for zigzag edge, armchair, and other edges.

3.4.1.2   Impact Due to the Angle of Twist

Bilayer graphene is a material where two monolayers of graphene are stacking vertically [27]. However, it is quite difficult to fabricate perfect stacking of two monolayers. As a result, a small twist is observed between two layers. Consequently, electrical property of bilayer graphene varies with twisting angle between two layers. Several methods are used to fabricate twisted bilayer graphene. Each method has its advantages and disadvantages. But the stacking of two layer graphene prepared by chemical vapor deposition is basically followed [28]. Here, vertical stacking of two graphene

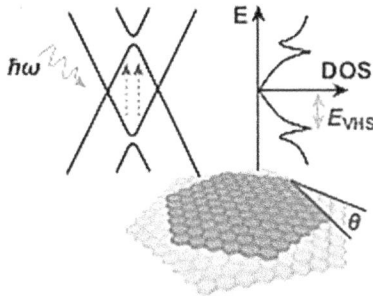

FIGURE 3.7   Schematic band structure and the equivalent density of state with van Hove singularity.

Source: Reproduced from ref.[31] with the permission from Springer Nature, copyright 2016

layers was used [27]. As a result, symmetry of bilayer graphene was broken, and this broken symmetry created inequivalent charge and electrostatic potential between two layers. But twisted bilayer graphene is not stacked bilayer graphene, wherein one layer rotates with respect to other layer. In Figure 3.7, schematic band structure and the equivalent density of state with van Hove singularity is shown. Evidently, the Dirac band dispersion changes dramatically with the twisting angle when the twisting angle between two layers is less than 5°. The Dirac cone of each layer of bilayer graphene intersects and creates a saddle point in the shared space of twisted bilayer graphene. In consequence, a van Hove singularity is created in the density of state. A small twisting angle between two layers creates a pattern on intercalated atoms and modifies its electronic properties. A twisted bilayer graphene with small twisting angle shows high mobility. Yu et al. discussed the interlayer conductance varies with twisting angle. Twisted bilayer graphene displays a high interlayer conductance when the twisting angle is small [29]. This means that high interlayer conductance can be observed at zero twisting angle. As a result, the interlayer conductance of 0° twisting angle of bilayer graphene is higher than the 30° twisting angle of twisted bilayer graphene. This indicates that the decoupling and coupling transition creates a twisting angle depended interlayer contact conductance in twisted bilayer graphene. The measurement of electrical transport for different twisted bilayer graphene devices with different twisting angles of 0.75° to 2° in room temperature has been observed by Polshyn et al. [30].

3.4.1.3   Effect of Impurity, Chemical Potential, Temperature, and Frequency

Similar to single-layer graphene, conductivity of bilayer graphene also depends on the chemical potential, frequency, temperature, and impurity level [32]. Both the parallel and perpendicular conductivities to the plane in bilayer graphene have been investigated. The parallel and perpendicular conductivity shows different variation with chemical potential, frequency, temperature, and impurity level. At zero temperature and zero frequency, the in-plane dc conductivity of bilayer graphene varies with chemical potential. For the different levels of impurity, this variation

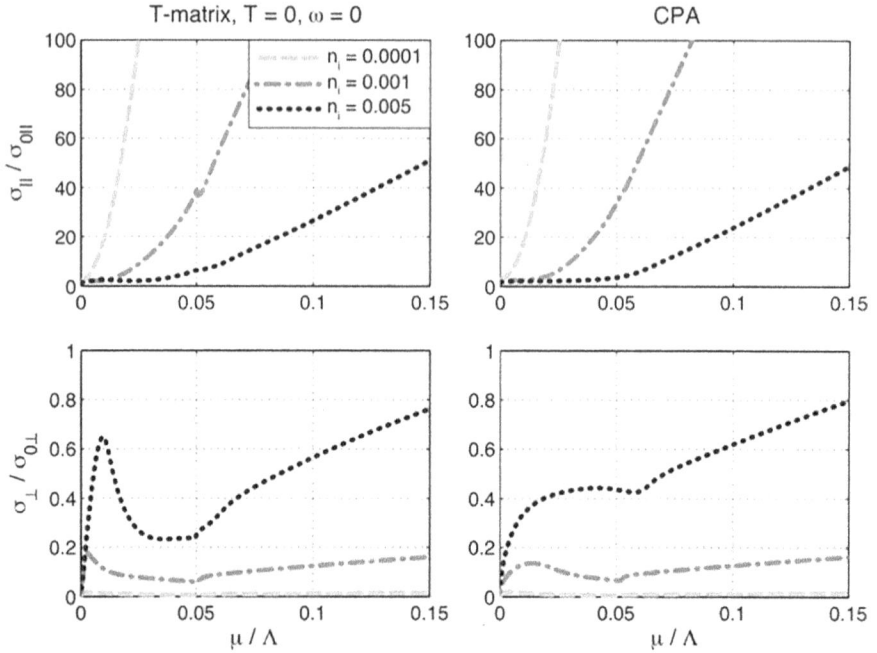

FIGURE 3.8 Variation in dc conductivity of bilayer graphene with chemical potential at T = 0 K, ω = 0.

Source: Reproduced from ref.[32] with the permission from American Physical Society, copyright 2008

would be different. Like for low impurity in bilayer graphene, the in-plane conductivity increases promptly with the increase of chemical potential. However, with the increase of impurity in bilayer graphene, this promptness will decrease gradually. In the case of perpendicular conductivity, the situation would be different. At zero temperature and zero frequency, the variation of perpendicular dc conductivity of bilayer graphene increases quickly for higher impurity. Even for higher impurity, an increase of perpendicular plane conductivity with chemical potential is greater than that of other lower impurity levels in bilayer graphene. For near zero impurity level in bilayer graphene, the perpendicular dc conductivity with chemical potential shows no variation. In Figure 3.8, the variation in dc conductivity of bilayer graphene with chemical potential has been shown at zero temperature and zero frequency. These figures have been drawn in t-matrix and coherent-potential approximation (CPA). In CPA, curves show no peak, and curves are smoother than the curve in t-matrix. At zero temperature and zero chemical potential, the variation in in-plane conductivity of bilayer graphene with frequency gets slight variation, and then suddenly it reaches at pick point for particular value of frequency. However, this pick point decreases with increase of impurity. For further increase of frequency, the value of in-plane conductivity would decrease with frequency and then shows flat line. In case of perpendicular conductivity, it increases irregularly with frequency, and suddenly it

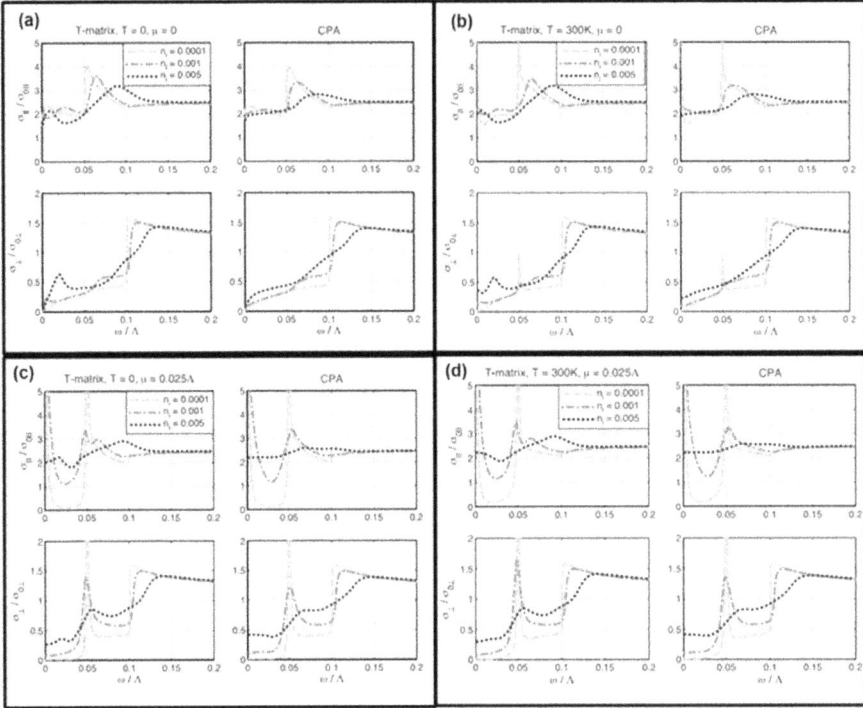

FIGURE 3.9 (a) Variation in conductivity of bilayer graphene with frequency at $\mu = 0$ Λ, T = 0 K. (b) Variation in conductivity of bilayer graphene with frequency at T = 300 K, $\mu = 0$ Λ. (c) Variation in conductivity of bilayer graphene with frequency at T = 0 K, $\mu = 0.025$ Λ. (d) Variation in conductivity bilayer graphene with frequency at $\mu = 0.025$ Λ and T = 300 K.

Source: Reproduced from ref.[32] with the permission from American Physical Society, 2008

reaches at a maximum value. After reaching its maximum value, it reaches saturation point. In Figure 3.9 (a), variation in conductivity of bilayer graphene with frequency at $\mu = 0$ Λ, T = 0 K is shown.

At 300 K temperature and zero chemical potential, the variation of in-plane conductivity and perpendicular conductivity shows the same variations as variation at zero temperature. But for low impurity levels, the parallel conductivity shows a pick point. In Figure 3.9 (b), the variation in conductivity of bilayer graphene with frequency at T = 300 K and $\mu = 0$ Λ is shown. At chemical potential $\mu = 0.025$ Λ and temperature T = 0 K, the in-plane conductivity of bilayer graphene dramatically varies with frequency. This variation decreases with increase of impurity levels in graphene. However, the value of in-plane conductivity reaches saturation level after a certain value of frequency. For a different impurity level, the variation would be different, but after a certain point, the in-plane conductivity reaches saturation point. In the case of perpendicular conductivity, the situation would be different. In Figure 3.9 (c) the variation in conductivity of bilayer graphene with frequency at $\mu = 0.025$ Λ

and temperature T = 0 K is shown. At μ = 0.025 Λ and temperature T = 0 K, perpendicular conductivity varies with frequency in an irregular manner. But after a certain value of frequency, the conductivity reaches saturation point. At T = 300 K and chemical potential μ = 0.025 Λ, both in-plane conductivity and perpendicular conductivity shows irregular variation with frequency for different impurity levels. In Figure 3.9 (d) the variation in conductivity of bilayer graphene with frequency at T = 300 K and μ = 0.025 Λ is shown. At μ = 0 and ω = 0, the variation in both direction varies with temperature for different levels of impurity. Only for a low level of impurity does the in-plane conductivity show variation. For a high value of impurity level, the in-plane dc conductivity maintains constant value with the variation of temperature. For higher impurity level, the perpendicular dc conductivity shows variation, and for lower impurity level, perpendicular conductivity maintains constant value with temperature. In Figure 3.10 (a) the variation in dc conductivity of bilayer graphene with temperature at μ = 0 and ω = 0 is shown. At μ = 0.025 Λ, ω = 0, both in-plane and perpendicular dc conductivity maintain nearly the same value with the increase of temperature. In Figure 3.10 (b) the variation in dc conductivity of bilayer graphene with temperature μ = 0.025 Λ, ω = 0 is shown.

3.4.2 MULTILAYER GRAPHENE

Multilayer graphene is a pile of monolayer graphene layers where more than two layers are stacked vertically. Monolayer fabricates due to the same reason bilayer graphene is used instead of monolayer graphene. Electrical properties of multilayer graphene also vary with different external parameters, like monolayer and bilayer graphene. The electrical property of multilayer graphene depends on twisting angle, chemical potential, impurities, temperature, etc.

3.4.2.1 Effect of Temperature and Twisting Angle of Multilayer Graphene

Some fascinating properties in twisted multilayer graphene (tMLG) originated from angular twists between the layers. Mogera et al. [33] described that the conductivity of the twisted multilayer graphene within the temperature range of 90 K to 273 K shows two different characteristics. The conductivity of each layer in MLG slowly increases with the increase of temperature and reaches a minimum value nearly at 180 K temperature and then linearly decreases up to the temperature 300 K. Two different natures (semiconducting and metallic) of MLG at different temperatures can be seen. In Figure 3.11, it shows the variation in conductivity of tMLG with temperature.

3.4.2.2 Impact of Chemical Potential, Frequency, Temperature, and Impurity Level

The conductivity of multilayer graphene depends on the chemical potential, frequency, temperature, and different impurity levels in multilayer graphene [34]. Individually, each parameter has influence over multilayer graphene, and collectively, all these parameters influence the electrical properties of graphene. For multilayer graphene, conductivity is being considered in two different directions,

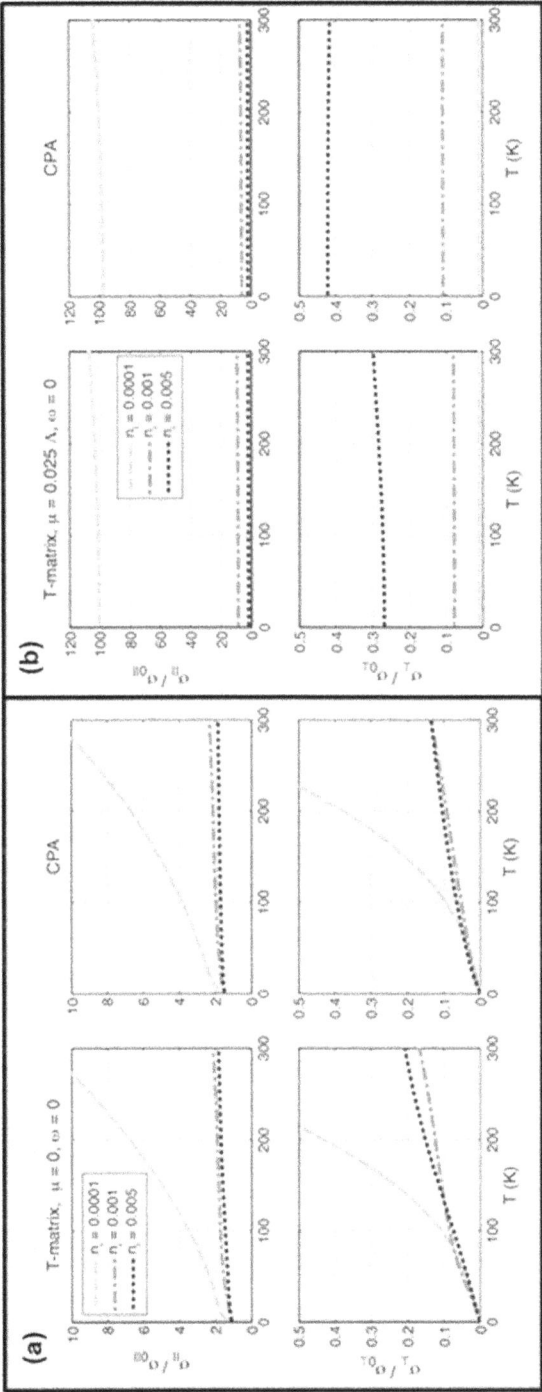

FIGURE 3.10 (a) Variation in dc conductivity of bilayer with temperature at $\mu = 0$ and $\omega = 0$. (b) Variation in dc conductivity of bilayer graphene with temperature, $\mu = 0.025$ Λ, $\omega = 0$.

Source: Reproduced from ref.[32] with the permission from American Physical Society, copyright 2008

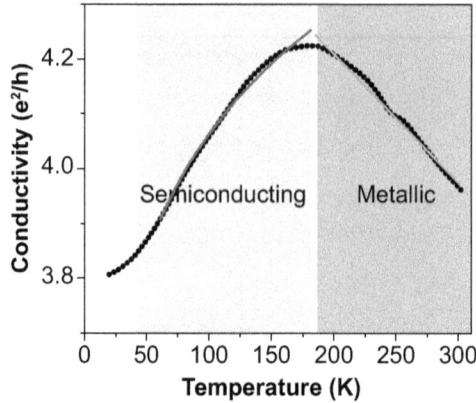

FIGURE 3.11   Variation in conductivity with temperature of twisted multilayer graphene.

Source: Reproduced from ref.[33] American Chemical Society, copyright 2017

like in-plane conductivity and perpendicular conductivity. Impacts of temperature, chemical potential, frequency, and impurity collectively on in-plane conductivity and the perpendicular conductivity are completely different. Because of that reason, in-plane conductivity and perpendicular conductivity of multilayer graphene show different variation [32]. In Figure 3.12 (a), it shows that at zero temperature and zero frequency, the in-plane conductivity of multilayer graphene varies with chemical potential. These figures have been drawn in t-matrix and coherent-potential approximation (CPA). For the different levels of impurity, this variation would be different. For instance, for low-level impurity in multilayer graphene, the in-plane dc conductivity increases promptly with the increase of chemical potential. However, with the increase of impurity in multilayer graphene, this promptness will decrease gradually. In the case of perpendicular dc conductivity, the situation is quite similar to the variation for in-plane dc conductivity with chemical potential. At zero temperature and zero frequency, the variation in perpendicular dc conductivity of multilayer graphene increases quickly for low impurity level, and this promptness decreases with the increase of impurity.

At zero temperature and zero chemical potential, the variation of in-plane conductivity of multilayer graphene with frequency gets variation. For higher-level impurity, the variation of in-plane conductivity is slightly greater than that of lower-level impurity. For lower impurity levels, the variation of in-plane conductivity is very little. After a certain value of frequency, the in-plane conductivity reaches saturation level for every level of impurity.

The variation in perpendicular conductivity of multilayer graphene with frequency is quite different than that of in-plane conductivity of multilayer graphene. The perpendicular conductivity continuously decreases with an increase in frequency and reaches a minimum value, then again it increases with frequency. After a certain value of frequency, it reaches a saturation level. For a higher level of impurity, the minimum value of multilayer graphene is slightly higher than that of the lower level

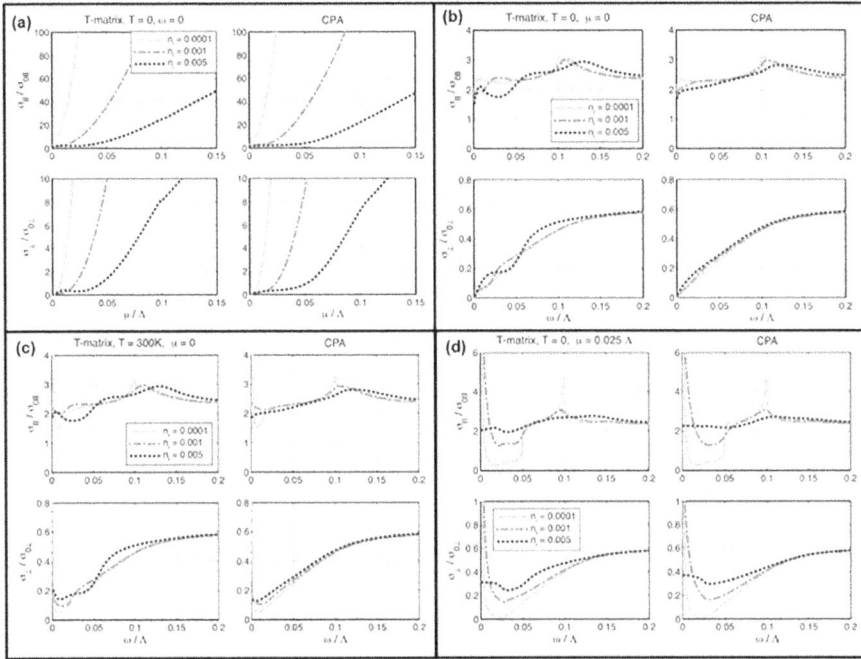

FIGURE 3.12   (a) Variation in dc conductivity of multilayer graphene with chemical potential, at T = 0 K, ω = 0 with frequency. (b) Variation in conductivity with frequency of multilayer graphene, at T = 0 K, μ = 0Λ. (c) Variation in conductivity with frequency of multilayer graphene, at T = 300 K, μ = 0Λ. (d) Variation in conductivity with frequency of multilayer graphene, at T = 300 K, μ = 0.025Λ.

Source: Reproduced from ref.[32] with the permission from American Physical Society, 2008

of impurity. In Figure 3.12 (b), the variation in conductance with frequency at μ = 0 and T = 0 K is shown.

At 300 K temperature and zero chemical potential, the in-plane conductivity of multilayer graphene variation with chemical potential is low. In the beginning, this value suddenly decreases, then it maintains low variation. At the same condition, the perpendicular plane conductivity of multilayer graphene suddenly decreases with frequency. After reaching a minimum value, it increases with frequency, and then the value of perpendicular conductivity reaches a saturation point. In Figure 3.12 (c), the variation in conductivity with frequency at T = 300 K and μ = 0 Λ is shown. At T = 0 K and μ = 0.025 Λ, the in-plane conductivity of multilayer graphene varies irregularly with frequency for different impurity levels. At the same condition, the perpendicular conductivity of multilayer graphene decreases with frequency, and after a certain value of frequency, the value of perpendicular conductivity of multilayer graphene increases, finally reaching saturation points. In Figure 3.12 (d), the variation in conductivity of graphene with frequency at T = 0 K and μ = 0.025 Λ is shown. At T = 300 K and μ = 0.025 Λ, the in-plane conductivity of multilayer

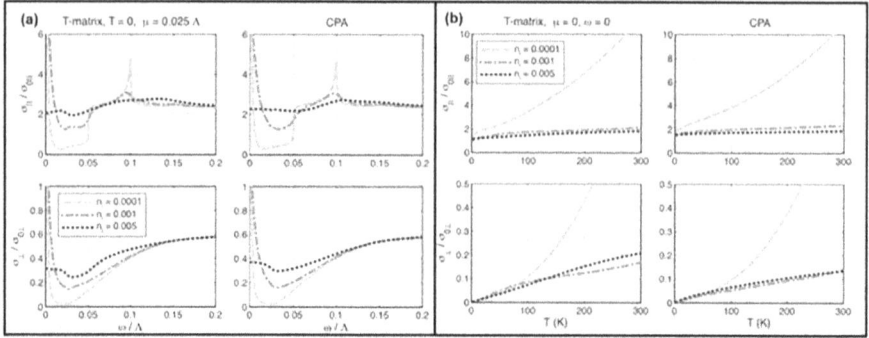

FIGURE 3.13 (a) Variation in dc conductivity of multilayer graphene with frequency at T = 300 K, μ = 0.025 Λ. (b) Variation in dc conductivity of multilayer graphene with temperature at μ = 0 Λ, ω = 0.

Source: Reproduced from ref.[32] with the permission from American Physical Society, 2008

graphene decreases frequency. After reaching the minimum value of conductivity, it will increase with frequency. Then it reaches a maximum value. With a further increase of frequency, the conductivity decreases again and reaches saturation level. The maximum value and minimum value of in-plane conductivity of multilayer graphene for a low level of impurity is greater than that of a higher level of impurity. At the same condition, the perpendicular conductivity of multilayer graphene decreases with frequency. After reaching the minimum value, it will increase, but after a certain value of frequency, it reaches a saturation point. In Figure 3.13 (a) the variation in conduction of multilayer graphene with frequency at T = 300 K and μ = 0.025 Λ is shown. At zero chemical potential and zero frequency, the in-plane dc conductivity for the lower level of impurity will increase with temperature. For a higher level of impurity, the temperature has a low impact on in-plane conductivity. In Figure 3.13 (b) the variation in conductance of multilayer graphene with temperature at μ = 0 Λ, ω = 0 is shown. However, the perpendicular dc conductivity of multilayer conductivity increases with temperature. From higher impurity level to lower impurity level, the temperature influence on outer plane conductivity increases. At μ = 0.025 Λ and zero frequency, the in-plane dc conductivity of multilayer conductivity shows weak temperature-dependent for different values of impurity level. For the lower level of impurity, the perpendicular conductivity shows weak temperature-dependent. But for higher impurity levels, it shows no variation. In Figure 3.14, the variation in dc conductivity with temperature at μ = 0.025 Λ, ω = 0 has been shown.

3.4.2.3   Variation in the Mobility with Layers Number

Graphene is described as an electron gas system having mobility near about 10,000 $cm^2v^{-1}s^{-1}$ [35] at room temperature. The electrical property of graphene on SiO_2 substrate varies from 2,000 to 20,000 $cm^2v^{-1}s^{-1}$ [36, 37]. Mobility in graphene depends on the layer numbers [38]. In Figure 3.15 (a) and 3.15 (b), the variation in conductivity

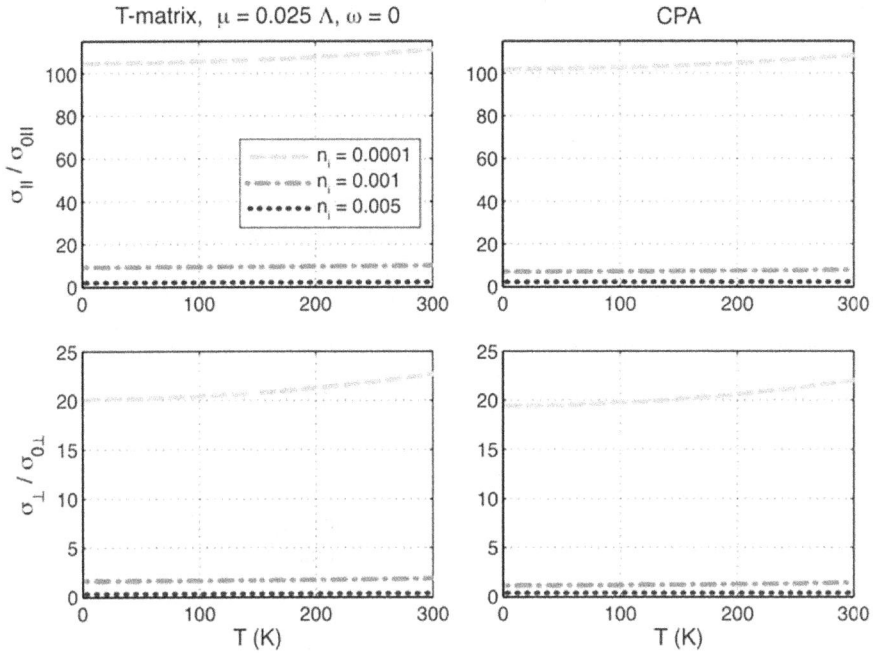

T-matrix, $\mu = 0.025\ \Lambda$, $\omega = 0$ **CPA**

$n_i = 0.0001$
$n_i = 0.001$
$n_i = 0.005$

FIGURE 3.14 Variation in dc conductivity of multilayer with temperature at $\mu = 0.025\ \Lambda$, $\omega = 0$.

Source: Reproduced from ref.[32] with the permission from American Physical Society, 2008

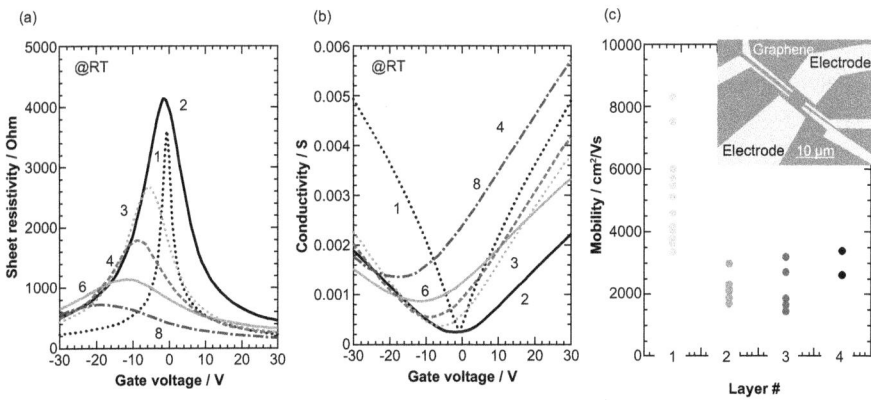

FIGURE 3.15 Variation in (a) conductivity and (b) resistivity of different layer graphene with gate voltage. (c) Variation in mobility of charge carrier in graphene with layer numbers.

Source: Reproduced from ref.[38] with permission from the Japan Society of Applied Physics, copyright 2009

and sheet resistance with gate voltage is shown. Here, the four-probe measurement technique was used (because graphene was not decayed by the deposition of metal electrode) [38]. The resistivity of graphene on SiO_2 substrate augments with the decrease in layer number. For monolayer, resistivity changes in a drastic manner. In Figure 3.15 (c), the variation in the mobility of charge carrier in graphene with layer number is shown. The carrier mobility in graphene varies with layer numbers. According to this figure, carrier mobility for monolayer graphene is greater than other layer numbers. But the transport of electrons in monolayer graphene is not ballistic. Electron conductivity of monolayer graphene is decreased by charge impurity scattering [39]. Reduction of interlayer scattering may enhance the current modulation with decreasing layer number. With the decrease of layer number from bilayer to monolayer, the mobility dramatically increases, owing to the variation from the quadratic relationship to linear dispersion relationship [38].

3.5 CONCLUSION

To be able to tune the graphene's electrical conductivity according to need is an astounding progress in the field of research. In this chapter, we have discussed graphene's structure. Theoretically, ideal graphene shows extraordinarily high in-plane electrical conductivity. The cause for such high conductivity has been discussed in brief. The electrical conductivity of graphene can be tuned in a number of different ways. The effect of different parameters, like temperature, pressure, external electric field, doping, defect, etc., on the electrical properties of graphene has been discussed. Ideal graphene is a zero bandgap material. However, through proper treatments, bandgap can be generated in graphene. This generated bandgap can also be tuned according to need. The process of generating bandgap has also been discussed. Finally, the electrical properties of bilayer graphene have been discussed. Proper engineering of this wonder material could cause a revolution in the field of electronics.

REFERENCES

1. K. S. Novoselov, A. K. Geim, and S. V. Morozov, "Electric field in atomically thin carbon films," *Science*, 306 (2004) 666–669. https://doi.org/10.1126/science.1102896
2. A. H. Castro Neto, F. Guinea, N. M. R. Peres, K. S. Novoselov, and A. K. Geim, "The electronic properties of graphene," *Rev. Mod. Phys.*, 81 (2009) 109–162. https://doi.org/10.1103/RevModPhys.81.109
3. B. Guo, L. Fang, B. Zhang, and J. Ru Gong, "Graphene doping: A review," *Insciences J.*, 1 (2011) 80–89. http://doi.org/10.5640/insc.010280
4. P. R. Wallace, "The band theory of graphite," *Phys. Rev.*, 71 (1947) 622–634. https://doi.org/10.1103/PhysRev.71.622
5. D. R. Cooper, B. D'Anjou, N. Ghattamaneni, B. Harack, M. Hilke, A. Horth, N. Majlis, M. Massicotte, L. Vandsburger, E. Whiteway, and V. Yu, "Experimental review of graphene," *ISRN Condens. Matter Phys.*, 2012 (2012) 56. https://doi.org/10.5402/2012/501686
6. S. Das Sarma, S. Adam, E. H. Hwang, and E. Rossi, "Electronic transport in two-dimensional graphene," *Rev. Mod. Phys.*, 83 (2011) 407. http://doi.org/10.1103/RevModPhys.83.407
7. M. I. Katsnelson, K. S. Novoselov, and A. K. Geim, "Chiral tunnelling and the Klein paradox in graphene," *Nat. Phys.*, 2 (2006) 620–625. http://doi.org/10.1038/nphys384

8. J. Schiefele, F. Sols, and F. Guinea, "Temperature dependence of the conductivity of graphene on boron nitride," *Phys. Rev.*, 85 (2012) 195420. http://doi.org/10.1103/PhysRevB.85.195420

9. C. H. Ahn, A. Bhattacharya, M. Di Ventra, J. N. Eckstein, C. Daniel Frisbie, M. E. Gershenson, A. M. Goldman, I. H. Inoue, J. Mannhart, A. J. Millis, A. F. Morpurgo, D. Natelson, and J.-M. Triscone, "Electrostatic modification of novel materials," *Rev. Mod. Phys.*, 78 (2006) 1185. http://doi.org/10.1103/RevModPhys.78.1185

10. E. Dorgan, M. Bae, and E. Pop, "Mobility and saturation velocity in graphene on SiO_2," *Appl. Phys. Lett.*, 97 (2010) 082112. http://doi.org/10.1063/1.3483130

11. K. S. Novoselov, V. I. Falko, L. Colombo, P. R. Gellert, M. G. Schwab, and K. Kim, "A roadmap for grapheme," *Nature*, 490 (2012) 192–200. http://doi.org/10.1038/nature11458

12. P. Ares, M. Pisarra, P. Segovia, C. Díaz, F. Martín, E. G. Michel, F. Zamora, C. Gómez-Navarro, and J. G. Herrero, "Tunable graphene electronics with local ultrahigh pressure," *Adv. Funct. Mater.*, 29 (2019) 1806715. https://doi.org/10.1002/adfm.201806715

13. L. Ci, Li Song, C. Jin, D. Jariwala, D. Wu, Y. Li, A. L. Srivastava, Z. F. Wang, K. Storr, L. Balicas, F. Liu, and P. M. Ajayan, "Atomic layers of hybridized boron nitride and graphene domains," *Nat. Mater.*, 9 (2010) 430–435. http://doi.org/10.1038/nmat2711

14. J. Kotakoski, A. V. Krasheninnikov, Y. Ma, A. S. Foster, K. Nordlund, and R. M. Nieminen, "B and N ion implantation into carbon nanotubes: Insight from atomistic simulations," *Phys. Rev. B*, 71 (2005) 205408. https://doi.org/10.1103/PhysRevB.71.205408

15. T. B. Martins, R. H. Miwa, A. Silva, and A. Fazzi, "Electronic and transport properties of boron-doped graphene nanoribbons," *Phys. Rev. Lett.*, 98 (2007) 196803. https://doi.org/10.1103/PhysRevLett.98.196803

16. P. A. Denis, "Band gap opening of monolayer and bilayer graphene doped with aluminium, silicon, phosphorus, and sulfur," *Phys. Lett.*, 492 (2010) 251–257. https://doi.org/10.1016/j.cplett.2010.04.038

17. O. Aktürk, and M. Tomak, "Bismuth doping of grapheme," *Appl. Phys. Lett.*, 96 (2010) 081914. https://doi.org/10.1063/1.3334723

18. S. Yu, W. Zheng, C. Wang, and Q. Jiang, "Nitrogen/boron doping position dependence of the electronic properties of a triangular graphene," *ACS Nano*, 4 (2010) 7619–7629. https://doi.org/10.1021/nn102369r

19. H. Tanaka, R. Arima, M. Fukumori, D. Tanaka, R. Negishi, Y. Kobayashi, S. Kasai, T. K. Yamada, and T. Ogawa, "Method for controlling electrical properties of single-layer graphene nanoribbons via adsorbed planar molecular nanoparticles," *Sci. Rep.*, 5 (2015) 12341. http://doi.org/10.1038/srep12341

20. T. Shimizu, J. Haruyama, D. C. Marcano, D. V. Kosinkin, J. M. Tour, K. Hirose, and K. Suenaga, "Large intrinsic energy bandgaps in annealed nanotube-derived graphene nanoribbons," *Nat. Nanotechnol.*, 6 (2011) 45–50. http://doi.org/10.1038/nnano.2010.249

21. F. Banhart, J. Kotakoski, and A. V. Krasheninnikov, "Structural defects in graphene," *Phys. Rev.*, 5 (2011) 26–41. http://doi.org/10.1021/nn102598m

22. K. Xu, and P. D. Ye, "Theoretical study of atomic layer deposition reaction mechanism and kinetics for aluminum oxide formation at graphene nanoribbon open edges," *J. Phys. Chem.*, 114 (2010) 10505–10511.

23. L. G. Cancado, M. A. Pimenta, B. R. A. Neves, M. S. S. Dantas, and A. Jorio, "Influence of the atomic structure on the raman spectra of graphite edges," *Phys. Rev. Lett.*, 93 (2004) 247401. https://doi.org/10.1103/PhysRevLett.93.247401

24. A. Orlof, J. Ruseckas, and I. V. Zozoulenko, "Effect of zigzag and armchair edges on the electronic transport in single-layer and bilayer graphene nanoribbons with defects," *Phys. Rev. B*, 88 (2013) 125409. http://doi.org/10.1103/PhysRevB.88.125409

25. K. Wakabayashi, M. Fujita, H. Ajiki, and M. Sigrist, "Electronic and magnetic properties of nanographite ribbons," *Phys. Rev.* 59 (1999) 8271. https://doi.org/10.1103/PhysRevB.59.8271

26. J. Li, I. Martin, M. Buttiker, and A. F. Morpurgo, "Topological origin of subgap conductance in insulating bilayer graphene," *Nat. Phys.*, 7 (2011) 38. http://doi.org/10.1038/nphys1822

27. A. Nimbalkar, and H. Kim, "Opportunities and challenges in twisted bilayer graphene," *Nano-Micro Lett.*, 12 (2020) 126. https://doi.org/10.1007/s40820-020-00464-8

28. J. T. Robinson, S. W. Schmucker, C. B. Diaconescu, J. P. Long, J. C. Culbertson, T. Ohta, A. L. Friedman, and T. E. Beechem, "Electronic hybridization of large-area stacked graphene films," *ACS Nano*, 7 (2013) 637–644. http://doi.org/10.1021/nn304834p

29. Z. Yu, A. Song, L. Sun, Y. Li, L. Gao, H. Peng, T. Ma, Z. Liu, and J. Luo, "Understanding interlayer contact conductance in twisted bilayer graphene," *Small* (2019) 1902844. https://doi.org/10.1002/smll.201902844

30. H. Polshyn, M. Yankowitz, S. W. Chen, Y. X. Zhang, K. Watanabe, T. Taniguchi, C. R. Dean, and A. F. Young, "Large linearin-temperature resistivity in twisted bilayer graphene," *Nat. Phys.*, 15 (2019) 1011–1016. https://doi.org/10.1038/s41567-019-0596-3

31. J. B. Yin, H. Wang, H. Peng, Z. J. Tan, and L. Liao, "Selectively enhanced photocurrent generation in twisted bilayer graphene with van Hove singularity," *Nat. Commun.*, 7 (2016) 10699. https://doi.org/10.1038/ncomms10699

32. J. Nilsson, A. H. Castro Neto, F. Guinea, and N. M. R. Peres, "Electronic properties of bilayer and multilayer graphene," *Physical Review B*, 78 (2008) 045405. https://doi.org/10.1103/PhysRevB.78.045405

33. U. Mogera, S. Walia, B. Bannur, M. Gedda, and G. U. Kulkarni, "Intrinsic nature of graphene revealed in temperature-dependent transport of twisted multilayer grapheme," *J. Phys. Chem. C*, 121 (2017) 13938–13943. https://doi.org/10.1021/acs. jpcc.7b04068

34. A. K. Geim, and K. S. Novoselov, "The rise of graphene," *Nat. Mater.*, 6 (2007) 183–191. http://doi.org/10.1038/nmat1849

35. Y. W. Tan, Y. Zhang, K. Bolotin, Y. Zhao, S. Adam, E. H. Hwang, S. Das Sarma, H. L. Stormer, and P. Kim, "Measurement of scattering rate and minimum conductivity in graphene," *Phys. Rev. Lett.*, 99 (2007) 246803. https://doi.org/10.1103/PhysRevLett.99.246803

36. C. Jozsa, M. Popinciuc, N. Tombros, H. T. Jonkman, and B. J. van Wees, "Electronic spin drift in graphene field-effect transistors," *Phys. Rev. Lett.*, 100 (2008) 236603. https://doi.org/10.1103/PhysRevLett.100.236603

37. K. Nagashio, T. Nishimura, K. Kita, and A. Toriumi, "Mobility variations in mono-and multi-layer graphene films," *Appl. Phys. Express*, 2 (2009) 025003 http://doi.org/10.1143/apex.2.025003

38. K. Nomura, and A. H. MacDonald, "Quantum hall ferromagnetism in graphene," *Phys. Rev. Lett.*, 96 (2006) 256602. https://doi.org/10.1103/PhysRevLett.96.256602

39. S. Adam, E. H. Hwang, V. M. Galitski, and S. Das Sarma, "A self-consistent theory for graphene transport," *PNAS*, 104 (2007) 18392. https://doi.org/10.1073/pnas.0704772104

4 Electrical Transport in Post-Graphene 2D Materials

Ghulam Dastgeer, Sobia Nisar, and Jonghwa Eom

CONTENTS

4.1 INTRODUCTION

Material science and condensed matter physics have disclosed the details about the astonishing applications of graphene as it was cleaved from the graphite chunk by a scotch tape. Graphene is considered an important catalyst in the field of two-dimensional (2D) materials having exceptional electronic features. Due to its extraordinary electronic spectrum, graphene has prompted the rise of another world-view of relativistic condensed matter physical science. Few of them are unobtrusive in high-energy physical science. In general, graphene is only a single atom thick, which introduces a new category of materials based on two-dimensional physics that provide a large variety of applications.[1,2] The study of 2D sheets of graphene has been ongoing for more than a half decade. However, in all those years, researchers were convinced that 2D crystals couldn't be realized due to their thermodynamic instability. The Peierls and Landau theory, which was further enhanced by Mermin, describes that in a low dimensional crystal lattice, due to the thermal fluctuations at

DOI: 10.1201/9781003247890-4

39

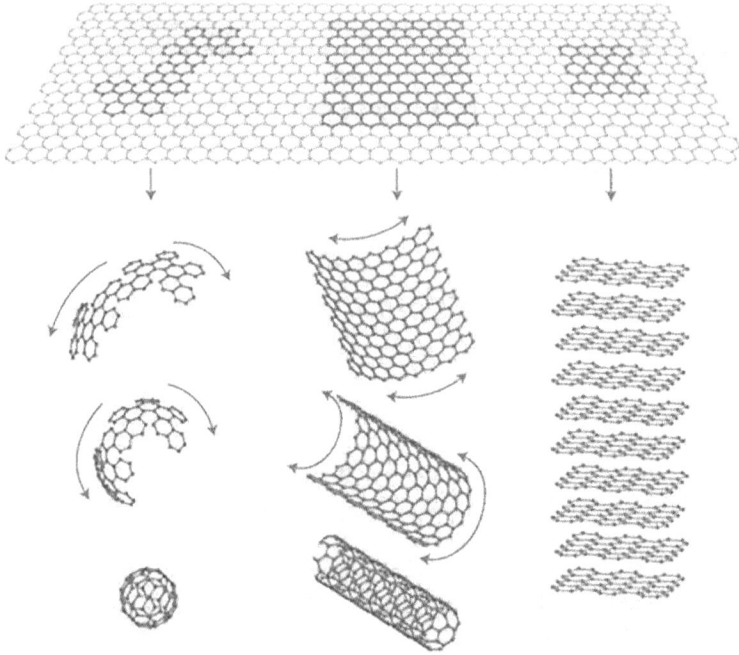

FIGURE 4.1   All dimensions of graphite materials are derived from the single atomic layer thick graphene. The rounded graphene is shaped to 0D buckyballs. The 1D nanorods/nanotubes are formed by rolling the 2D graphene sheets while stacking several layers over one another to form 3D bulk graphite.

Source: (ref.[2]) Copyright 2007, Nature Publishing Group

any finite temperature, the atom displacement became comparable to interatomic distance.[3] Moreover, as the thickness declines, the melting temperature of the thin film quickly reduces, which makes the film unstable.[4,5] However, in 2004, graphene was discovered by tearing the graphite pieces over and over again and managed to generate solitary thick atom flakes. Now, we know the ripples in graphene play an important role in the realization of truly 2D materials. Graphene is the wonderful masterpiece of carbon, having a flat monolayer of carbon atoms strongly packed in the 2D honeycomb lattice and is the fundamental building block of graphitic material. It can be enfolded into 0D, rolled into 1D nanotubes, or piled up into 3D graphite.[6] In general, all the dimensions of a graphite are based on the single atomic layer of graphene, as shown in Figure 4.1.

Yet before exploring the previous work on 2D materials, it is valuable to characterize what 2D crystals are. The 2D crystals are based on a single atomic plane, in which atoms are attached with each other by covalent bonds. In electronics, the number of layers extensively made an impact on the electronic structure. Approximately, hundred layers are favorable to make the thin film of 3D material. However, in the case of graphene (2D crystals) composed of one or two atomic layers, electronic spectra are quite simple. Graphene is narrated as zero energy bandgap or zero-overlap semimetals. For layers three or more, the spectra become complex because the conduction and valance band begin to overlap. Also, various charge carriers become visible. Therefore, graphene one, two, and few (three to less than ten) layers discriminate three diverse types of 2D crystals.[7–9] Toward the electronic applications, the most practical and doable way is the epitaxial development of graphene and, with so much in question, fast advancement toward this path is anticipated.

Graphene-supported electronic applications grab the attention of various organizations. Intel and IBM centered their research in this direction to keep an eye on possible development. Graphene's golden era is not unanticipated because now Si-based technology is outdated. Therefore, there must be any new material that can take over the position of Si, and for that, graphene seems to have a potential option. The most instant and pressing application of graphene is found in the utilization of composite material. In electric batteries, the use of graphene powder is another exciting likelihood for the graphite market. It will help in upgrading the effectiveness of batteries. Carbon nanotubes can also be used for this application, but due to its low price of production and being economical, graphene powder has an advantage over the carbon nanotube.[10] Moreover, graphene powder can offer even more promising emitting properties, and thin graphite flakes were utilized in the commercial prototype (plasma display). Graphene also provides outstanding material for solid-state gas sensors.[11] In addition, hysteretic magnetoresistance,[12] and substantial bipolar supercurrents[13] are considered as major applications of graphene.

The main reason for putting all theoretical and experimental to graphene is due to its unique nature of charge carriers and excellent incomparable electronic traits in isolated form. Schrodinger equation to explain electronic properties of the material is favorable in condensed matter physics. On the other hand, graphene is different, which elucidates with the Dirac equation rather than Schrodinger equation because their charge carrier resembles the relativistic particles.[14,15] The researcher, from experience, knows that new exciting dimensions of physics will be revealed after high-quality samples. That's why the entire focus of the research community is on the graphene. After a series of experiments, it is revealed that the zero-bandgap of the graphene limits its potential for the logic devices, switching applications and photovoltaic devices. This problem has led the researcher to investigate some new class of flexible and bandgap 2D semiconductor materials that could be cleaved in thin sheets like graphene.

4.1.1 Transition Metal Dichalcogenides (TMDs)

The new class of semiconductor 2D materials was the transition metal dichalcogenides (TMDs), which are also atomically thin in nature like graphene. Generally, the TMDs are denoted by a formula of MX_2. Where "M" denotes the transition metals like (Mo, W, etc.) and "X" denotes the chalcogen atoms as (S, Se, or Te). In their basic structure, one layer of "M" atoms is surrounded by two layers of "X" atoms. Some TMDs have distinctive structural phases due to transition metallic atoms having dissimilar coordination spheres. Usually, the 2H phase is hexagonal, 1T phase is octahedral, and 1T′ phase is known as trigonal prismatic phase. These most regular depicted structural stages are shown in Figure 4.2.

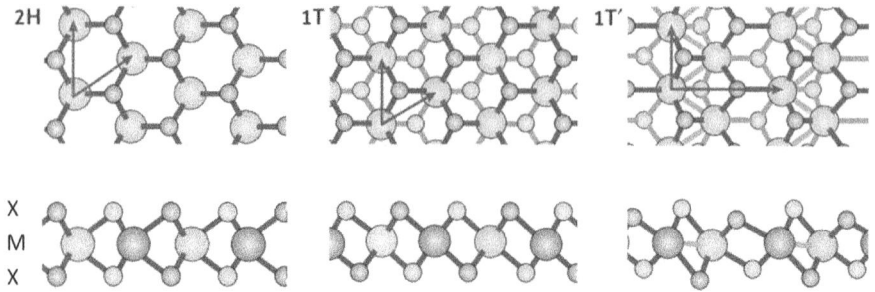

FIGURE 4.2   The general crystal structure of the single-layer TMDs 2D materials with hexagonal (2H), octahedral (1T), and dimerized (1T′) phases. The in-plane intra-atomic bonding of transition metal and chalcogen atoms is presented for each 2H, 1T, and 1T′ phases.

Source: (ref.[16]) Copyright 2017, Nature Publishing Group

The stacking order of these three atoms (X-M-X) defines the crystal structure and its phases because the 2H and 1T phase are considered thermodynamically more stable as compared to 1T′ metastable phase.[16] The 2D TMDC semiconductor that gains inclusive focus among all TMDs semiconductor materials is MoS_2. It appears as a new class material having astonishing properties.[17] MoS_2 is one of the most stable 2D TMDC material with a high electron mobility. Each molybdenum atom (Mo) is surrounded by two sulfur atoms (S), and its crystals exist either in 2H phase or 1T phase, as show in Figure 4.3 (a–b). The 2H phase of MoS_2 is an excellent and stable n-type semiconductor owing to a large electron density, while the 1T phase is octahedral and shows metallic behavior. MoS_2 is a non-centrosymmetric material having a direct energy bandgap of 1.8 eV in its monolayer, while its indirect bandgap is 1.2 eV in case of multilayer MoS_2.[18–20] The structure of MoS_2 is the trigonal prismatic configuration of Mo atoms and hexagonal planes of S. Furthermore, the Raman spectra of MoS_2 flakes of different layers are illustrated in Figure 4.3 (c), showing in-plane and out-of-plane modes of vibrations. Strong bands at ~386 and ~404 cm^{-1} are the reason for the ascription of E_{2g}^1 (in-plane) and A_g^1 (out-of-plane) SL vibrational methods. SL to ML expansion in the number of layers causes the movement of E_{2g}^1 band toward the lesser wave numbers. In contrast, the band that moves to a higher wave number is A_g^1. At ~384 cm^{-1}, the E_{2g}^1 and at ~406 cm^{-1}, the A_g^1, BL bands are seen. The ML peaks E_{2g}^1 and A_g^1 are noticed at ~384 and ~408 cm^{-1}, respectively.

The electric transport in few-layer MoS_2 device with metal electrodes is presented in Figure 4.3 (d–f). The transfer curve is showing a clear threshold current as the top gate is swept from negative to positive gate voltages, while the inset image is output characteristics of the monolayer MoS_2 FET with gold electrodes. The linear trend of the I_{ds}-V_{ds} lines are demonstrating the Ohmic behavior at various gate voltages.[23] The electron mobility, which is defined as "how quickly an electron can move in a semiconductor or metal" under the applied electric field and is mathematically formulated as follows:

$$\mu_{FE} = \frac{L}{W}\left(\frac{dI_{ds}}{dV_{bg}}\right)\frac{1}{C_g V_{ds}} \tag{1.1}$$

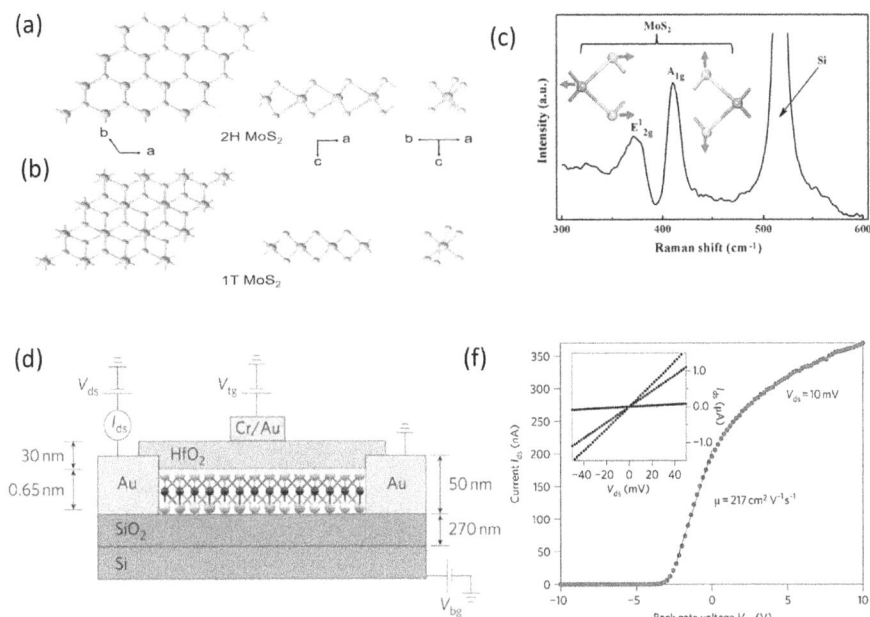

FIGURE 4.3 (A–B) The schematic illustration is showing the crystal structure of 2H and 1T phase of MoS_2 crystals, respectively. The Mo atoms are presented with cyan color, while the S atoms are presented in yellow color. (c) The two Raman peaks of MoS_2 are showing the in-plane and out-of-plane modes of vibrations, over the Si/SiO_2 substrate. (d) The schematic image of the HfO_2 covered MoS_2 device with gold metal electrodes. (e–f) The back gate dependent transfer curve of monolayer MoS_2 sheet and inset image is showing the gate dependent output (I_{ds}-V_{ds}) characteristics.

Source: (a–b) (Ref[21]) copyright 2020, Inorganic Chemistry Communications; (c) (Ref.[22]) copyright 2015, *Journal of Alloys and Compounds*; (e–f) (Ref[23]) Copyright 2011, Nature Publishing Group

Where "μ_{FE}" is field effect mobility, "L" is the channel length, "W" is the channel width, "C_g" is the gating material/dielectric capacitance, and "V_{ds}" is source-drain voltage and "$\dfrac{dI_{ds}}{dV_{bg}}$" is the slope of linear fit in the accumulation region. The linear trend of the output characteristics at various gates is showing the pure ohmic behavior. This ohmic behavior is attributed to the work function of the Ni-metal and the capping of the MoS_2 sheet with tunnel barrier h-BN, which reduces the Schottky barrier height.[24,25]

4.1.2 BLACK PHOSPHORUS (BP)

Black phosphorus (BP) is another attractive 2D semiconductor material with a small bandgap. It consists of a single layer of phosphorus atoms and considered a crystalline material. The atoms are contented with each other to form an armchair structure,[26] as shown in Figure 4.4 (a). The high-resolution transmission electron microscopic

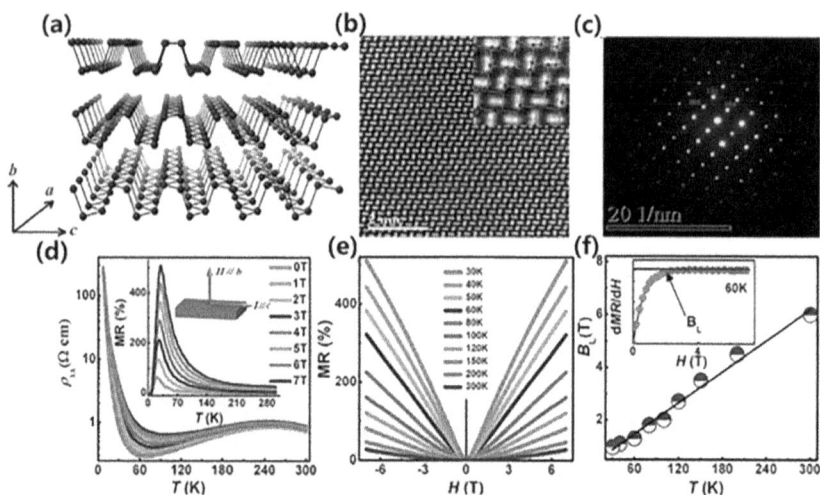

FIGURE 4.4 (a) Crystal structure of BP sheet demonstrating the armchair structure. The distance between two consecutive layers is equal to ~5.3 A°. (b) High-resolution transmission electron microscopic image of few-layers thin sheet of BP is showing the atomic planes clearly. (c) The electron diffraction pattern of the BP flake. (d) The resistivity ρ_{xx} of the BP is plotted as a function of temperatures at different values of the applied magnetic fields in perpendicular direction, ranging from 300 K to 10 K. (e) The normalized magneto resistance of BP as a function of magnetic field H at various temperatures. (f) The temperature versus crossover magnetic field plot. The inset graph shows the variation in MR as a function of magnetic field.

Source: (ref[30]) Copyright 2016, Nature Publishing Group

HTEM scan of a few-layers thin sheet of BP and its electron diffraction pattern is presented in Figure 4.4 (b–c). Intrinsically, BP is a *p*-type semiconductor material in nature whose bandgap varies from 0.3 eV ~ 0.2 eV, which depends upon its sheet thickness.[27–29] In monolayer BP, large hole mobility (10,000 cm²/Vs) is predicted[29] theoretically, which makes it more promising to use in the electronic device.

The anisotropic magneto-transport through the BP plays a vital role in future magneto-electronic devices. The sheet dependent resistivity ρ_{xx} of BP is plotted as a function of temperature at several magnetic fields. The temperature was elevated from 10 K to 300 K, while the values of perpendicular magnetic field were changed from 0T to 7T, as shown in Figure 4.4 (d). It is observed that the resistivity decreases abruptly up to 60 K, while above this temperature, BP showed the metallic behavior. This transition is attributed to the thermal activation of the impurities that act like donors. These impurity donors freeze as the temperature is decreased further, and BP shows a significant variation in its resistivity below 60 K. The inset of Figure 4.4 (d) is showing the variation in magneto-resistance (MR) as a function of temperature at various magnetic fields. Generally, the MR is described as $[\rho_{(H)} - \rho_{(0)}] / \rho_{(0)} \times 100\%$, where $\rho_{(0)}$ and $\rho_{(H)}$ are the values of resistivity at zero field and H field, respectively. The highest value of MR was observed up to 510% at a temperature of 30 K under the effect of 7T magnetic field.[30] Moreover, the field-dependent MR is plotted at various temperatures, as shown in Figure 4.4 (e–f). Under a threshold value of the applied magnetic field B_L (6 T) at room temperature, the MR is demonstrating a quadratic increase, as shown in Figure 4.4 (f), and then shows a linearly increasing trend with the increasing magnetic field without saturation.

4.1.3 MXENES AS 2D MATERIAL

A newly emerging and attention seeking material has been discovered from the group of transition metal carbides, nitrides, and carbonitrides, known as MXene. Generally, the MXenes are synthesized from MAX phases comprising of layered ternary carbides, represented with the formula $M_{n+1}AX_n$. Here M is denoting an early transition metal (Sc, Ti, Nb, Mo, V, Cr, Zr, Hf, or Ta), A is denoting an element that belongs to groups 12–16 (Cd, Al, Ga, Ge, Si, P, S, As, In, Sn, Tl, Pb, or S), and X could be carbon and/or nitrogen. MXenes are usually produced by the top-down synthesis methodology by careful etching of A elements from the MAX phase structure, using reactive acids and exfoliation technique. In the subsequent $M_{n+1}AX_n$ phases, the powerful bonding between M and A elements is substituted by hydrogen bond (e.g., OH, F, or O after replacement of element A), at room temperature. The aqueous solution of HF is used as an etching agent.[31–33] All established MAX phases having P6$_3$/mmc symmetry are almost layered hexagonal, and the M layers are considered almost closely packed while the X atoms occupied the octahedral vacancies. In general, all the MAX phases are similar to the 2D layered materials, in which carbides and/or nitrides are bonded with an element A, as shown in Figure 4.5. The M-X bond could be a covalent/metallic/ionic bond much stronger, though the M-A bond is metallic.[31,34]

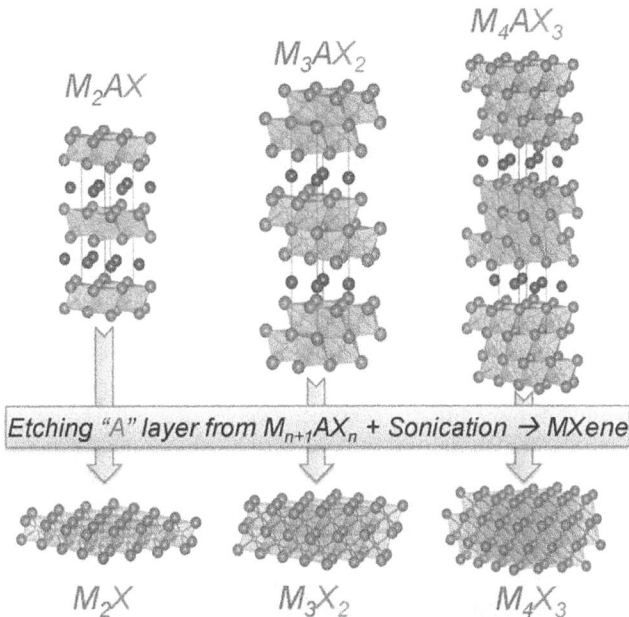

FIGURE 4.5 Schematic demonstration of the structure of various MAX phases and the subsequent MXenes.

Source: Reprinted with permission from ref[31], copyright 2014, Wiley

The band structure of the resultant MXenes calculated by first-principles density-functional theory is typically different than its parent MAX phase. The functionalization of MXenes with F, O, and OH is the promising way to shift it from metallic phase to semiconductor. The work function and bandgap of MXenes could be modulated by surface engineering technique (e.g., the work function of the $Ti_3C_2T_x$ MXene could be tuned from 6.15 eV to 1.81 eV by terminating the entire O or OH from the surface, respectively). The tunable bandgap and its phase shift from metallic to semiconductor makes MXene a promising candidate for electric and optoelectronic devices. Its sheet resistance varies from 0.1 to 8 KΩ per square with a transmittance efficiency up to 90%, which are considered quite good for electric and optoelectronic applications.[35,36] This high transmittance and flexible nature of the MXene sheets can play a vital role to fabricate the high efficiency solar cells, photodetectors, light emitter diodes, and liquid crystal displays (LCDs).[37] After modulating the work function of MXenes, it has been also deployed to form electrodes in thin film transistors, FETs, and logic devices.[38,39]

Including a variety of applications, recently MXenes have obtained substantial interest for energy harvesting, energy conversion, and energy storage tools, such as batteries, supercapacitors, and water splitting, because of its excellent physicochemical features of high electric conductivity, high melting point, hardness, and high thermal conductivity and oxidation resistance as well.[32,33,40,41]

4.2 OVERALL VIEW OF POST-GRAPHENE 2D MATERIALS

The presence of 15 elemental main group 2D materials has been tentatively confirmed or hypothetically anticipated until this point in time, with an assortment of captivating and valuable applications. Only the chemical composition of elements is not enough to describe the material properties of 2D materials, but allotropes are also considered as a strong ally to dictate the optical, electronic, and thermal properties. The electronic attributes of 2D elements in Group IV resemble carbon, the lightest elements of the group. Moreover, Si, Ge, and Sn show the hybridization state between sp^2 and sp^3, which contains buckled crystal lattice structure like graphene instead of a planner form.[42] Group III elements containing 2D aluminum, which is known as aluminene, and 2D boron that is recognized as borophene have various allotropes, in theory, with a bit complicated multi-atomic unit cell. In Group III elements, a lower number of valance electrons might be linked with this complication, ensuing their exclusive electronic band structure.[43]

The transition from metallic to semimetal or semiconducting actions of monolayer 2D materials is experienced as we move from Group III to Group VI in a periodic table. For example, Group III elements like aluminene, gallenene, and borophene are considered metals. Similarly, Group IV 2D materials attain Dirac cone semimetal band structure like graphene. Moreover, Group IV 2D analogues have spin-orbit coupling, which, with the increase of atomic number, creates a small bandgap around the Dirac cone.[44,45] In Group IV, V, and VI, elements like plumbene, bismuthene, selenene, and tellurene also illustrate large spin-orbit coupling.[46–49] 2D materials from Group IV elements also have several magnetic characteristics that are theoretically forecasted. For example, at Curie temperature, T_c = 122 K and 145 k semi-hydrogenated silicone and germanene show the behavior of ferromagnetic semiconductors.[50]

In nanoscale materials, bandgaps assume an essential part that plays a key role in optoelectronic properties and directs their possible future applications. Therefore, several techniques have been adopted in earlier work, like the electric field, strain, edge, doping, surface passivation, and functionalization to modulate the bandgap of elemental 2D materials and to bring new and thrilling applications. 2D materials illustrate the unique and outstanding properties in the field of electronics and sensing, power, and energy. Also, it provides future advancements in photonic and optoelectronic applications.

4.2.1 ELECTRONICS AND SENSING

The novel and outstanding properties of elemental 2D materials are ready to empower future headways in electronics and sensing for silicon gadgets. Large numbers of the elemental 2D materials show profitable properties, including high electron and hole mobilities and on/off ratios, topological insulating conduct, material adaptability, and high sensitivities to adsorbed particles that can develop the present state of the art system.

Fabrication and material measurement during FETs operation are the conventional means to assess the electrical properties of 2D materials. Until this point in time, active elemental 2D materials with field-effect transistor (FET) is synthesized using silicene, selenene, phosphorene, and tellurene. All these elements contain a large range of on/off ratios and mobilities. In 2017, exfoliated crystals were utilized to fabricate the transistors of tellurene and selenene. Selenene gadgets with 16 nm thickness showed generally low mobilities in contrast with other basic 2D materials (0.26 cm^2 V^{-1} s^{-1}) and low on current (20 mA mm^{-1}), however a high on/off proportion $\approx 10^6$ at room temperature.[51] Where the on/off ratio is essential, for logic-type apparatus, this ratio might be significant to guarantee low current leakage. Thickness-based electrical execution of tellurene crystals, in contrast, a notable and extensive electronic performance is displayed. Moreover, thickness-based optical properties are not shown by typical layered 2D materials.[52] Until today, in 2D materials, the future of the electronic and sensing platform is based on three main areas of concentration (flexible electronics, sensors, and spintronics).

2D materials seem to be a strong candidate for future electronics and sensors as they display various stimulating properties; moreover, these materials can show further improved behavior if these materials combine in the form of heterostructure instead of being individual. For example, borophene/perylenetetra carboxylic dianhydride (PTCDA) recently was used for the experimental lateral heterostructure; moreover, abrupt borophene/polymer interface is the reason why these heterostructures exhibit metal/semiconductor conduct and resistive switching.[53] In borophene/PTCDA heterostructure, the evolution in the density of states happens at the molecular length scale due to these sharp interfaces.

4.2.2 Photonic and Optoelectronic Applications

The plasmonic, photodetector, and ultrafast lasers are key areas of photonics and optoelectronics where 2D materials are deeply explored. Plasmons are aggregate charge oscillations that happen in solids through Coulomb interactions. While in semiconducting 2D material like graphene and phosphorene, plasmons show up in the mid/close infrared frequency range, plasmons in borophene can reach out into the visible range because of their higher carrier concentrations.[43] In an anisotropic plasmonic gadget wherein the appliance reaction strain or light polarization induce charges, phosphorene-based devices are used.[54] Another rising application of 2D material involving plasmons is thermoplasmonics, where localized hotspots can create using surface plasmons coupled with light irradiation. This technique is highly useful in the medical field and nowadays successfully utilizing for removing tumors in mice.[55]

Photodetection is one of the earliest applications of optoelectronics for 2D materials. Bandgaps of 2D semiconducting materials going from visible to infrared frequencies, making them unmistakably appropriate for broadband photodetection.[56–58] Current photodetection innovation is fundamentally founded on silicon with its elite performance, maturity, low-cost, and high level of integration with electronics. Perhaps the most intensely read natural materials for photodetectors is phosphorene. Because of the trouble in synthesizing elemental 2D material, for example, borophene, silicene, germanene, what's more, stanene, their adequacy as photodetector materials remains to be considered and further investigated.

In ultrafast laser (fiber-based), continuous wavelength laser is being converted into pulsed laser using mode-locking technology. This is accomplished by adopting SA (saturable absorber), a nonlinear optical material whose light absorption is inversely proportional to intensity. Whereas, the ideal SA has good absorption properties. Phosphorene, one of the primary elemental 2D material, was studied as an SA.

4.2.3 Power and Energy

Global communities desire strategies that are reliable, vigorous, and exceedingly proficient for the deliverance, storage, and transmission of power in upcoming years. Batteries are the key component in future electronic devices as they enhance gravimetric and volumetric storage abilities. From in power perspective, 2D materials are considered as a strong entity for next-generation energy devices. Among 2D materials, borophene holds the maximum theoretical capacity for Li.[59] For energy storage, rechargeable batteries are the highest priority because of their high energy densities. However, lower power density also occurs due to slow charging and discharging time and short life cycle. Therefore, as an alternate, ultracapacitor or supercapacitor offers super lifetimes, fast charge/discharge rate, and higher power densities. Moreover, phosphorene was considered as one of the prior materials among 2D materials for utilization in supercapacitors.[60] Also, elemental 2D materials present a significant role in the creation and formation of novel thermoelectric gadgets. However, after loads of applications of 2D materials in the field of electronics until this point in time, still, technological uprising requires a considerable amount of study and exploration of 2D materials in the coming years.

4.3 BAND STRUCTURE

Nanodevices that are based on 2D materials instead of conventional semiconductors have been broadly researched and show extremely competitive performance. However, there are extraordinary difficulties to be conquered before figuring out its practical applications. According to the physics of semiconductors, the Fermi level in *n*-type 2D semiconductor materials exists near the conduction band, while in *p*-type materials, it lies near the valance band. In 2D gadgets, one of the vital components that direct the charge carrier from metal to 2D channel is the "contact" among metal electrode and semiconductor.[61] In deciding the Schottky barrier (SB) that prompts non-negligible contact opposition and accordingly restricts gadget performance, the major and leading part is played by the energy level alignments (ELAs) at the 2D metal surface.[62] For instance, a huge SB or contact resistance can essentially break down the performance of a 2D FET, for example, current on/off proportion, field-impact mobility, and subthreshold swing (SS). Therefore, it is significant to create successful contact designing ways to form ohmic contacts in 2D apparatus.

Conventional devices that are Si supported to face the unwanted interfacial impacts, for example, Fermi level pinning (FLP) due to the existence of significant dangling bonds on the surface of Si and hence control the configuration of ohmic contacts.[63] 2D materials with an atomically flat surface unlike bulk Si give an extraordinary chance to form ohmic contacts with metals. However, the crystal structure of 2D materials with the thin atomic nature can be harmed during standard device fabrication procedure, for example, e-beam lithography (EBL) and physical vapor deposition (PVD) of metals, which brings about huge SB or interfacial states at 2D/metal contacts. Furthermore, for accomplishing ohmic contacts, the substitutional doping methods are broadly utilized. For instance, thermal diffusion and ion implantation are not relevant to 2D material because of the introduction of critical defects and annihilation of the crystal lattice. Different methodologies have been adopted to set up van der Waals (vdWs) 2D/metal contacts; for example, to preserve the 2D crystal, mechanical transfer of metal layers was processed to eliminate interfacial residue during device fabrication, thus making a deformity-free contact and removing interfacial states or impacts.

In 2D materials, top contact and edge contact are two types of contact geometry. By placing metals on top of 2D material, top contact is created. On the other hand, due to the limited edge area, it is difficult to attain good edge contact. Heterostructure, according to band alignment, is separated into three major categories recognized as symmetric, staggered, and broken as type I, type II, and type III, respectively, as shown in Figure 4.6 (a–c).

In type I, the valance and conduction bands of one material lie within the range of other semiconductor material, while in type II, the bands are shifted within bandgap size. In type III heterostructures, the energy bands of both materials exist out of range with respect to each other. Each band arrangement has specific applications to empower various assortments of gadgets. In optical devices such as LEDs

(a)   Type I   (b)   Type II   (c)   Type III

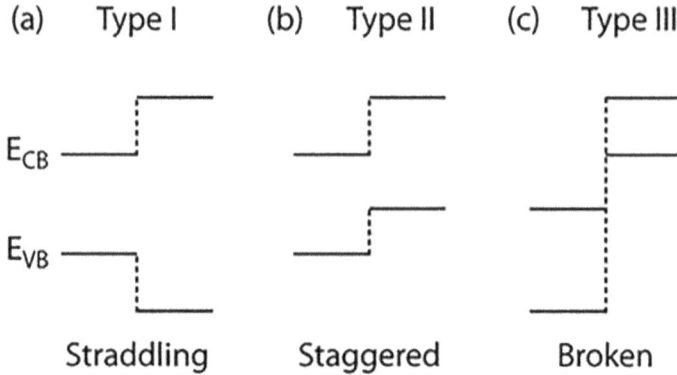

FIGURE 4.6   Three basic types of the energy band diagrams based on various heterostructures. (a) The straddling band alignment is also known as type I. (b) The staggered band alignment is called type II. (c) The broken bandgap, or Type III band alignment.

Source: (ref[64]) Copyright 2013, *Chemical Society Reviews*

(light-emitting diodes), the most generally used band alignment is type I. Moreover,
type I band alignment is utilized in lasers as they give away to spatially restrict electrons and holes, so effective recombination can happen. The capacity to create single
and multiple quantum wells has additionally improved the performance of lasers and
LED appliances.[65] For unipolar electronic tool applications, band alignment, which
is considered very valuable, is type II because it has exceptionally strong carrier
confinement as they let larger offset on one side (either conduction or valance band).
An excellent example of strong carrier confinement in type II is the HEMTs (high
electron mobility transistors) based InAs/AlSb quantum.[66] In bipolar transistors, as
well in quantum well resonant tunneling bipolar transistors,[67] conduction band levels can be adopted as hot electron injectors. To design the conduction to valance
band transition energy, the heterostructure of type II and type III are additionally
helpful. This is especially significant in TFETs (tunneling field-effect transistors) to
upgrade the tunneling current density,[68] also plays an important role in wavelength
photodetector[69] and infrared inter-sub band superlattice lasers. Probably the greatest
gadgets empowered by semiconductor heterostructures are quantum cascade lasers,
which utilize complex heterostructure stacks to design minibands and intra-sub band
changes, which permit proficient light emanation from the mid-infrared to the tera-
hertz systems.[70] Utilizing 2D materials, numerous gadget ideas have additionally
been proposed or acknowledged. To display current rectification and compilation of
photoexcited carriers, recently, the exhibition of atomically thin p-n diodes of type II
heterostructure has been done.[71] The latter can additionally be accomplished utilizing graphene/TMD/graphene heterostructures. Type I heterostructure that contains
LEDs with the TMD sandwiched among BN and graphene as contacts can also be
designed across a wide spectral range.[72] In the recent past, type II heterostructures
have likewise been investigated as a platform for attaching long-lived interlayer (or
indirect) excitons.[73] Moreover, for different applications, lateral heterostructures of

TMDs with different classes of monolayer materials can be utilized. Therefore, in semiconductor heterostructures, the relative band alignment is one of the paramount design parameters because it discovers the type of the heterostructure.

4.4 DOPING, DEFECTS, ALLOYS

Intrinsically the semiconductor materials are either *p*-type (having holes as majority carriers) or *n*-type (having electrons as majority charge carriers), but these semiconductors could be modified according to our desires to develop the desired charge transport. In case of silicon, the phosphorus doping make it *n*-type semiconductor, while the addition of boron atoms doped the silicon as a *p*-type semiconductor. The doping of few atoms of magnesium to gallium nitride was also used to build an LED-emitting blue light.[74] The discovery of graphene and other bandgaped 2D materials opened the door to develop the smart electric and optoelectronic devices even for the quantum applications. A lot of techniques had been introduced to enhance the performance of the devices composed of 2D materials (e.g., mobility, carrier density and bandgap modulation by surface passivation, high k-dielectric substrate, different metal contacts, and doping).[27,76,77] There are various doping techniques of the 2D materials, such as photoinduced doping by DUV light in different gas environments, using dissimilar metals of high or low work functions to dop them as a *p*-type or *n*-type semiconductors, and chemical doping by different chemicals.[75, 77–79]

One of the effective techniques to enhance the device and modulate the charge carrier density from electrons to holes and vice versa is chemical doping. This chemical doping of 2D materials could change the Fermi level and may also increase or decrease the bandgap of the pristine materials. Usually, the *p*-type doping of the 2D materials shifts their Fermi level near the valance band maxima, while *n*-type doping shifts the Fermi level near the conduction band maxima. Hua-Min Li et al. introduced the chemical doping of few-layer thick 2D sheets of MoS_2 to fabricate a vertical p-n junction with symmetric metal contacts.[75] The one surface of the MoS_2 sheet was doped to *n*-type semiconductor by introducing the benzyl-viologen (BV), while the other surface was doped to *p*-type semiconductor by using gold chloride ($AuCl_3$), as shown in Figure 4.7 (a–d).

The electric transport through this chemically doped vertical p-n junction of MoS_2 sheets was controlled by back-gate voltage, and an ideality factor of 1.6 was extracted at the positive gate voltage of +60V. Additionally, the photo response and tunneling behavior of chemically doped MoS_2 p-n junction was also studied with varying its layer thickness. These characteristics of the chemically doped MoS_2 p-n junction were showing a remarkable effect of chemical doping of the 2D materials to build a highly efficient logic devices, p-n junctions, solar cells, and photodetectors.

Furthermore, the performance of the 2D materials based devices may be affected because of defects in 2D TMDs sheets. These defects modulate the optical, chemical, magnetic, and electrical properties of the 2D materials and their devices.[80] Typically, these defects could occur during the growth of 2D materials, chemical doping, environmental effects, or stability of the 2D materials. The 2D materials have two possible polytypes 2H phase, and 1T semimetallic phase. This 1T phase is not as stable as 2H, and it is not considered as a semiconductor. Secondly, the surface defects

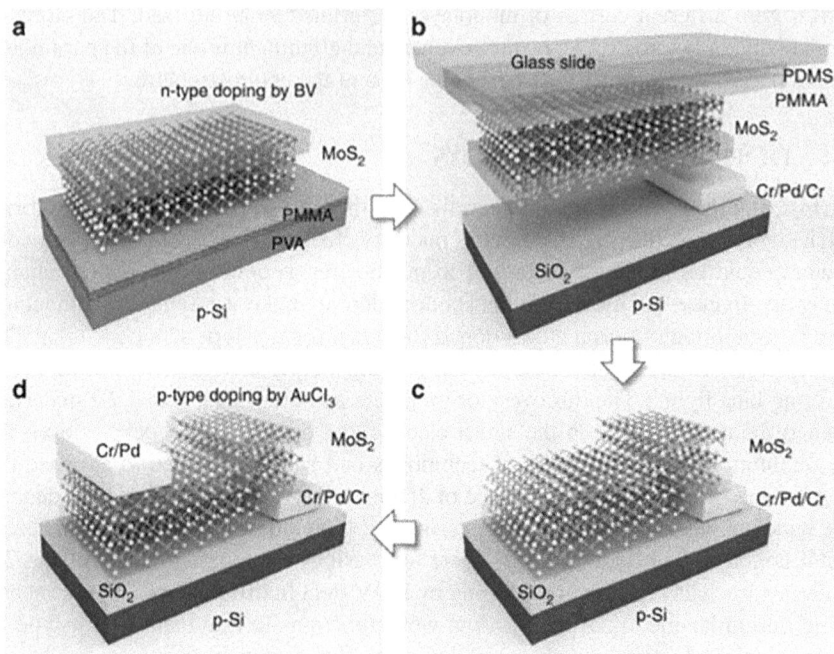

FIGURE 4.7 Chemically doped few-layers thick MoS$_2$ sheet to fabricate a vertical p-n junction. (a) The one side of the MoS$_2$ sheet was transferred over the PMMA/PVA/Si substrate, and it is n-doped by BV. (b) The n-doped sheet of MoS$_2$ was transferred to a PDMS/glass stamp. (c) The BV doped MoS$_2$ sheet was stacked over the Cr/Pd/Cr electrode and SiO$_2$/Si substrate. (d) The top surface of MoS$_2$ sheet was doped to p-type semiconductor with AuCl$_3$, and a Cr/Pd electrode was deposed over it as a top electrode.[75]

Source: Copyright 2015, Nature Publishing Group

in the sheets of 2D materials may also arise because of oxidation or inappropriate growth.[81,82] It is also known as zero defect, in which some vacancies could be observed within the atomic planes. These defects could degrade the performance of the 2D material for the electric or optoelectronic applications. One of the most environmentally unstable material is black phosphorus (BP), which is oxidized rapidly. Its monolayer flakes could not be deal in air or ambient environment. After exposing the thick flakes of BP in ambient environment, it is observed that its layers are etched gradually up to monolayer and reduced the device performance significantly.[27,84,85] These defects generated because oxidation could be controlled by passivation of some insulator oxide films or 2D sheets of h-BN.[27,85]

4.5   ELECTRICAL TRANSPORT

In 2D semiconductors, Schottky barrier height (SBH) is the key component that charge inoculation highly depends on. To optimize the procedure, the familiarity of its value and information about how to amend it is necessary to know. Schottky-Mott rule is referred to as the difference $q\varphi_{B0} = q(\varphi_m - \chi)$ among metal's work function ($q\varphi_m$) and electron affinity ($q\chi$) of semiconductor. This is how, in the best-case scenario, the SBH ($q\varphi_{B0}$) between metal and semiconductor is decided. However, in a real situation, the pinning of Fermi level at the metal/semiconductor interface occurs. This can be verified by checking the dependence of Schottky barrier height on φ_m, given by $S = d\varphi_{B0}/d\varphi_m$, with $S = 1$ equivalent to the best case or Schottky limit, and $S \approx 0$ related to that of a pinned Fermi level. The existence of metal-induced gap states is the reason behind the metal/bulk semiconductor pinning.[63]

In thermionic emission, the measurement of activation energy is the generic methodology of extracting the Schottky barrier height. At the source and drain, two Schottky barriers are linked back-to-back, respectively, in the geometry of SB FET. However, the consumption of voltage drop goes to the maximum in the reverse-biased contact. Therefore, transistor behavior dominates. The following equation represents the current density injection through a reverse-biased SB:

$$J = A^* T^\alpha exp\left[-\frac{q\varphi_{B0}}{k_B T}\right]\left[1 - exp\left(-\frac{qV}{k_B T}\right)\right] \tag{1}$$

Where A^* represents the Richardson constant that can be derived for semiconductors of 2D and 3D. SBH shows with the $q\varphi_{B0}$, exponential characterized by α which is equal to 2 and 3/2 for SCs and 2D SCs, respectively, T is temperature, V stands for applied bias, and k_B is the Boltzmann constant. Moreover, for $qV \gg k_B T$, equation (1) is modified to:

$$J = A^* T^\alpha exp\left[-\frac{q\varphi_{B0}}{k_B T}\right] \tag{2}$$

If the transistor conduct in the subthreshold regime (small gate voltage V_g). Therefore, the previous equation again is simplified as follows:

$$J = A^* T^\alpha exp\left[-\frac{E_A}{k_B T}\right] \tag{3}$$

Where $E_A = q\varphi_{B0} + E_C^\infty - E_C^O$ is the total activation energy that charge carriers must trounce to access the channel, whereas $E_C^\infty - E_C^O$ is the difference between the conduction band minimum in the bulk and at the interface.

Schottky barrier and activation energy are not a similar thing. Unlike all values of V_g, when the flat band state is met, it is possible to identify SB and activation energy. The channel opposition plays an important role in the observed temperature reliance. With decrease of temperature, the resistance goes down when the MoS_2 channel is gated further than the metal-insulator transition. Also, the overall opposition (channel plus contact) can then decrease, but it cannot misconstrue as a negative SB. The SBH is required to rely unequivocally upon the number of layers because the bandgap rises as the thickness is reduced in the restriction of one to three layers. From 2 nm to multilayers, there is no huge reliance on the SBH on MoS_2 thickness. It is experienced that if the Fermi level is pinned to around 100 meV beneath the

TABLE 4.1
Performance Summary of Different Metal
Electrodes Fabricated over MoS$_2$ Sheet[87]

Metals	Work Function, φ_M (eV)	Schottky Barrier Height, $\varphi_B = \varphi_M - X$ (eV)
Al	4.5	0.5
W	5.1	1.1
Au	5.4	1.4
Pt	5.7	1.7

conduction band, then the SBH relies feebly upon the metal work (S = 0.09). This is in inconsistency with early outcomes from X-beam photoelectron spectroscopy estimations, which showed that the SBH for bulk MoS$_2$ is in the Schottky limit (S = 1).[86]

Four diverse metals were carefully considered with varying work functions to investigate the properties of metal when it comes in contact with nanostructured films of 2D MoS$_2$ nanoflakes. These metals include Al, Pt, W, Au. Comparative to the X = 4.0 eV that is electron affinity of MoS$_2$, Au, and Pt with $\left(\varphi_{Au} = 5.40\text{eV}\right)$ and $\left(\varphi_{Pt} = 5.70\text{eV}\right)$ were opted as a high work function while as a low function, metals were Al and W with $\left(\varphi_{Al} = 4.54\text{eV}\right)$ and $\left(\varphi_W = 5.09\text{eV}\right)$. Under identical bias conditions, Al/MoS$_2$ and Pt/MoS$_2$ interface reveals the highest and lowest current, respectively.

The experimental outcome attained utilizing various metal contacts with varying functions are recapitulated in Table 4.1. It indicates that Al (the lowest work function metal) has the highest bias current, representing its better ability of charge injection at room temperature for an interface with MoS$_2$ in contrast to Au and Pt. The estimated SBH by the Schottky-Mott affiliation is also publicized in Table 4.1.[87]

For infusion of electron (*n*-type) and hole (*p*-type), ohmic contacts are acknowledged through *n+/p/n+* or *p+/n/p+* substitutional/debasement doping profiles, respectively, in traditional Si FETs. In any case, early showings of 2D FETs without such controllable and supportable doping plans depended on the utilization of elemental metals with various work capacities for carrier infusion into the respective band of the 2D channel. There has been some advancement in doping 2D materials; however, these procedures are generally surface electrostatic doping, where a charged species are generated on a superficial level outcome in the gathering of electrons or hole openings in the semiconductor

4.5.1 HETEROSTRUCTURES

In previous few years, the novelty and versatility of TMDCs (WS$_2$, MoS$_2$, MoSe$_2$, etc.) and 2D materials like hexagonal boron nitrides (h-BN) and graphene have attracted considerable attention. Figure 4.8 (a) illustrates the evolution from semimetallic graphene to insulator h-BN 2D materials, which are possibly cleaved to get the required layer numbers. The graphene is a 2D material with zero bandgap while the

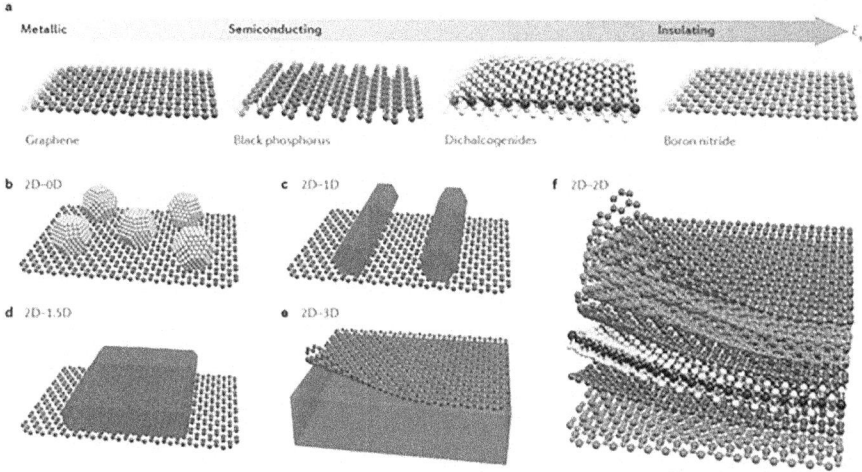

FIGURE 4.8 (a) The evolution of 2D materials from semi metallic graphene to insulator h-BN. (b–f) Schematic representation of 2D materials and various type of their heterostructures with 0D, 1D, 2D, and 3D materials.

Source: (ref.[89]). Copyright 2016, Nature Publishing Group

2D semiconductors have a reasonable bandgap varying from 0.3 eV to 2eV and h-BN is an insulator 2D material with a large bandgap of 6.08 eV.[1,27,28,88,89]

The modified 2D layers and their integration into lateral and 2D heterostructure provides many more opportunities to fabricate switching devices, photo detectors, photovoltaic cells, and high frequency devices.[27,79,90] The hybrid structure of the operating artificial 2D provides a paragon of innovative applications and properties beyond their component 2D atomic crystals. Therefore, this is considered as a novel and exciting field of progressive research. The 2D materials based heterostructure could be categorized to, e.g., 0D quantum dots, 1D nano rods/nanowires, 2D sheets, and 3D stacking of bulk materials over the 2D material presented in Figure 4.8 (b–f). The 0D-2D heterostructure are usually based on the 2D materials decorated with nanoparticles or quantum dots, while the 1D-2D heterostructures are based on 2D material and nano-rods/nanowires. These two formerly mentioned heterostructures are considered very promising to fabricate the fast and broadband photodetectors and atomically thin FETs with switching applications. Each heterostructure has its unique characteristic that can be used for various applications (e.g., power conversion efficiency, p-n diodes, photovoltaic applications, energy harvesting, and data storage).[27,91–93]

The most intriguing applications could be attained by 2D-2D atomically thin van der Waals heterostructure.[94,95] In optoelectronics and electronics, a comprehensive scope of bandgap-engineered applications has been provided by this heterostructure strategy.[96] In vertical stacking, weak interlayer van der Waals forces and strong intralayer covalent bonding has been present in layered materials.[97] To make atomic-layered 2D materials, several schemes have been anticipated thus far. In experimental and theoretical work, many vertical heterostructure such as TMDC/TMDC stack, h-BN/Gr stack, Gr/h-BN stack, and Gr/TMDC stack has been investigated. Most of stacks shows a huge potential for various applications.[98-100] Different technologies were recommended for the formation of artificial 2D layers, from manipulation, transfer, and stacking of layers. Moreover, to achieve a high standard vertical heterostructure, mechanical exfoliated 2D layers could be employed. The 2D-2D van der Waals heterostructures are getting attention to fabricate the p-n junctions, homo junctions, n-p-n (p-n-p) transistors, photovoltaic devices, and solar cells.[27,28] Recently, TMDs and graphene-based heterostructures are also showing promising results to fabricate the spintronic devices. The stacking of few layers of TMDs over the graphene increases the spin orbit interaction of graphene up to several millielectron volts, which makes it more efficient to control the spin transport through it.[101,102]

4.5.2   PN DIODES AND HOMOJUNCTIONS DEVICES

The major building block of integrated circuits (ICs) is the p-n junction diode, which consists of two different types of semiconductor materials. The combination of a *p*-type semiconductor material and *n*-type semiconductor material establishes a p-n junction, which is also known as a diode. These diodes are used as rectifiers for the rectification purposes, logic switches, high frequency, and logic devices and many more applications. As the thin and flexible 2D materials (e.g., WS_2, WSe_2 MoS_2, MoTe2, BP, etc.) are discovered, the van der Waals heterojunction attained a huge attention to develop the p-n junctions, photovoltaic cells, solar cells, and the optoelectronic devices. A thin sheet of the *p*-type 2D material is stacked over the *n*-type 2D material, which results in a van der Waals p-n heterojunction. The diodes composed of thin sheets of 2D materials are classified into three major groups. The first class is known as lateral p-n junctions, in which a *p*-type TMDC sheet is attached with the *n*-type TMDS sheet or *n*-type thin oxide film in a plan. Recently, Deng, Y. et al. reported[103] a lateral van der Waals diode composed of *n*-type monolayer MoS_2 sheet and *p*-type black phosphorus sheet. Another combination of thin 2D materials and semiconducting oxide films were also used for the fabrication of p-n diode, as shown in Figure 4.9 (a). A thin sheet of 2D semiconducting *p*-type black phosphorus and *n*-type metal oxide indium gallium zinc oxide (n-IGZO) was combined together to form a van der Waals p-n junction. The electric transport through such diodes may arise because of the sharp interface between *p*-type and *n*-type materials, metal electrodes, carrier concentration, and electrostatic gating.[79,104] Using a gate voltage, the carrier transport and rectification ratio could be modulated through the lateral p-n junctions, as shown in Figure 4.9 (b–c). This lateral van der Waals p-n junctions are considered best candidates to get a large rectification ratio up to 10^6, and for the efficient photovoltaic applications.[27,28,103]

The second class of the diodes is the vertical diodes. The vertical diodes are composed of the single or multilayer sheets of 2D TMDS materials of different nature, sandwiched between two metal contacts. The p-i-n diodes also belong to this class of diodes, as shown in Figure 4.10 (a–b). The WSe_2 sheets of various thickness varying from monolayer to 72 layers were inserted in between Gd metal and Pt metal electrodes. The Gd is a low work function metal ($\Phi_{Gd} \approx 2.9$ eV) which doped the

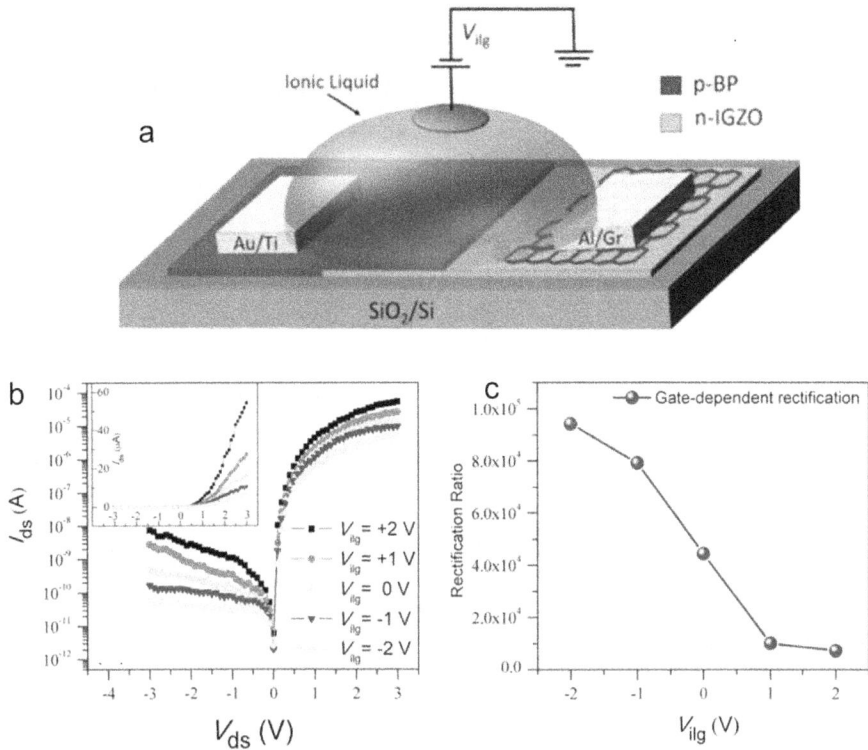

FIGURE 4.9 (a) The schematic diagram of IGZO/p-BP van der Waals diode. (b) The ionic liquid gate dependent logarithmic I-V curves are plotted as a function of source-drain biasing voltage. Inset is showing the linear I-V curves at various gate voltages. (c) The gate-dependent rectification in IGZO/p-BP van der Waals diode.

Source: (adapted from[27]) Copyright 2019, American Chemical Society

few layers of WSe$_2$ to more *n*-type, while because of the high work of Pt ($\Phi_{Pt} \approx 5.6$ eV), it doped WSe$_2$ to more *p*-type. Such asymmetric metal electrodes are providing an excellent choice to make vertical p-i-n diodes, in which rectification ratio could be modulated by changing their thickness. The thickness dependent I-V curves and rectification ratio are depicted in Figure 4.10 (c–d). The electric transport through the vertical diodes based on the tunneling through the 2D material, its thickness, and metal electrodes of different work functions.[27,79] Increasing the thickness of the 2D semiconducting sheets, the rectification ratio increases. Even in vertical diodes, a series of quantum tunneling occur such as direct tunneling (DT), Fowler-Nordheim tunneling (FN), and Schottky emission tunneling (SE), and these tunneling also play a major role to enhance the rectification ratio. The band alignment and various types of the tunneling are presented in Figure 4.10 (a–c).

The latterly mentioned two effects (DT and FT) are because of electrons and holes tunneling through the metal and 2D semiconductor sheets junction, while the

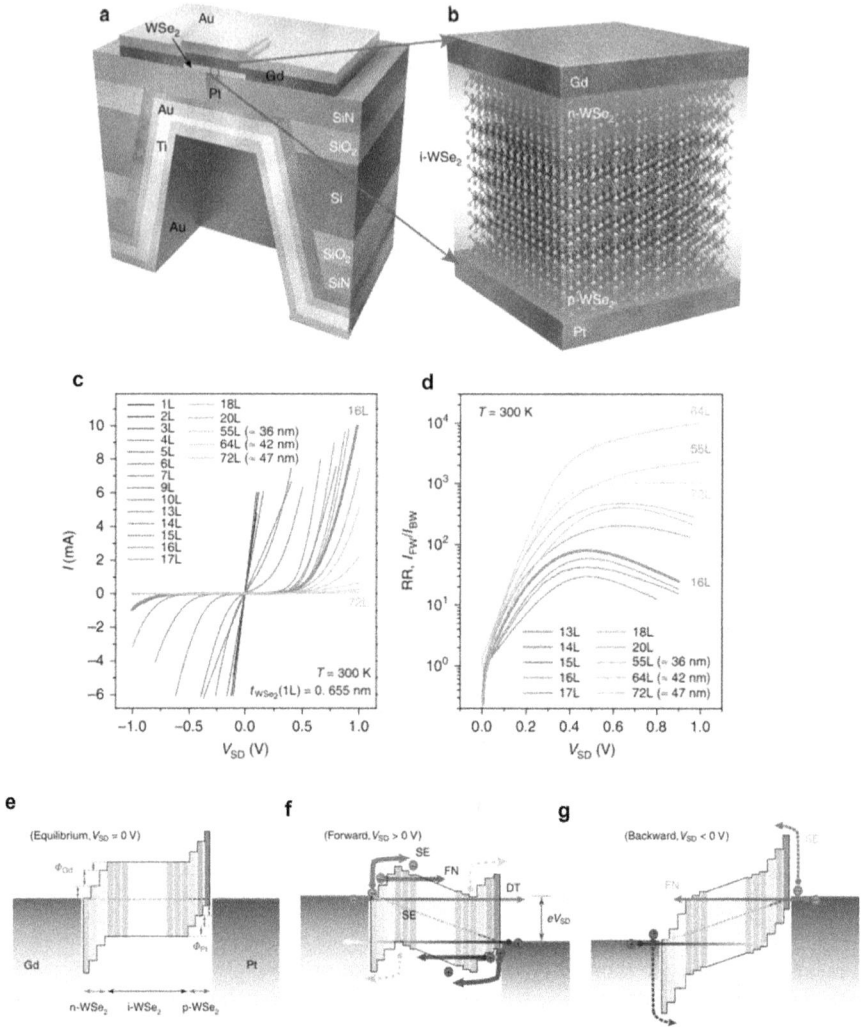

FIGURE 4.10 (a–b) The schematic images of the vertically stacked Gd/WSe2/Pt device. (c) The output characteristics of the vertical p-i-n diode with varying the thickness of WSe$_2$ sheet from monolayer to 72 layers. (d) At room temperature, the highest rectification ratio was attained with the 64 layers thick WSe$_2$ sheet, sandwiched between Gd and Pt metal electrodes. (e) The band alignment and electric charge transport because of a series of quantum tunneling is presented through a Gd/WSe$_2$/Pt based p-i-n heterojunctions. The energy band alignment is presented at equilibrium (e) with zero biased voltage $V_{SD} = 0V$. (f) Tunneling mechanism and charge transport during the forward biasing and (g) during the reverse biasing when $V_{SD} < 0V$.

Source: (ref.[79]) Copyright 2018, Nature Publishing Group

former tunneling (SE) mechanism is responsible of the energy bands of the insulating layers. In contrast, when the vertical diodes are reverse biased, then the electron and holes transport via diodes are entirely dependent on the DT and FN tunneling, but the SE tunneling will be negligible because of the large Schottky barriers. The vertical diodes are used for the high frequency devices, logic switching, and tunneling applications.

The third class of the p-n diodes is the homojunctions p-n diodes. In homojunction devices, one-half of the 2D semiconductor material's sheet is doped as a *p*-type semiconductor, while the rest of the sheet is doped as an *n*-type semiconductor.[105] The same 2D material possessed both types of the charge carriers (holes and electrons). This doping concentration can also be controlled by electrostatic gating.[106] Recently, Andreas, P. et al. reported a WSe_2 monolayer p-n diode[106] in which charge transportation was controlled by using two bottom metal gates. A monolayer WSe_2 sheet was transferred over the silicon nitride, having two bottom metal electrodes as gating electrodes. Finally, they fixed the one bottom gate at $V_{G1} = -40$ V, and the other bottom gate was fixed to $V_{G2} = +40$ V. This gating technique ($V_{G1} < 0$V, $V_{G2} > 0$ V) tuned the one half of the monolayer WSe_2 to *p*-type by inducing the holes as majority carriers, and the other half part of monolayer WSe_2 was doped to *n*-type semiconductor by inducing electrons as majority carriers, as shown in Figure 4.11 (a–c).

The half part of the WSe_2 sheet over the $V_{G1}<0$V will behave as a *p*-type WSe_2, while the other half part of WSe_2 sheet, which is over $V_{G2}>0$V, will behave as *n*-type WSe_2. In this technique, the same flake is turned into a homojunction p-n diode, and the carrier transport is modulated by gate voltages.

In contrast, there are several reports based on chemical doping techniques to tune the charge carriers in semiconductors (e.g., Li et al. stated vertical MoS_2 p-n junction by $AuCl_3$ to doped *p*-type and benzyl viologen (BV) as n-dopant to make an ultimate thin MoS_2 p-n junction diode).[107] However, in such chemical doped devices, the rectification ratio is limited.[107] That's why electrostatic doping is considered most effective technique in such van der Waals diodes.

4.5.3 HIGH-FREQUENCY DEVICES

The load transient response, bandwidth, and speed of the electronic devices could be improved through the high-frequency devices. The large bandwidth and high speed are only achievable if a device is operating at a low voltage drop and minimum power loss to provide a higher switching frequency.[28,108] For this purpose, using different metal contacts the organic semiconductors, or metal oxides were used to fabricate the Schottky diodes over the silicon chips or glass.[109–114] These Schottky diodes could operate up to the maximum frequency range of 900 MHz but could not be used for the flexible or wearable devices because of the lack of flexibility. Wearable and flexible electronic devices have been expected because of the suddenly growing thin films and graphene devices, but to date, no reasonable devices have been introduced yet. J. Zhang et al., fabricated a Schottky diode based on the indium-gallium-zinc-oxide (IGZO) thin film, which could operate up to a high frequency of 2.45 GHz, but the IGZO-thin film growth over the flexible substrate is still challenging.[90]

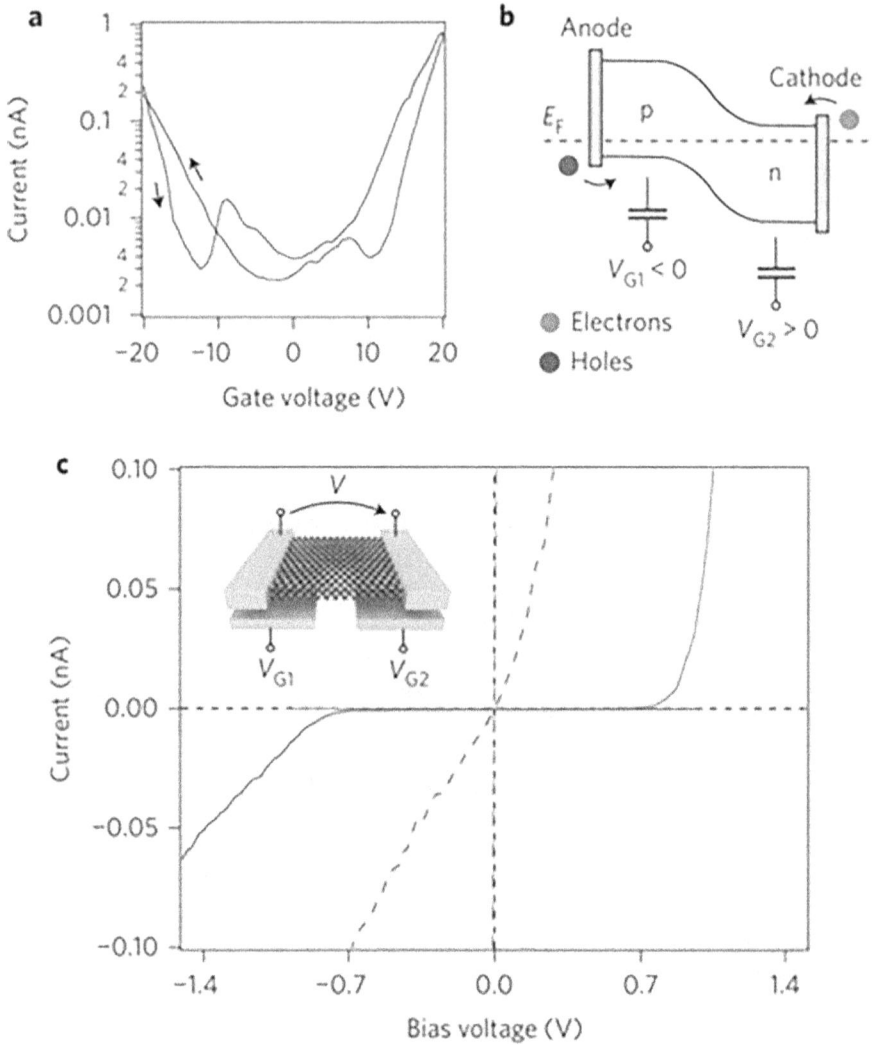

FIGURE 4.11   (a) The transfer characteristics of the monolayer WSe_2 device at a fix biasing voltage of 0.2V. The I_{ds} versus V_g cyclic-curve is extracted after sweeping the gate voltage from −20V to +20V. (b) Band diagram of the monolayer WSe_2 device while operating as a p-n diode. (c) The output characteristics of the device are plotted as I-V curves. The inset is showing how to fix the metal gates. The dashed green line is for n-n at fixed $V_{G1} = V_{G2} = 40$ V, p-p is presented in dashed blue line for $V_{G1} = V_{G2} = −40$ V, the solid green line is for p-n at fixed $V_{G1} = −40$ V, and $V_{G2} = 40$ V, and the solid blue line is for n-p at fixed $V_{G1} = 40$ V, $V_{G2} = −40$ V.

Source: (ref.[106]) Copyright 2014, Nature Publishing Group

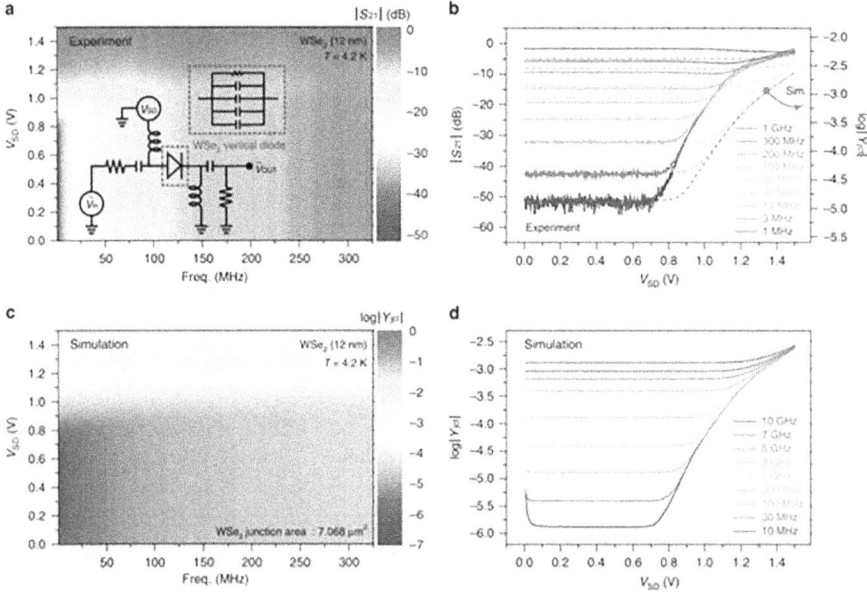

FIGURE 4.12 The vertically stacked p-i-n diode composed of Gd/WSe$_2$/Pt is showing a high-frequency operating limit. (a) The transmission coefficient (|S21|) is plotted at T = 4.2 K as V$_{SD}$, and driving frequency is changed. (b) High-frequency switching (|S21|) at differing frequencies. (c–d) The line profile and two-dimensional display, (c) line profiles, (d) the simulated junction admittances under an active Gd/WSe$_2$/Pt junction area of 7.068 μm², showing that the 2D vdW vertical diodes could be used for high-frequency switching applications.

Source: (ref.[79]) Copyright 2018, Nature Publishing Group

To overcome these challenges, 2D materials are considered the best candidates over the 3D bulk materials to fabricate the high-frequency smart devices.[27,88,115,116] Recently, a fast and ultrahigh frequency switching of 10 GHz is reported using few nanometers thick sheet of tungsten di-selenide (WSe$_2$).[79] The WSe$_2$ sheet was sandwiched between the Pt (used as a high work function metal) and Gd (used as a low work function metal) to fabricate a tunneling device, as shown in Figure 4.10 (a–b). The Gd is a low work function metal (Φ = 2.9 eV), and it doped the top surface of the WSe$_2$ to n-type, while the high work function of Pt (Φ = 5.6 eV) generates more hole on the bottom surface of WSe$_2$ and make it p-type. The middle layers of the WSe$_2$ remain intrinsically insulator. This 2D TMDs-based p-i-n device can operate up to 10GHz frequency.

Moreover, the frequency limits could be improved by limiting the capacitive loss through the substrate because these capacitive losses also cause to decrease the switching frequency. Such vertical diodes based on the thin sheets of the 2D TMDs semiconductor materials may provide a rout to fabricate the high-speed, flexible electronic devices. The mechanically flexible LCDs, mobiles, and wireless devices could be developed by growing a large area thin sheets of the 2D TMDs semiconductor materials.[109]

4.6 CHALLENGES AND OPPORTUNITIES OF POST-GRAPHENE 2D MATERIALS

Detection and preparation of a single layer are the earlier challenges in single-layer characterization. From bulk crystal, mechanical exfoliation is the most uncomplicated and effortless method for moving 2D sheets of layered van der Waals solids like graphene to any substrate. Flakes of size 10 μm are normally fabricated from exfoliation, which varies in thickness from single to multilayer. To identify single and multiple layer flakes, the preliminary methodology with the most influential elevated throughput is optical microscopy. Moreover, to establish the thickness of the layer, the highly potent procedure with 5% of precision is AFM. However, inconsistencies emerged from the difference in the associations of the tip with the sample and substrate. For instance, estimating the height of the second layer on a first layer is the better determination of the solitary layer thickness, as opposed to the tallness between a solitary layer and the substrate because in the previous case the tip layer interaction is consistent. To exclude any hysteretic artifacts, the optimization of a tip surface distance is necessary to attain a highly precise height profile of a single layer.

The astonishing applications of elemental 2D substances occur in the field of electronics and sensing, which allow superb technological development in future devices. The valuable properties that exhibit in 2D materials include high sensitivity to adsorbed molecule, high on/off ratio and high electron-hole mobility, material flexibility, and insulating attitude. For several gases, B, N, and BN-doped stanine[117] work as a sensor. Moreover, 2D materials having buckled or puckered structure direct the path to the anisotropy in the optical conductivity because new electronics and optoelectronics designing of gadgets are made possible. Plasmonic, ultrafast laser, and photodetectors are the major areas of electronics and optoelectronics, where amazing applications lie. Also, the technical revolution in the power and energy sector depends on such strategies that are highly robust and reliable. 2D materials play an important role in this direction, providing next-generation devices for energy storage that provide lightweight energy storage gadgets without sacrificing performance. 2D materials also upgrade batteries and supercapacitors to offer various benefits in the formation of new and novel thermoelectric devices.

Hi-tech uprising worked around basic 2D materials will require a lot of exploration, designing, also, improvement in the coming years. In the first place, utilization for basic 2D materials should be better perceived at a fundamental level and developed to guarantee reliability during production at a large scale. Future working procedures should be customized to the application of interest with lower cost and more versatile strategies for applications, like plasmonic, photodetectors, and spintronics. Moreover, researchers must focus on the properties of 2D materials that are predicted but neglected, such as ferroelectricity, because it might be vital to opening another range of future applications. Therefore, it can be concluded that despite the huge difficulties ahead, there is no uncertainty that the outstanding properties of these exciting materials will impact the future innovation space in applications, for example, nanoelectronics, spintronics, detecting and sensing, thermoelectric, photonics, and energy frameworks.[71,79,88,115,116]

REFERENCES

1. Geim, A. K. & Novoselov, K. S. *Nanoscience and technology: A collection of reviews from nature journals* 11–19 (World Scientific, 2010).
2. Geim, A. K. & Novoselov, K. S. The rise of graphene. *Nature Materials* **6**, 183–191, http://doi.org/10.1038/nmat1849 (2007).
3. Peierls, R. E. Annales de l'institut Henri Poincare. *Quelques proprietes typiques des corpses solides*, **5**, 177 (1935).
4. Venables, J. & Spiller, G. Nucleation and growth of thin films. *Surface Mobilities on Solid Materials*, 341–404 (1983).
5. Evans, J., Thiel, P. & Bartelt, M. C. Morphological evolution during epitaxial thin film growth: Formation of 2D islands and 3D mounds. *Surface Science Reports* **61**, 1–128 (2006).
6. Novoselov, K. S. *et al.* Two-dimensional atomic crystals. *Proceedings of the National Academy of Sciences* **102**, 10451–10453 (2005).
7. Novoselov, K. S. *et al.* Electric field effect in atomically thin carbon films. *Science* **306**, 666–669 (2004).
8. Partoens, B. & Peeters, F. From graphene to graphite: Electronic structure around the K point. *Physical Review B* **74**, 075404 (2006).
9. Morozov, S. *et al.* Two-dimensional electron and hole gases at the surface of graphite. *Physical Review B* **72**, 201401 (2005).
10. Stankovich, S. *et al.* Graphene-based composite materials. *Nature* **442**, 282–286 (2006).
11. Schedin, F. *et al.* Detection of individual gas molecules adsorbed on graphene. *Nature Materials* **6**, 652–655 (2007).
12. Hill, E. W., Geim, A. K., Novoselov, K., Schedin, F. & Blake, P. Graphene spin valve devices. *IEEE Transactions on Magnetics* **42**, 2694–2696 (2006).
13. Heersche, H. B., Jarillo-Herrero, P., Oostinga, J. B., Vandersypen, L. M. & Morpurgo, A. F. Bipolar supercurrent in graphene. *Nature* **446**, 56–59 (2007).
14. Schakel, A. M. Relativistic quantum Hall effect. *Physical Review D* **43**, 1428 (1991).
15. González, J., Guinea, F. & Vozmediano, M. Unconventional quasiparticle lifetime in graphite. *Physical Review Letters* **77**, 3589 (1996).
16. Manzeli, S., Ovchinnikov, D., Pasquier, D., Yazyev, O. V. & Kis, A. 2D transition metal dichalcogenides. *Nature Reviews Materials* **2**, 17033, http://doi.org/10.1038/natrev-mats.2017.33 (2017).
17. Novoselov, K. S. Nobel lecture: Graphene: Materials in the Flatland. *Reviews of Modern Physics* **83**, 837–849, http://doi.org/10.1103/RevModPhys.83.837 (2011).
18. Gusakova, J. *et al.* Electronic properties of bulk and monolayer TMDs: Theoretical study within DFT framework (GVJ-2e method). *Physica Status Solidi (a)* **214**, 1700218, https://doi.org/10.1002/pssa.201700218 (2017).
19. Li, X. D., Wu, S. Q. & Zhu, Z. Z. Band gap control and transformation of monolayer-MoS_2-based hetero-bilayers. *Journal of Materials Chemistry C* **3**, 9403–9411, http://doi.org/10.1039/C5TC01584G (2015).
20. Conley, H. J. *et al.* Bandgap engineering of strained monolayer and bilayer MoS_2. *Nano Letters* **13**, 3626–3630, http://doi.org/10.1021/nl4014748 (2013).
21. Gupta, D., Chauhan, V. & Kumar, R. A comprehensive review on synthesis and applications of molybdenum disulfide (MoS_2) material: Past and recent developments. *Inorganic Chemistry Communications* **121**, 108200, https://doi.org/10.1016/j.inoche.2020.108200 (2020).
22. Liu, Y. *et al.* Electrical characterization and ammonia sensing properties of MoS_2/Si p-n junction. *Journal of Alloys and Compounds* **631**, 105–110, https://doi.org/10.1016/j.jallcom.2015.01.111 (2015).

23. Radisavljevic, B., Radenovic, A., Brivio, J., Giacometti, V. & Kis, A. Single-layer MoS_2 transistors. *Nature Nanotechnology* **6**, 147–150, http://doi.org/10.1038/nnano.2010.279 (2011).

24. Wang, J. *et al.* High mobility MoS_2 transistor with low Schottky barrier contact by using atomic thick h-BN as a tunneling layer. *Advanced Materials* **28**, 8302–8308, https://doi.org/10.1002/adma.201602757 (2016).

25. Lee, S., Tang, A., Aloni, S. & Philip Wong, H. S. Statistical study on the Schottky barrier reduction of tunneling contacts to CVD synthesized MoS_2. *Nano Letters* **16**, 276–281, http://doi.org/10.1021/acs.nanolett.5b03727 (2016).

26. Coleman, P. Lending an iron hand to spintronics. *Physics* **2**, 6 (2009).

27. Dastgeer, G. *et al.* Black phosphorus-igzo van der waals diode with low-resistivity metal contacts. *ACS Applied Materials & Interfaces* **11**, 10959–10966, http://doi.org/10.1021/acsami.8b20231 (2019).

28. Dastgeer, G. *et al.* Temperature-dependent and gate-tunable rectification in a black phosphorus/WS2 van der Waals heterojunction diode. *ACS Applied Materials & Interfaces* **10**, 13150–13157, http://doi.org/10.1021/acsami.8b00058 (2018).

29. Li, Q. F., Wang, H. F., Yang, C. H., Li, Q. Q. & Rao, W. F. Theoretical prediction of high carrier mobility in single-walled black phosphorus nanotubes. *Applied Surface Science* **441**, 1079–1085, https://doi.org/10.1016/j.apsusc.2018.01.208 (2018).

30. Hou, Z. *et al.* Large and anisotropic linear magnetoresistance in single crystals of black phosphorus arising from mobility fluctuations. *Scientific Reports* **6**, 23807, http://doi.org/10.1038/srep23807 (2016).

31. Naguib, M., Mochalin, V. N., Barsoum, M. W. & Gogotsi, Y. 25th anniversary article: MXenes: A new family of two-dimensional materials. *Advanced Materials* **26**, 992–1005, https://doi.org/10.1002/adma.201304138 (2014).

32. Anasori, B., Lukatskaya, M. R. & Gogotsi, Y. 2D metal carbides and nitrides (MXenes) for energy storage. *Nature Reviews Materials* **2**, 16098, https://doi.org/10.1038/natrevmats.2016.98 (2017).

33. Seh, Z. W. *et al.* Two-dimensional molybdenum carbide (MXene) as an efficient electrocatalyst for hydrogen evolution. *ACS Energy Letters* **1**, 589–594, http://doi.org/10.1021/acsenergylett.6b00247 (2016).

34. Chaudhari, N. K. *et al.* MXene: An emerging two-dimensional material for future energy conversion and storage applications. *Journal of Materials Chemistry A* **5**, 24564–24579, http://doi.org/10.1039/C7TA09094C (2017).

35. Hantanasirisakul, K. *et al.* Fabrication of Ti3C2Tx MXene transparent thin films with tunable optoelectronic properties. *Advanced Electronic Materials* **2**, 1600050, https://doi.org/10.1002/aelm.201600050 (2016).

36. VahidMohammadi, A., Rosen, J. & Gogotsi, Y. The world of two-dimensional carbides and nitrides (MXenes). *Science* **372**, eabf1581, https://doi.org/10.1126/science.abf1581 (2021).

37. Agresti, A. *et al.* Titanium-carbide MXenes for work function and interface engineering in perovskite solar cells. *Nature Materials* **18**, 1228–1234, https://doi.org/10.1038/s41563-019-0478-1 (2019).

38. Lyu, B. *et al.* Large-area MXene electrode array for flexible electronics. *ACS Nano* **13**, 11392–11400, https://doi.org/10.1021/acsnano.9b04731 (2019).

39. Wang, Z., Kim, H. & Alshareef, H. N. Oxide thin-film electronics using all-MXene electrical contacts. *Advanced Materials* **30**, 1706656, https://doi.org/10.1002/adma.201706656 (2018).

40. Ma, T. Y., Cao, J. L., Jaroniec, M. & Qiao, S. Z. Interacting carbon nitride and titanium carbide nanosheets for high-performance oxygen evolution. *Angewandte Chemie International Edition* **55**, 1138–1142, https://doi.org/10.1002/anie.201509758 (2016).

41. Abel, M., Clair, S., Ourdjini, O., Mossoyan, M. & Porte, L. Single layer of polymeric Fe-phthalocyanine: An organometallic sheet on metal and thin insulating film. *Journal of the American Chemical Society* **133**, 1203–1205, http://doi.org/10.1021/ja108628r (2011).
42. El Bachra, M., Zaari, H., Benyoussef, A., El Kenz, A. & El Hachimi, A. First-principles calculations of van der Waals and spin orbit effects on the two-dimensional topological insulator stanene and stanene on ge (111) substrate. *Journal of Superconductivity and Novel Magnetism* **31**, 2579–2588 (2018).
43. Huang, Y., Shirodkar, S. N. & Yakobson, B. I. Two-dimensional boron polymorphs for visible range plasmonics: A first-principles exploration. *Journal of the American Chemical Society* **139**, 17181–17185 (2017).
44. Fu, B., Abid, M. & Liu, C.-C. Systematic study on stanene bulk states and the edge states of its zigzag nanoribbon. *New Journal of Physics* **19**, 103040 (2017).
45. Matthes, L., Pulci, O. & Bechstedt, F. Massive Dirac quasiparticles in the optical absorbance of graphene, silicene, germanene, and tinene. *Journal of Physics: Condensed Matter* **25**, 395305 (2013).
46. Reis, F. *et al.* Bismuthene on a SiC substrate: A candidate for a high-temperature quantum spin Hall material. *Science* **357**, 287–290 (2017).
47. Yuhara, J., He, B., Matsunami, N., Nakatake, M. & Le Lay, G. Graphene's latest cousin: Plumbene epitaxial growth on a "nano watercube". *Advanced Materials* **31**, 1901017 (2019).
48. Aktürk, E., Aktürk, O. Ü. & Ciraci, S. Single and bilayer bismuthene: Stability at high temperature and mechanical and electronic properties. *Physical Review B* **94**, 014115 (2016).
49. Xu, Y. *et al.* Large-gap quantum spin Hall insulators in tin films. *Physical Review Letters* **111**, 136804 (2013).
50. Gillgren, N. *et al.* Gate tunable quantum oscillations in air-stable and high mobility few-layer phosphorene heterostructures. *2D Materials* **2**, 011001 (2014).
51. Ji, Q., Li, J., Xiong, Z. & Lai, B. Enhanced reactivity of microscale Fe/Cu bimetallic particles (mFe/Cu) with persulfate (PS) for p-nitrophenol (PNP) removal in aqueous solution. *Chemosphere* **172**, 10–20, http://doi.org/10.1016/j.chemosphere.2016.12.128 (2017).
52. Qin, J. *et al.* Controlled growth of a large-size 2D selenium nanosheet and its electronic and optoelectronic applications. *ACS Nano* **11**, 10222–10229 (2017).
53. Liu, X. *et al.* Self-assembly of electronically abrupt borophene/organic lateral heterostructures. *Science Advances* **3**, e1602356 (2017).
54. Liu, Z. & Aydin, K. Localized surface plasmons in nanostructured monolayer black phosphorus. *Nano Letters* **16**, 3457–3462 (2016).
55. Sun, C. *et al.* One-pot solventless preparation of PEGylated black phosphorus nanoparticles for photoacoustic imaging and photothermal therapy of cancer. *Biomaterials* **91**, 81–89 (2016).
56. Wang, G. *et al.* Two dimensional materials based photodetectors. *Infrared Physics & Technology* **88**, 149–173 (2018).
57. Konstantatos, G. Current status and technological prospect of photodetectors based on two-dimensional materials. *Nature Communications* **9**, 1–3 (2018).
58. Wang, F. *et al.* 2D library beyond graphene and transition metal dichalcogenides: A focus on photodetection. *Chemical Society Reviews* **47**, 6296–6341 (2018).
59. Jiang, H., Lu, Z., Wu, M., Ciucci, F. & Zhao, T. Borophene: A promising anode material offering high specific capacity and high rate capability for lithium-ion batteries. *Nano Energy* **23**, 97–104 (2016).
60. Hao, C. *et al.* Flexible all-solid-state supercapacitors based on liquid-exfoliated black-phosphorus nanoflakes. *Advanced Materials* **28**, 3194–3201 (2016).

61. Sze, S. M., Li, Y. & Ng, K. K. *Physics of semiconductor devices* (John Wiley & Sons, 2021).

62. Tersoff, J. Schottky barrier heights and the continuum of gap states. *Physical Review Letters* **52**, 465 (1984).

63. Heine, V. Theory of surface states. *Physical Review* **138**, A1689 (1965).

64. Selinsky, R. S., Ding, Q., Faber, M. S., Wright, J. C. & Jin, S. Quantum dot nanoscale heterostructures for solar energy conversion. *Chemical Society Reviews* **42**, 2963–2985, http://doi.org/10.1039/C2CS35374A (2013).

65. Kash, K. Optical properties of III-V semiconductor quantum wires and dots. *Journal of luminescence* **46**, 69–82 (1990).

66. Werking, J. D. *et al.* High-transconductance InAs/AlSb heterojunction field-effect transistors with delta-doped AlSb upper barriers. *IEEE Electron Device Letters* **13**, 164–166 (1992).

67. Capasso, F., Mohammed, K. & Cho, A. Y. Electronic transport and depletion of quantum wells by tunneling through deep levels in semiconductor superlattices. *Physical Review Letters* **57**, 2303 (1986).

68. Pop, E. Energy dissipation and transport in nanoscale devices. *Nano Research* **3**, 147–169 (2010).

69. Yu, Y. *et al.* High-gain visible-blind UV photodetectors based on chlorine-doped n-type ZnS nanoribbons with tunable optoelectronic properties. *Journal of Materials Chemistry* **21**, 12632–12638 (2011).

70. Faist, J. *et al.* Quantum cascade laser. *Science* **264**, 553–556 (1994).

71. Lee, C.-H. *et al.* Atomically thin p-n junctions with van der Waals heterointerfaces. *Nature Nanotechnology* **9**, 676–681 (2014).

72. Lee, J. Y., Shin, J. H., Lee, G. H. & Lee, C. H. Two-dimensional semiconductor optoelectronics based on van der Waals heterostructures. *Nanomaterials (Basel, Switzerland)* **6**, http://doi.org/10.3390/nano6110193 (2016).

73. Rivera, P. *et al.* Observation of long-lived interlayer excitons in monolayer MoSe 2–WSe 2 heterostructures. *Nature Communications* **6**, 1–6 (2015).

74. Li, J., Wei, Z. & Kang, J. *Two-dimensional semiconductors: Synthesis, physical properties and applications* (John Wiley & Sons, 2020).

75. Li, H.-M. *et al.* Ultimate thin vertical p-n junction composed of two-dimensional layered molybdenum disulfide. *Nature Communications* **6**, 6564, http://doi.org/10.1038/ncomms7564 (2015).

76. Dev, D. *et al.* High quality gate dielectric/MoS$_2$ interfaces probed by the conductance method. *Applied Physics Letters* **112**, 232101, http://doi.org/10.1063/1.5028404 (2018).

77. Chaves, A. *et al.* Bandgap engineering of two-dimensional semiconductor materials. *NPJ 2D Materials and Applications* **4**, 29, http://doi.org/10.1038/s41699-020-00162-4 (2020).

78. Zhang, K. & Robinson, J. Doping of two-dimensional semiconductors: A rapid review and outlook. *MRS Advances* **4**, 2743–2757, http://doi.org/10.1557/adv.2019.391 (2019).

79. Nazir, G. *et al.* Ultimate limit in size and performance of WSe2 vertical diodes. *Nature Communications* **9**, 5371, http://doi.org/10.1038/s41467-018-07820-8 (2018).

80. Lin, Z. *et al.* Defect engineering of two-dimensional transition metal dichalcogenides. *2D Materials* **3**, 022002, http://doi.org/10.1088/2053-1583/3/2/022002 (2016).

81. Kim, D., Lee, R., Kim, S. & Kim, T. Two-dimensional phase-engineered 1T′- and 2H-MoTe2-based near-infrared photodetectors with ultra-fast response. *Journal of Alloys and Compounds* **789**, 960–965, https://doi.org/10.1016/j.jallcom.2019.03.121 (2019).

82. Gan, X. *et al.* 2H/1T phase transition of multilayer MoS$_2$ by electrochemical incorporation of S vacancies. *ACS Applied Energy Materials* **1**, 4754–4765, http://doi.org/10.1021/acsaem.8b00875 (2018).

83. Island, J. O., Steele, G. A., Zant, H. S. J. V. d. & Castellanos-Gomez, A. Environmental instability of few-layer black phosphorus. *2D Materials* **2**, 011002, http://doi.org/10.1088/2053-1583/2/1/011002 (2015).

84. Xu, Y., Shi, Z., Shi, X., Zhang, K. & Zhang, H. Recent progress in black phosphorus and black-phosphorus-analogue materials: Properties, synthesis and applications. *Nanoscale* **11**, 14491–14527, http://doi.org/10.1039/C9NR04348A (2019).

85. Miao, J., Zhang, L. & Wang, C. Black phosphorus electronic and optoelectronic devices. *2D Materials* **6**, 032003, http://doi.org/10.1088/2053-1583/ab1ebd (2019).

86. Lince, J. R., Carré, D. J. & Fleischauer, P. D. Schottky-barrier formation on a covalent semiconductor without Fermi-level pinning: The metal-MoS 2 (0001) interface. *Physical Review B* **36**, 1647 (1987).

87. Walia, S. *et al.* Characterization of metal contacts for two-dimensional MoS$_2$ nanoflakes. *Applied Physics Letters* **103**, 232105 (2013).

88. Dastgeer, G., Abbas, H., Kim, D. Y., Eom, J. & Choi, C. Synaptic characteristics of an ultrathin hexagonal boron nitride (h-BN) diffusive memristor. *Physica Status Solidi (RRL)—Rapid Research Letters* **n/a**, 2000473, https://doi.org/10.1002/pssr.202000473.

89. Liu, Y. *et al.* Van der Waals heterostructures and devices. *Nature Reviews Materials* **1**, 16042, http://doi.org/10.1038/natrevmats.2016.42 (2016).

90. Zhang, J. *et al.* Flexible indium-gallium-zinc-oxide Schottky diode operating beyond 2.45 GHz. *Nature Communications* **6**, 7561, http://doi.org/10.1038/ncomms8561 (2015).

91. Jeon, P. J. *et al.* Black phosphorus–zinc oxide nanomaterial heterojunction for p-n diode and junction field-effect transistor. *Nano Letters* **16**, 1293–1298, http://doi.org/10.1021/acs.nanolett.5b04664 (2016).

92. Sokolov, P. M. *et al.* Graphene-quantum dot hybrid nanostructures with controlled optical and photoelectric properties for solar cell applications. *Russian Chemical Reviews* **88**, 370–386, http://doi.org/10.1070/rcr4859 (2019).

93. Afzal, A. M., Dastgeer, G., Iqbal, M. Z., Gautam, P. & Faisal, M. M. High-performance p-BP/n-PdSe2 near-infrared photodiodes with a fast and gate-tunable photoresponse. *ACS Applied Materials & Interfaces* **12**, 19625–19634, http://doi.org/10.1021/acsami.9b22898 (2020).

94. Fan, X. F., Shen, Z. X., Liu, A. Q. & Kuo, J. L. Band gap opening of graphene by doping small boron nitride domains. *Nanoscale* **4**, 2157–2165, http://doi.org/10.1039/c2nr11728b (2012).

95. Muchharla, B. *et al.* Tunable electronics in large-area atomic layers of boron-nitrogen-carbon. *Nano Letters* **13**, 3476–3481, http://doi.org/10.1021/nl400721y (2013).

96. Fiori, G., Betti, A., Bruzzone, S. & Iannaccone, G. Lateral graphene-hBCN heterostructures as a platform for fully two-dimensional transistors. *Acs Nano* **6**, 2642–2648, http://doi.org/10.1021/nn300019b (2012).

97. Miao, J. *et al.* Vertically stacked and self-encapsulated van der Waals heterojunction diodes using two-dimensional layered semiconductors. *ACS Nano* **11**, 10472–10479, http://doi.org/10.1021/acsnano.7b05755 (2017).

98. Sachs, B., Wehling, T. O., Katsnelson, M. I. & Lichtenstein, A. I. Adhesion and electronic structure of graphene on hexagonal boron nitride substrates. *Physical Review B* **84**, http://doi.org/10.1103/Physrevb.84.195414 (2011).

99. Gopalan, D. P. *et al.* Formation of hexagonal boron nitride on graphene-covered copper surfaces. *Journal of Materials Research* **31**, 945–958, http://doi.org/10.1557/jmr.2016.82 (2016).

100. Li, C. *et al.* WSe2/MoS$_2$ and MoTe2/SnSe2 van der Waals heterostructure transistors with different band alignment. *Nanotechnology* **28**, http://doi.org/10.1088/1361-6528/Aa810f (2017).

101. Safeer, C. K. *et al.* Room-temperature spin hall effect in graphene/MoS$_2$ van der Waals heterostructures. *Nano Letters* **19**, 1074–1082, http://doi.org/10.1021/acs.nanolett.8b04368 (2019).

102. Ghiasi, T. S., Kaverzin, A. A., Blah, P. J. & van Wees, B. J. Charge-to-spin conversion by the rashba-Edelstein effect in two-dimensional van der Waals heterostructures up to room temperature. *Nano Letters* **19**, 5959–5966, http://doi.org/10.1021/acs.nanolett.9b01611 (2019).

103. Deng, Y. *et al.* Black phosphorus–Monolayer MoS$_2$ van der Waals heterojunction p-n diode. *ACS Nano* **8**, 8292–8299, http://doi.org/10.1021/nn5027388 (2014).

104. Li, X., Xiong, X. & Wu, Y. Toward high-performance two-dimensional black phosphorus electronic and optoelectronic devices. *Chinese Physics B* **26**, 037307, http://doi.org/10.1088/1674–1056/26/3/037307 (2017).

105. Chen, J.-W. *et al.* A gate-free monolayer WSe2 pn diode. *Nature Communications* **9**, 3143, http://doi.org/10.1038/s41467-018-05326-x (2018).

106. Pospischil, A., Furchi, M. M. & Mueller, T. Solar-energy conversion and light emission in an atomic monolayer p-n diode. *Nature Nanotechnology* **9**, 257–261, http://doi.org/10.1038/nnano.2014.14 (2014).

107. Li, H.-M. *et al.* Ultimate thin vertical p-n junction composed of two-dimensional layered molybdenum disulfide. *Nature Communications* **6**, 6564, http://doi.org/10.1038/ncomms7564. www.nature.com/articles/ncomms7564#supplementary-information (2015).

108. Afzal, A. M., Iqbal, M. Z., Dastgeer, G., Ahmad, A. U. & Park, B. Highly sensitive, ultrafast, and broadband photo-detecting field-effect transistor with transition-metal dichalcogenide van der Waals heterostructures of MoTe2 and PdSe2. *Advanced Science*, 2003713 (2021).

109. Lin, C.-Y. *et al.* High-frequency polymer diode rectifiers for flexible wireless power-transmission sheets. *Organic Electronics* **12**, 1777–1782, https://doi.org/10.1016/j.orgel.2011.07.006 (2011).

110. Cvetkovic, N. V. *et al.* Organic half-wave rectifier fabricated by stencil lithography on flexible substrate. *Microelectronic Engineering* **100**, 47–50, https://doi.org/10.1016/j.mee.2012.07.110 (2012).

111. Pal, B. N. *et al.* Pentacene-zinc oxide vertical diode with compatible grains and 15-MHz rectification. *Advanced Materials* **20**, 1023–1028, https://doi.org/10.1002/adma.200701550 (2008).

112. Chen, W.-C. *et al.* Room-temperature-processed flexible n-InGaZnO/p-Cu2O heterojunction diodes and high-frequency diode rectifiers. *Journal of Physics D: Applied Physics* **47**, 365101, http://doi.org/10.1088/0022–3727/47/36/365101 (2014).

113. Im, D., Moon, H., Shin, M., Kim, J. & Yoo, S. Towards gigahertz operation: Ultrafast low turn-on organic diodes and rectifiers based on C60 and tungsten oxide. *Advanced Materials* **23**, 644–648, https://doi.org/10.1002/adma.201002246 (2011).

114. Lilja, K. E., Bäcklund, T. G., Lupo, D., Hassinen, T. & Joutsenoja, T. Gravure printed organic rectifying diodes operating at high frequencies. *Organic Electronics* **10**, 1011–1014, https://doi.org/10.1016/j.orgel.2009.04.008 (2009).

115. Dastgeer, G. *et al.* Surface spin accumulation due to the inverse spin Hall effect in WS 2 crystals. *2D Materials* **6**, 011007, http://doi.org/10.1088/2053-1583/aae7e8 (2018).

116. Dastgeer, G., Shehzad, M. A. & Eom, J. Distinct detection of thermally induced spin voltage in Pt/WS2/Ni81Fe19 by the inverse spin hall effect. *ACS Applied Materials & Interfaces* **11**, 48533–48539, http://doi.org/10.1021/acsami.9b16476 (2019).

117. Molle, A. *et al.* Buckled two-dimensional Xene sheets. *Nature Materials* **16**, 163–169, http://doi.org/10.1038/nmat4802 (2017).

5 2D Material Photonics and Optoelectronics

Muhammad Farooq Khan, H. M. Waseem Khalil, Shania Rehman, Muhammad Asghar Khan, and Jonghwa Eom

CONTENTS

5.1 BROADBAND PHOTODETECTORS

Broadband photodetectors have a vital role in the new era technology of medical sciences, wireless communication, telecom engineering, etc. The broadband light having wavelengths (λ) larger than the energy bandgap (eV) of material will work for any photodetector. To detect light, the photoelectric effect or photothermal effect is often employed for successful experiments. Photodetector follows the mechanism of photoelectric effect, where an electron is ejected from the valance band by an incident photon of a light source to the conduction band. Recently 2D materials such as graphene TMDCs achieved gigantic consideration from researchers in optoelectronics devices because of their impressive optical and remarkable electrical properties [1–7]. Usually, thin-film semiconductors for broadband light detection have narrow bandgaps of < 1eV. Principally, the excess carriers are excited by incident photons, attributed to the photoconductance, which increases linearly with time until the recombination process starts. The production of electron-hole pairs is normally separated because of the energy bandgap of the 2D material, application of gate voltage, geometry of devices, and barrier heights formed at semiconductors and metals interfaces. However, the thick TMDC semiconductors have an indirect energy bandgap, whereas the monolayer of TMDC semiconductors owes a direct energy bandgap. Particularly, photoluminance (PL) has a great impact on the layer number of TMDCs. In PL the peak intensity is increased significantly when decreasing the thickness of TMDCs

DOI: 10.1201/9781003247890-5

(a)   (b)   (c)

(d)   (e)   (f)

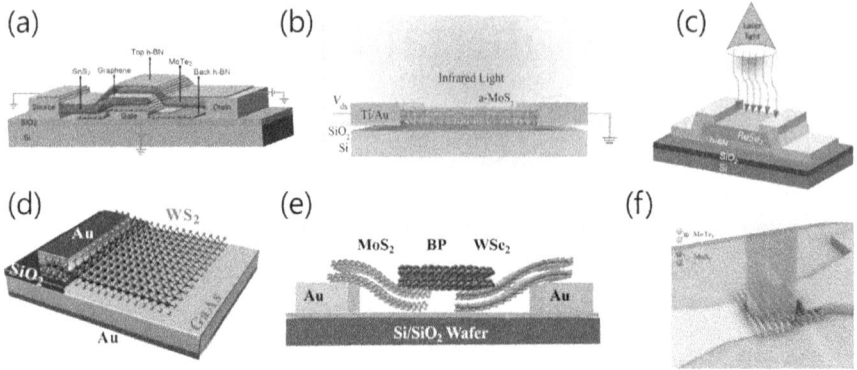

FIGURE 5.1   (a) Cross-sectional view of MoTe$_2$/Graphene/SnS$_2$ P-G-N junctions. (b) The layout of ultra-broadband amorphous MoS$_2$ photodetector. (c) The schematic representation of ReSe$_2$ broadband photodetector on h-BN with Sc/Au electrodes. (d) A diagrammatic illustration of self-driven 2D-WS$_2$/GaAs broadband photodetector. (e) Depiction of WSe$_2$/BP/MoS$_2$ heterojunction-based high-performance broadband floating-base bipolar phototransistor. (f) Self-Powered and multiband detection 2D MoTe$_2$/MoS$_2$ photodiode.

Source: (a) [19] Copyright 2018 Wiley. (b) [20] Copyright 2019 of American Chemical Society. (c) [6] IOP. (d) [21] Royal Society of Chemistry 2020. (e) [22] Copyright 2017 of American Chemical Society. (f) [23] Copyright 2020 of American Chemical Society.

up to monolayer, which ascribes the effect of strong quantum confinement in the atomically thin layer. Such kind of property is well compatible with TMDCs in photoelectronic [8–12]. Surprisingly, monolayers of TMD semiconductors can efficiently absorb and emit the incident photons through its basic bandgap with tremendous PL and electroluminescence traits [13]. During electron-hole pair generation, the excitons or trions are tightly bound because of their configurational constraints and weak dielectric screening. While from their PL spectra at room temperature, it is evident that their charges could be altered by electric field or chemical doping [14, 15]. Normally, thick layers of TMDC semiconductors demonstrate a large photocurrent as compared to a thin layer, which is possibly due to optical absorption. Nevertheless, the performance of photodetectors can be improved by various techniques, such as doping, defect engineering, graphene/TMDCs structure, and 2D-2D heterostructure materials [5, 16–18]. Such schemes can boost up photogeneration, photoresponsivity, and quantum efficiency without increasing the active area of the devices. However, there are many unique broadband photodetectors established based on TMDCs and their heterostructures. A few of them are shown in Figure 5.1 [6, 19–23].

Elahi et al. fabricated the vertical graphene/ReSe$_2$/graphene photodetector (UV-Vis-NIR) and attained a remarkable photoresponsivity (~1.5 × 10^7 A W^{-1}) and external quantum efficiency (EQE) of ~64% at wavelength λ = 220 nm [24]. Also, this photodetector is exposed under several power density values to reveal the mechanism of the recombination process at shallow and deep levels. Similarly, Hu and the group analyzed the black phosphorus (BP) ink that is compatible with inkjet printers and encourages scalable progress in optoelectronic and photonic devices. In this work, it is also determined that the printed BP can stably work as a switch under powerful irradiation for laser applications and also visible to NIR photodetector, having large photoresponse [25].

5.1.1 Photovoltaic Devices Based on 2D Materials

Previously, in photovoltaic devices, few issues like cost-effectiveness, small operating voltage, high efficiency, and low power consumption remained in discussion to explore the alternative source of energy. However, the exponential increase of energy consumption and environmental crises by using fossil fuels have created an alarming situation, and renewable energy has become the fundamental need for the world. Sun is the abundant source of renewable energy. To use this energy pragmatically, devices that convert solar energy into useful energy like electric energy are needed. Photovoltaics, the devices which convert solar energy into electrical energy, became the prime solution to the demand. But it is necessary to find inexpensive, efficient, biocompatible, and environmentally-friendly photovoltaic devices to overcome such challenges [26]. Universally, it is essential to develop inexpensive, wearable, and eco-friendly semiconducting materials that can easily support the scalable, facile, and economical process to convert solar energy [27]. Meanwhile, for a broad range of applications, the electronic devices should be envisioned with society and technology developments to uplift the future of the next generation [28]. Therefore, to accomplish these proposed applications by using conventional rigid semiconducting materials seems challenging [29, 30]. A honeycomb-structured and monolayer carbon atom (graphene) was discovered in 2004. In addition, a vast number of 2D TMDC semiconducting materials in layered forms have appeared on the research horizon. Also, their energy bandgaps vary from graphene (gapless, 0 eV) to hexagonal boron (insulator, 6 eV), and the electrical properties from metallic (graphene) to the semiconducting (TMDcs) and dielectric (h-BN) changes, as shown in Figure 5.2 [31]. Generally, the atoms are bound by a strong covalent bond, while the integration of bulk form (3D structure) is formed by its atomic monolayers with weak forces (van der Waals) which are stacked vertically to the 2D plane. In comparison with typical 3D semiconducting materials, the 2D TMDC demonstrates numerous extraordinary properties that are tempting for novel photovoltaic devices [32, 33].

Usually, the process of photovoltaic happens within the p-n junction or Schottky junction (SK-junction) having an internally built electric field. When the devices are illuminated, e^--h^+ pairs are generated by incoming photons on the device area, which have equivalent or more energy than the energy bandgap of the semiconductor.

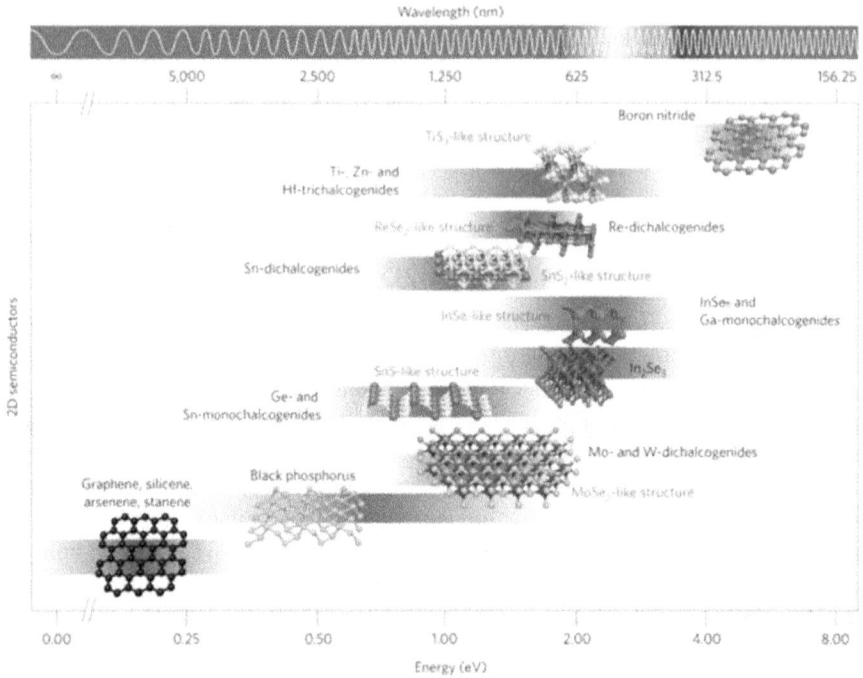

FIGURE 5.2 A comparative study of energy bandgap values of various 2D layered semi-conductor materials groups. The band structure of materials is also shown to highlight the resemblances and dissimilarities between these groups. The horizontal gray bars designate the possible bandgap ranges.

Source: Used with the permission [31]. Copyright 2018, Wiley

Thus, they are split due to junction internal built-in electric field and then finally collected by electrodes, which ultimately generate the electric power. In addition, Jun Ma et al. reported a high-performance SK-junction solar cell (SC) based on graphene/MoS_2/Si heterostructure [34]. In this article, they optimized the thickness of MoS_2 and achieved 15.6% efficiency after coating TiO_2 as an antireflection layer. See in Figure 5.3.

However, in Gr/Si and TMDCs/Si SK SCs, the flexibility of devices is ignored. In contrast, all TMDCs-based p-n junction brings novel promises for their conduction engineering in SCs. However, such schemes are relatively wonderfully implemented in power conversion devices due to their flexibility and high optical absorption. Recently, various 2D-2D p-n junction and graphene flexible photovoltaic devices have been established for energy-harvesting purposes, shown in Figure 5.4.

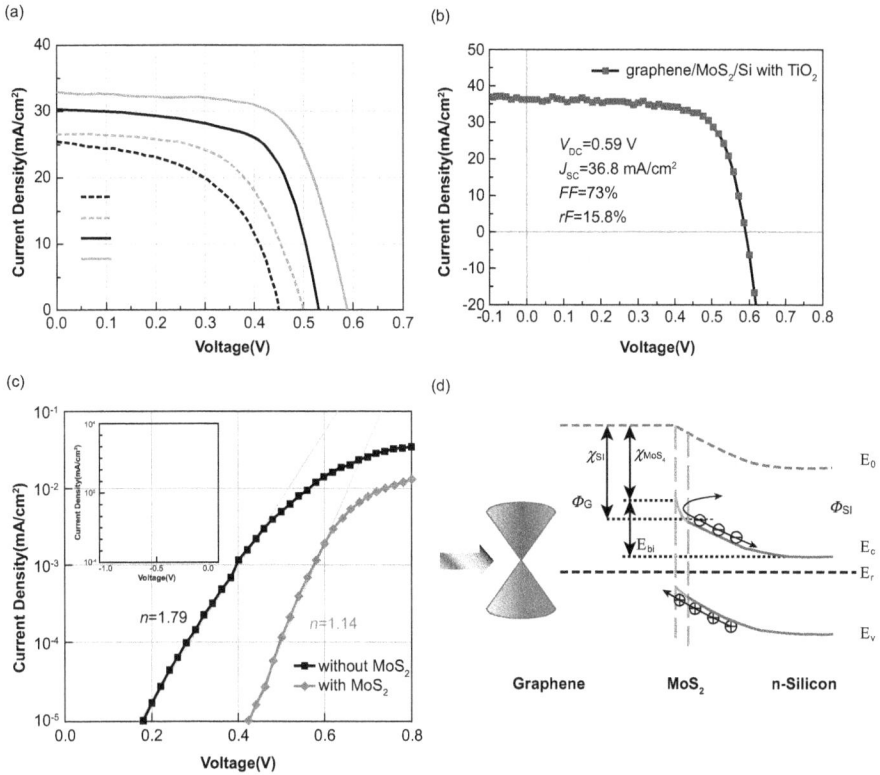

FIGURE 5.3   J-V curves of (a) Gr/MoS$_2$/Si SK-junction SCs. (b) Gr/MoS$_2$/Si SKjunction SCs with a TiO$_2$ anti-reflective coating. (c) The Dark J-V curves of graphene/Si SCs having and without having MoS$_2$ interlayers. (d) Diagrammatic representation of band diagrams for Gr/MoS$_2$/Si SK-junction SCs.

Source: Reprinted with permission [34], copyright 2018, Elsevier

5.2   2D OPTOELECTRONIC MEMORY DEVICES

Memory-based data storage systems are the keystone of today's electronic information industries. Optoelectronic memory devices have a pivotal role in the advancement of digital data computing and optical communication. The numerous opportunities of multifunctional incorporation for optical sensing, data storage, and processing have been established by optoelectronic random-access memories (ORAMs) in a single device. The single-atom-thick 2D TMDCs have been deliberated as auspicious contenders for flexible, wearable, and transparent optoelectronics memory applications [39, 40]. This superb aspect attributed to the 2D materials is its high optical absorption and strong light-matter interactions within the material. The electronic memories based on TMDCs materials present promising designs and concepts for ORAM structures based on 2D materials. The basic standard of state-of-the-art ORAMs is

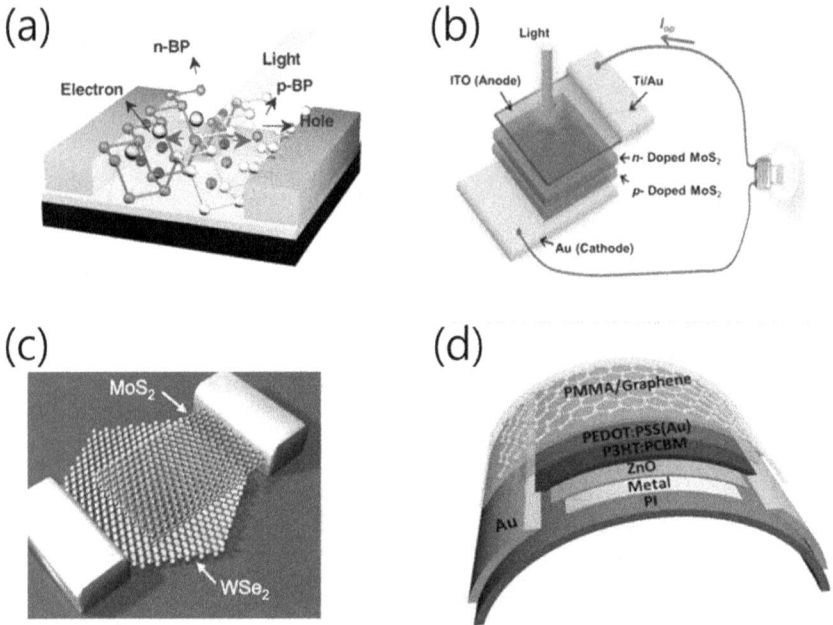

FIGURE 5.4 (a) The diagrammatical representation of the n-BP/p-BP homo-junction photovoltaic device. (b) Graphic design of an n-MoS$_2$/p-MoS$_2$ based p-n junction SC treated by CHF$_3$ plasma. (c) Schematic illustration of a MoS$_2$/WSe$_2$ heterojunction SC under 514 and 633 nm laser illumination. (d) Schematic view of an OPV with the inverted structure: PI/Metal/ ZnO/P$_3$HT:CBM/PEDOT:PSS(Au)/graphene/PMMA.

Source: Reprinted with permission from (a) [35], copyright 2017, Wiley. (b and c) [36] [37], copyright 2014 of American Chemical Society. (d) [38], copyright 2013 Wiley.

usually based upon the trapping of charges in FETs and light-tunable filamentary switching devices. In 2D materials and their heterostructure, the integration of floating gate memories has shown great potential, having a high ON/OFF ratio with long, extended retention time. Because 2D materials have demonstrated a broad spectral response from UV to NIR light with high responsivity and robust response, so by designing an appropriate structure, it is likely to develop photo-sensing and information storage devices, which can lead to potential applications, including image sensors, logic gates, and artificial intelligence. The description of ORAMs is divided into two schemes according to their device configuration: ORAM manifested by 2D materials and ORAM manifested by 2D heterostructures. It will be explained in sections 5.3.1 and 5.3.2, respectively.

In both ORAMs and customary photodetectors, the light signal is converted into an electrical signal. But the scheme of ORAMs is quite different from the photodetectors. The photogenerated carriers are immediately recombined in the photodetectors when light is switched off, and the photocurrent is also quickly dropped from I_P (photocurrent) to I_D (dark current). The photoresponsivity "R" of the photodetector describes the proficiency of conversion from optical to electrical, which can be estimated by Eq. (1).

$$R \ (A/W) = (I_P - I_D)/P \ x \ S \tag{1}$$

In which "P" represents the light power density, and the active channel area of the device is represented by "S."

Conversely, in ORAM, the optical data/information can be stored and sustained for a long time even after removing the light pulse, corresponding to a persistent and retained photocurrent I_S. In ORAMs, the photocurrent I_P usually decreases gradually when the light is switched off and reaches up to a stable current storage I_S. Therefore, we can describe a particular nonvolatile responsivity (NR) for ORAM, which can be calculated as follows:

$$NR \ (A/W) = (I_S - I_D)/P \ x \ S \tag{2}$$

Where "I_S" is the storage current.

5.2.1 ORAMs Manifested by 2D Materials

The 2D materials based ORAMs have been performing subsequently due to their active charge trapping site. Photogenerated charge carrier electrons/holes can be stored proficiently even after switching off the light. Although, in this scenario, the ORAM can store the optical data because the number of photogenerated carriers is estimated by the value of intensity of light, which is normally associated with the presence of charge storing or interfacial physics. While tuning the intensity of light provides the multiple resistance states. Furthermore, the light pulse of different wavelengths provides a variety of optical excitation energies and optical absorption, resulting in dynamic charge trapping/de-trapping physics and numerous resistance states. Recently, Lie et al. established an optoelectronic memory array by atomically thin $CuIn_7Se_{11}$ for image sensing [41]. Further, this work proceeds with InSe and MoS_2 materials, in which photogenerated charge carriers are trapped and de-trapped in the barriers by applying a gate voltage to 2D material, as shown in Figure 5.5. The optoelectronic memory composed of atomically thin layered semiconductors can store the photon-generated carriers during the exposure of light, and then carriers can be read out during the data processing and long-lasting storage.

In addition, the researcher also demonstrated the multilevel, high ON/OFF ratio ($\sim 10^7$), long retention time ($\sim 10^4$s), and highly responsive (8×10^3 A/W) nonvolatile 2D ORAM and a highly responsive nonvolatile 2D ORAM by incorporating Au nanoparticles on SiO_2 surface to trap charges, as shown in Figure 5.6 (a) [42]. Similarly, the artificially charged trapping states are induced at the MoS_2/SiO_2 interface by using oxygen plasma on SiO_2 before the growth of monolayer

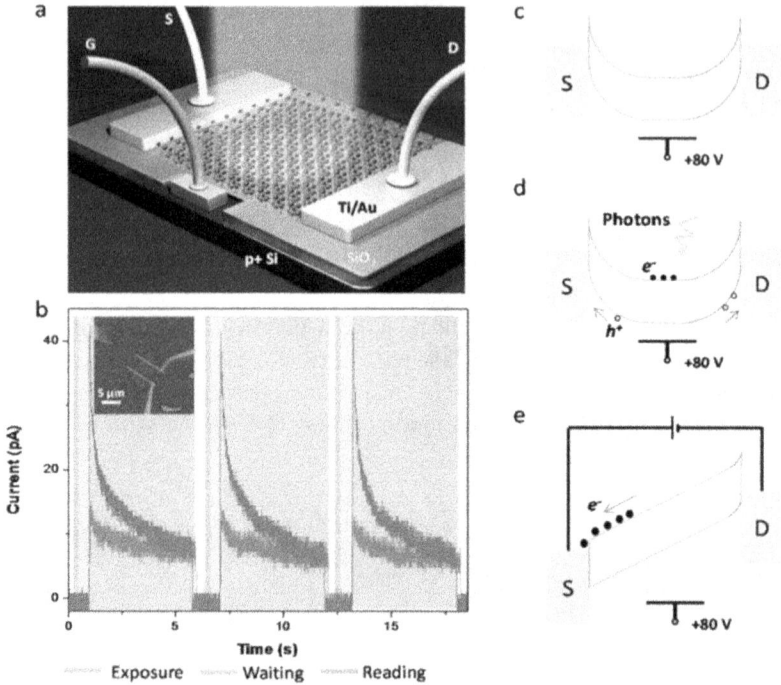

FIGURE 5.5 (a) Schematic of a CIS-based transistor. (b) Working arrangement and read-out current from the CIS test structure shown in the inset. (c) At the +80 V gate, layered CIS creates SK barriers with Ti/Au electrodes. When the light is OFF, the SK barrier puts the potential well empty. (d) At +80 V gate voltage, the SK barrier forms a potential well that traps photoelectrons. When light is ON, e⁻-h⁺ pairs are generated. The electrons gets trapped at the potential wall when the holes dispatched from the CIS moves toward the electrodes. (e) After illumination, a bias is applied to the device, which opens the potential well from one side, and the trapped electrons are released.

Source: Reprinted permission is taken from [41], copyright 2015 of American Chemical Society.

MoS_2, see Figure 5.6 (b, c). Deliberately, the formation of charge traps of the MoS_2 ORAM device revealed an enhanced 4700 ON/OFF ratio with 6,000 s retention time [42].

5.2.2 ORAMs Manifested by 2D Materials' Heterostructures

The ORAMs based on 2D materials are usually sensitive to surrounding environments because virtually all memory devices give great response to charge trapping states. Whether these are intrinsic or artificially induced at the interface of materials. The principle of charge trapping mechanism typically encourages the short retention time in ORAMs, which deters its application. However, the heterostructure of 2D

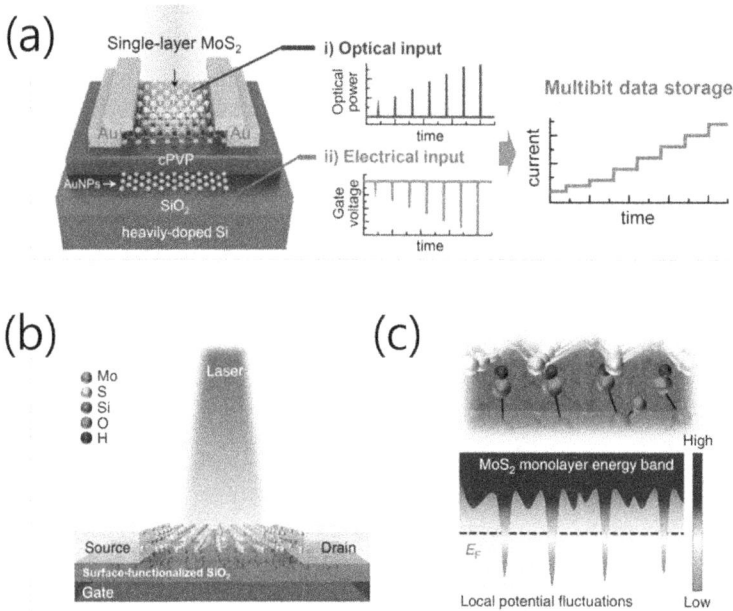

FIGURE 5.6 (a) Schematic diagram of MoS_2 ORAM with a layer of Au nanoparticles between MoS_2 and SiO_2 as charge trapping layer. (b) Schematic design of MoS_2 ORAM with functionalized plasma-treated MoS_2/SiO_2 interface. (c) Representation of the extended atomic structure of MoS_2 having the functional groups. The summarized energy band structure demonstrates the local potential fluctuations encouraged by intentionally generated artificial trap sites.

Source: Reproduced with the permissions (a) from Wiley [42] and (b) and (c) from Nature [39]

materials or integration of 2D materials along with the different semiconductors can extend the scope and performance of environmental stability, light absorption range, nonvolatile properties, and data storage capacity in ORAMs.

Liu and the group reported a multibit nonvolatile optoelectronic memory of single-layer WS_2 and few-layer thick h-BN, as shown in Figure 5.7 (c). The WSe_2/h-BN memory that demonstrates a memory switching ratio near $\sim 1.1 \times 10^6$ confirms about 128 (7 bit) discrete storage states. This memory displays robustness in the retention time, around $\sim 4.5 \times 10^4$ s. Also, the advantage of wideband light spectrum empowers its usage in filter-free color image sensors [43]. Similarly, Huang et al. in 2019 claimed a nonvolatile multibit optoelectronic memory manifested by MoS_2, h-BN, and graphene by using a top-floating-gated structure, see Figure 5.7 (d). Moreover, after electric programming, laser irradiation can be erased by these devices with a broadband spectrum. For the multilevel storage property, the data states are controlled by different electrical and laser pulse illuminations [44]. Furthermore, recently Mukherjee et al. designed a laser-assisted multilevel nonvolatile memory of few-layer thick ReS_2/h-BN/graphene heterostructures, see Figure 5.7 (e, f) [45]. The

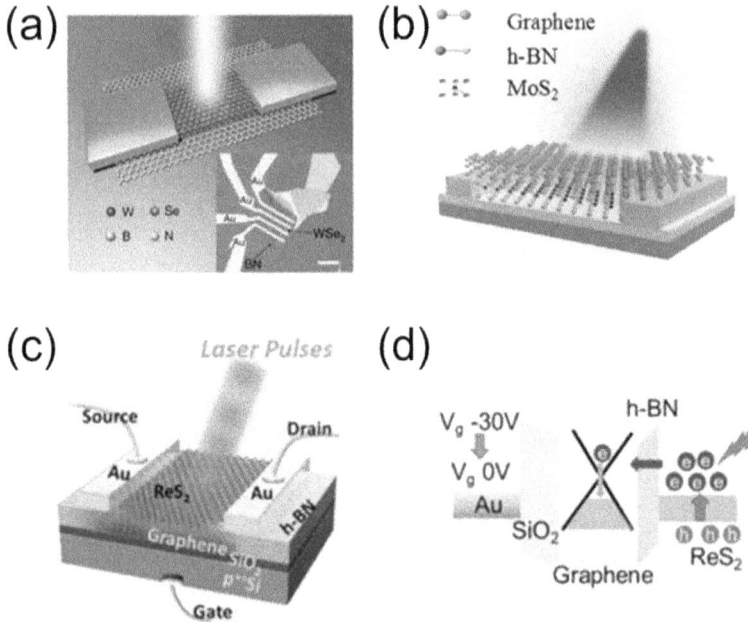

FIGURE 5.7 (a) Diagrammatic representation of graphene/h-BN/MoS$_2$ ORAM with Ti/ Au contacts. (b) The dynamic response for photon writing (high current) and electrical erasing (low current) states. (c) Schematic drawing of the optoelectronic memory made by WSe$_2$ flake on h-BN. (d) Schematic of memory structure designed with MoS$_2$ as a light-sensitive channel (bottom), h-BN as a dielectric (middle), and a Gr as a floating gate (top). (e) Schematic illustration of the memory device composed with graphene, h-BN, and ReS$_2$ layers stack. (f) Band structure demonstrating charge accumulation procedure where the amount of stored charge in the graphene is controlled by light and voltage pulses.

Source: Reused with the permission of (a) [43], copyrights from Nature; (b) [44], copyrights from Wiley; (c) and (d) [45], copyrights from Wiley.

direct bandgap of ReS$_2$ fulfills several requirements as an active channel for optical memory, which makes it a potential candidate in photo-assisted optoelectronic applications. Such nonvolatile memory described a high current ON/OFF ratio, a large memory window, good endurance (> 1,000 cycles), and stable retention (> 10^4 s). At 10 ms gate pulses of +10 V and −10 V, a successive program and erase states are observed, respectively.

5.3   BIO-PHOTONICS OF 2D MATERIALS

The growing demands of diagnostics diseases and biomedical applications have significantly encouraged the rational structure and synthesis of an inclusive range of functional nanomaterials. The association of 2D nanomaterials with the recognition of biomolecular events establishes the sensing and diagnostic investigations.

2D materials are among the most studied materials in the twenty-first century of nanotechnology field. Materials such as graphene and TMCs attracted a plethora of attention because of their wide range of properties leading to numerous applications in biosensing and bio-photonics [46–50]. Because of their exceptional optical properties, 2D materials will emerge in biomedical photonic applications once their bio-compatibility is satisfactory, depending upon their chemical and physical modifications. We address the latest advancement in bio-photonics of 2D materials and their imaging applications in this section. A biosensor is a diagnostic device that is used to locate biochemical substances and the concentration or existence of different kinds of biological analytes, like antibodies, DNA, enzymes, proteins, or antigens. There are three components of a biosensor [51].

1. The probe
2. The transducer component
3. The electronic system

The function of the probe is to identify the analyte that is of main concern. The transducer transforms the event of bio-recognition into a quantifiable signal, and the electronic system that reads the output is often used in conjunction with the transducer [52].

Mainly, optical biosensors of 2D materials are focused. In 2D materials, it is widely known that a nanosheet is exceedingly susceptible to the differences on its surface. For example, an optical biosensor based on graphene exhibited single-cell flow detection with excellent resolution and sensitivity [53]. Down to single cells, this sensor can differentiate between tiny normal lymphocyte cells and bigger malignant, cancerous cells. Furthermore, the dynamic range is wide and easily adjustable by modifying the angle of incidence and power of incidence. Recently, MoS_2 has been utilized in a photo-electrochemical sensor to detect miRNA. The photoanode comprises Au NPs and MoS_2 NSs heterojunction on ITO [51]. The photoresponsive material is the MoS_2-Au NPs heterojunction, which improves the visible light-harvesting efficiency of MoS_2. The MoS_2 sensor on ITO with Au NPs displays high sensitivity for miRNA detection, according to photocurrent measurements, with the capacity to differentiate interrelated and single-base mismatched miRNA.

For a moment, the graphene-based optical refractive index sensor showed effectively good resolution (1.7×10^{-8}) and enhanced sensitivity of (4.3×10^7) mV/RIU with extensive dynamic range, see Figure 5.8 [54]. This sensor can precisely detect a minor amount of cancer cells among other normal cells at the single-cell level, which provides a true statistical distribution of cancer and normal cells by fewer cells.

Further, it is a matter of great importance to detect specific DNA or peptide sequences for several biomedical applications, including disease diagnostics, biomolecular analysis, gene therapy, environmental forecast, etc. In 2014, Loan and the group designed a device structure of graphene/MoS_2 to use for selective and label-free detection of DNA hybridization, see Figure 5.9 (a–b) [50]. Recently, Shorie et al. made a nanohybrid mediated SERS substrate, which was established by incorporating the Au NPs on exfoliated NSs of WS_2 to make plasmonic concentrations. High selectivity for cardiac marker myoglobin was achieved by functionalizing the

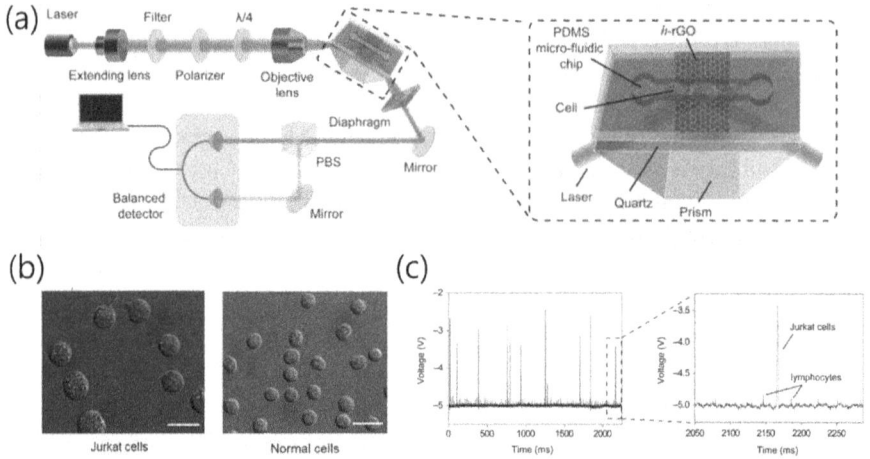

FIGURE 5.8 (a) Flow-sensing system set up for a single cell. (b) Micrographs of Jurkat cells and lymphocytes with scale bar of 15 μm (left and right side, respectively). (c) Distinct time-dependent variation in voltage linked with mixed lymphocytes and Jurkat cells.

Source: Reprinted with the permission of [54], copyright from American Chemical Society

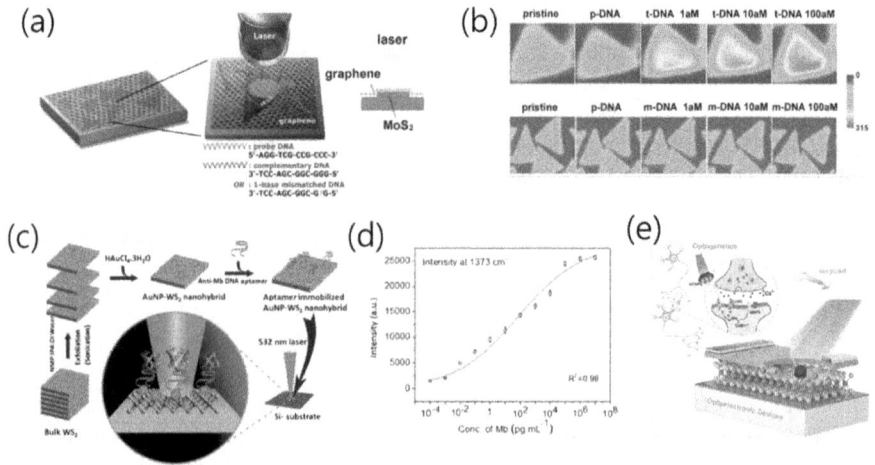

FIGURE 5.9 (a) Schematic drawing of DNA detection technique by microscope and structure of sensor composed by graphene/MoS$_2$ heterostructure along with (b) the photoluminescence (PL) peak area mappings. (c) Schematic representation of SERS device. (d) Calibration curve of Mb (Myoglobin) with related Raman intensity. (e) Schematic structure of as-prepared NIR phototransistor.

Source: Reused with the permission of (a) and (b) from [50], copyrights from Wiley, (c) and (d) [55], copyrights from Springer, and (e) [56], copyrights from Elsevier

surface of the nanohybrid with specific aptamers, Figure 5.9 (c, d) [55]. This investigation demonstrates a paragon to synergistically manipulate the distinct electromagnetic and chemical characteristics of Au NPs and WS_2 for many-fold augmentation of SERS signals. In addition, Zhai et al. designed an up-conversion optogenetics-inspired NIR phototransistor synapse, manifested by up-converting nanoparticles sensitized MoS_2 [56]. In this device, MoS_2 could be used as light-sensitive ion channels to reabsorb the visible light emitted from up-converting nanoparticles under near-IR illumination, see Figure 5.9 (e). This idea of bringing biology, optics, and electronics close together can open new routes for emerging optogenetics-inspired neuromorphic technology in bio-photonics.

5.4 LIGHT-EMITTING DIODES (LEDs) OF 2D MATERIALS

The LEDs are considered to be the most valuable optoelectronic innovation in the past few decades, even winning the 2014 race for the Nobel Prize in Physics. LED is a semiconductor device that emits light (visible or invisible) when energized. In detail, when a p-n junction is forward biased, recombination of charges (electrons and holes) takes place within the structure and mostly near to the p-n junction. This recombination process requires that the energy possessed by the free electron be released or transferred in the form of heat or in the form of photons. The color or wavelength of the released light depends on the bandgap of the semiconducting material used for the making of LEDs. LEDs have several advantages over an incandescent light source, including brighter light, longer lifetime, less combustion, and smaller size. Various types of LEDs have been studied and commercialized in recent years.

The two-dimensional materials have appealed to great attention in recent times due to their captivating optical, mechanical, and electrical properties, which are originated from their unique band structure and morphology [12, 57–59]. Graphene, a monolayer of carbon atoms hexagonally packed like a 2D honeycomb lattice structure, has excellent carrier movement and broadband light absorption [57, 60]. Graphene has a commendably modifiable charge carrier density, which makes it a promising contender for electrically tunable devices. However, with the lack of bandgap in graphene, 2D TMDs (MoS_2, WS_2, $MoSe_2$, $MoTe_2$, WSe_2, etc.) have become promising layered semiconductors, having a bandgap in the range of 1.57 to 2eV [61], which is more pertinent for making LEDs. Besides the layered materials, there is PbS, a non-layered 2D material having a direct bandgap of around 0.4 eV, which has great potential in a broad range of wavelength detection from visible region to IR [62, 63]. The ultrahigh carrier mobility, a wide range of light emission, and powerful light-matter interaction of 2D materials make them suitable contenders in optoelectronics, such as LEDs and photodetectors.

It is important to find single-layer TMDs with direct bandgap structure and large light-emitting proficiency. The properties of light emission in 2D TMDs depend on the excitation effect of the materials [14, 64]. Strong Coulombic force interaction and charged carriers generated electrically or optically in TMDs remarkably give rise to a multiple-particle phenomena such as the generation of a different kind of exciton (e-h pair), including trion (two h^+s with an e^- or vice versa) and bi-exciton (two h^+s with two e^-s) [65]. In these formations, trion exhibits a charged feature, whereas

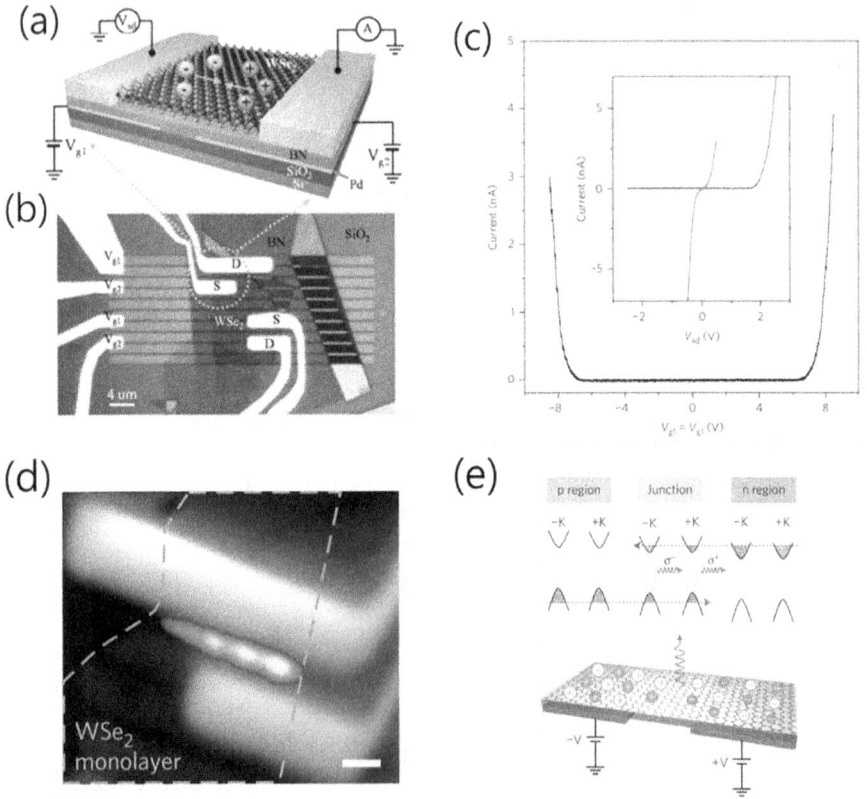

FIGURE 5.10 (a) Schematic of single-layer WSe_2 p-n junction devices. (b) Optical image of the multiple monolayer WSe_2 p-n junction devices. The scale bar is 4 μm. (c) The variation of current values. (d) The electroluminescence image (red) placed over the device image (gray-scale). (e) Band diagram and device schematic drawing the electroluminescence generation from valley excitons.

Source: Reproduced with the permission of [74], copyrights from Nature

exciton and bi-exciton express a charge-neutral feature [66]. In 2D materials, there are several ways to create p-n junctions for LEDs, such as chemical doing, electrostatic gating, van der Waals heterostructures, and single photons emitters [67–71].

The most common technique used to build LEDs is electrostatic gating, which uses two local gates to create a p-n junction within the 2D material sheets [72, 73]. Since monolayer WSe_2 is one of the promising TMD materials, it has been widely used to create LEDs based on the p-n junction with a large photon emission rate (~16 million s^{-1}) at an applied current of 35 nA, as shown in Figure 5.10 [74].

Despite all that mentioned in the preceding section, the previous study reveals that ambipolar transport behavior can be encouraged in a field-effect transistor with conventional gate dielectrics that may have serious issues of dielectric breakdown [73, 75]. To overcome the dielectric breakdown problem, an ionic liquid gated structure

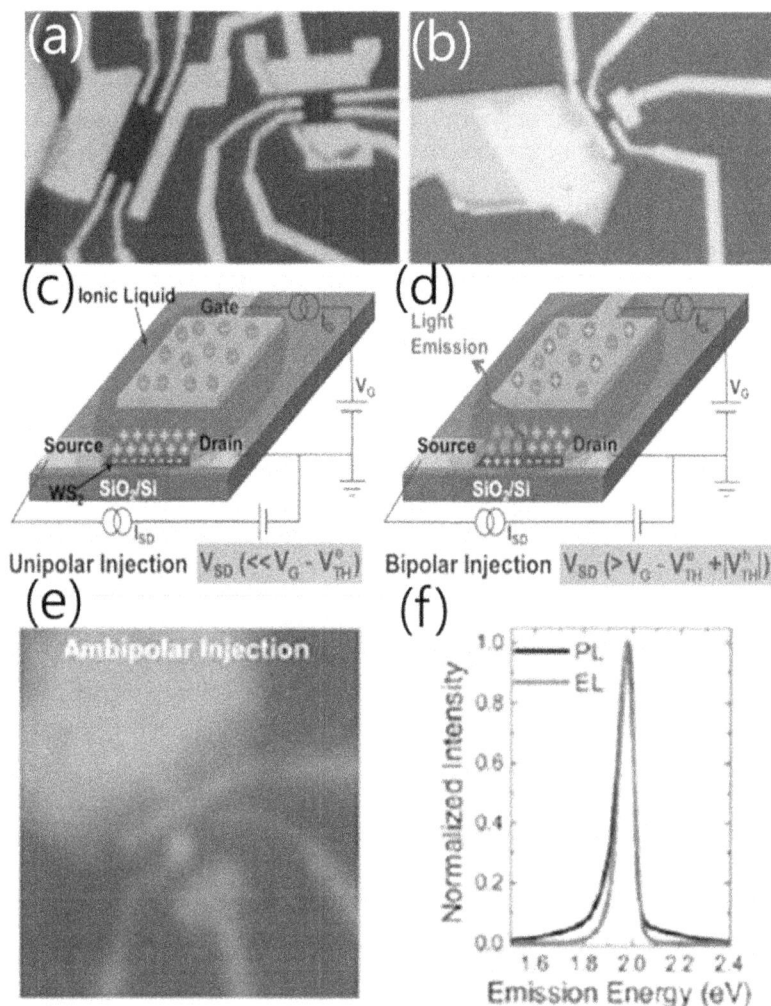

FIGURE 5.11 The optical microscope images of final devices after Au contacts for (a) mono- (ML) and bilayer (BL) and (b) only ML flakes. (c) Schematic drawing of the ionic liquid gated FET of monolayer WS_2 devices in unipolar injection regime and (d) bipolar injection regime. (e) The optical image of the monolayer WS_2 device with visible emitted light spot due to electroluminescence. (f) A comparison of the normalized electro- and photoluminescence spectra of monolayer WS_2.

Source: Reprinted with the permission of [76], copyrights from the American Chemical Society

is used, which demonstrates unipolar and bipolar injection regimes, see Figure 5.11 [76]. In this work, Jo et al. designed an LED of a monolayer WS_2 with ionic liquid gated FET.

Under bias conditions, this device shows an emitted light spot because of electroluminance with high visibility. The position of the emission spot in an ambipolar regime can be easily shifted by changing the gate and source-drain voltage. The effective electrons and holes are accumulated in the FET, execute the operation in an ambipolar injection regime, with electrons and holes injected at the opposite electrodes of the device simultaneously, which leads to light emission. The electroluminescence and photoluminescence spectra for the monolayer device are similar to each other, as expected [76].

Notably, van der Waals heterostructures of 2D semiconductors are the backbone of solid state physics and allows a variety of 2D materials with different crystal axes orientations and lattice constant to be combined [77]. Recently, Ye et al. studied the microscopic investigation of the electroluminescence from a diode manifested by the assembly of heavily p-type doped silicon and monolayer MoS_2 [78]. Therefore, to check the device performance, a voltage is applied to the MoS_2/silicon heterojunction, as shown in Figure 5.12. In this work, the injection of holes from silicon across the junction gives rise to a recombination process due to the direct bandgap of monolayer MoS_2. The electroluminescence spectrum of the device is found at room and low temperatures under different current values. The electroluminescence spectra exhibit a current threshold of about 15 µA; however, the basic features of the excitation emission (AX and DX) are observed at currents higher than threshold.

Similarly, in 2017 Ross et al. designed a lateral p-n junctions of WSe_2 and $MoSe_2$ hetero-bilayer, which electrostatically determined the interlayer exciton optoelectronics, see Figure 5.13 [79]. Since the application of a forward bias facilitates the early detection of electroluminescence from interlayer excitons. The p-n junction at the neutral bias performs as the most sensitive photodetector, where the wavelength-dependent photocurrent measurement permits the direct observation of resonant optical excitation of the interlayer exciton.

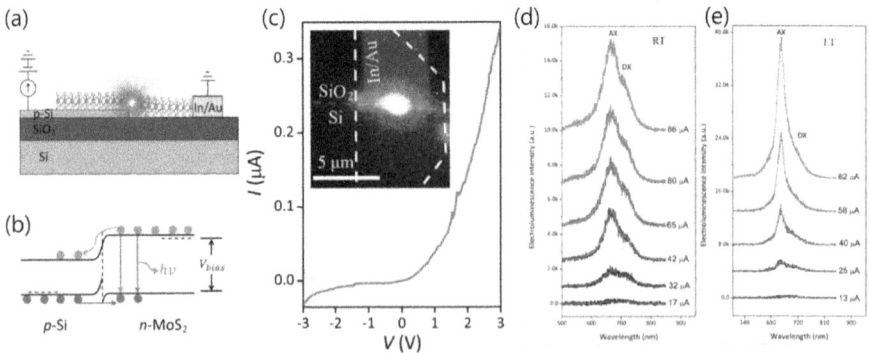

FIGURE 5.12   (a) Diagrammatic representation of the monolayer MoS_2/Si heterojunction. (b) Model band structure of the MoS_2 diode under a forward bias voltage. (c) Electrical characteristics of the monolayer MoS_2 diode. Inset: surface plot of the electroluminescent emission. (d) At room temperature and (e) at low temperature, the electroluminescence (EL) spectra of a MoS_2 diode were recorded at different current magnitudes.

Source: Reproduced with the permission of [78], copyrights from the American Institute of Physics

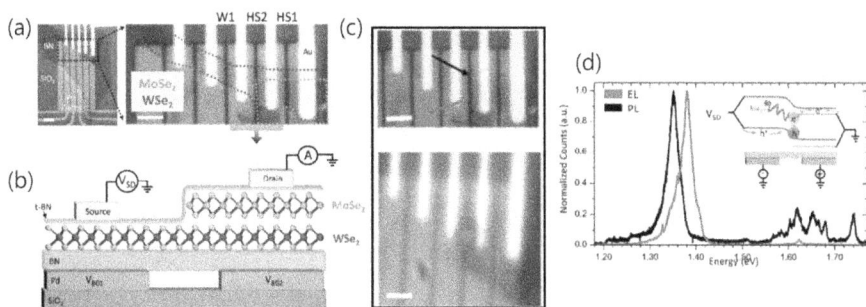

FIGURE 5.13   (a) Top view of a microscope image of the device. The scale bar is 2 μm. (b) Drawing of device region (HS2) cross-section. (c) Micrograph and electroluminescence image of the device area (top and bottom, respectively). Scale bars are of 2 μm (d) spectra of EL (red) and photoluminescence (black) for the HS2 region.

Source: Reprinted with the permission of [79], copyrights from the American Chemical Society

The interlayer exciton has about 200 times lower photocurrent amplitude as compared to that of the resonant excitation. This shows that the spatial separation of holes and electrons affects the oscillator strength of the interlayer exciton to a level of two orders of magnitude.

5.5   PHOTODETECTORS OF MXene

Based on their phenomenal physical and chemical characteristics, MXenes have gained much attraction as a group of 2D layered materials [80, 81]. The general chemical formula of MXene is $M_{n+1}X_nT_x$, where n = 1, 2, or 3. M represents transition metal (Ta, Mo, Cr, Hg, Zr, and Ti); X denotes carbon or nitrogen; and T_x is a surface functional group (hydroxyl, oxygen, chlorine, or fluorine) [82, 83]. Several types of MXenes have been obtained by experimentation, and a lot more have been predicted theoretically [84–86]. Owing to their stochiometric properties and various surface functionalization, MXenes have shown tunable properties. In particular, $Ti_3C_2T_x$ (type of MXene) has been verified to be a high optical transmittance (~93%) and conductivity (~5736 cm^{-1}) when its thickness is ~4 nm [87, 88]. Another exciting feature of MXenes is the tunable work function (~2.14 eV to ~5.65 eV) that covers a wide range of materials [89]. Meanwhile, unique optical and electrical properties can be finely tuned by transition metals, surface termination, thickness, electrochemical behavior, etc. [90]. Subsequently, these tunable optoelectronic properties make MXenes a promising candidate for many future applications, like transparent conductors, photodetectors, light-emitting diodes, photovoltaic devices, energy storage devices [85, 91–93].

However, a photodetector is considered to be an essential optoelectronic device and can convert incident light into current or voltage, having a wide range of applications, such as image sensing, night visions, and communications. Despite their rich optical and electrical properties, 2D MXenes photodetectors are not well explored in the research community. However, the recent development in 2D MXenes based

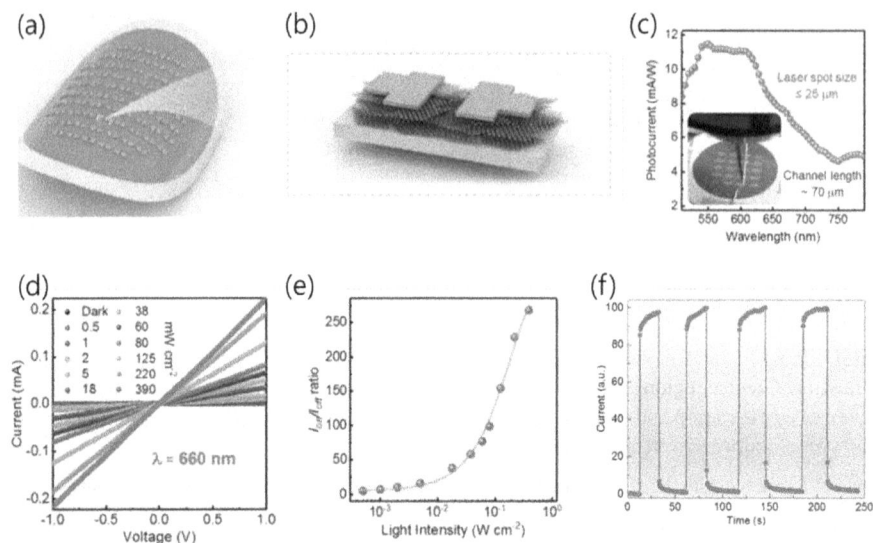

FIGURE 5.14 (a) Schematic of design flexible photodetectors, parallel-type thin-film Mo_2CT_x. (b) Zoomed-in view of one device from an array. (c) Photo-response of a Mo_2CT_x photodetector at 0.7 V bias. Inset: Photograph of the photocurrent arrangement. (d) I-V characteristics of the Mo_2CT_x photodetector in the dark and under different light intensities of a wavelength of 660 nm. (e) The ratios of the photocurrent to the dark current of the Mo_2CT_x thin-film photodetector versus different light intensities. (f) Photo-response of Mo_2CT_x photodetector, the light-switch ON and OFF cycles were ~500 ms at a wavelength of 660 nm.

Source: Reprinted with the permission of [94], copyrights from the Wiley

photodetectors over the past few years cannot be ignored. There are several reports on the MXenes based photodetectors, including simple photodetector, plasmon-enhanced photodetector, and self-driven photodetector [94, 95][16].

Photodetectors based on five different MXenes nanosheets (Mo_2CT_x, T_2CT_x, T_2CT_x, $Ti_3C_2T_x$, $Ti_3C_2T_x$, Nb_2CT_x, and Nb_2CT_x) are fabricated, among which Mo_2CT_x shows promising results [94]. In 2019, Velusamy et al. deposited thin films on paper substrates, which demonstrated high-level responsivity and detectivity (9 A W^{-1} and ~5 × 10^{11} Jones, respectively) in a broad range of 400–800 nm with consistent photoswitching characteristics at λ = 660 nm. Moreover, Mo_2CT_x thin-film devices are found to be resistive to change in the natural environment along with a prolonged exposure to light and mechanical forces, making them useful in the visible range of light spectrum, see Figure 5.14.

Recently, Deng et al. demonstrated a photodetector of combined 2D material and MXene. This photodetector is manifested by an all-sprayed-processable method having a large area on common paper by 2D $CsPbBr_3$ and conductive Ti_3C_2Tx (MXene) [96]. This is a flexible photodetector and successfully revealed the photo communication with an on/off current ratio up to 2.3 × 10^3 and an outstandingly fast response to light exposure (~18 ms), see Figure 5.15. Also, after bending the 1,500 cycles,

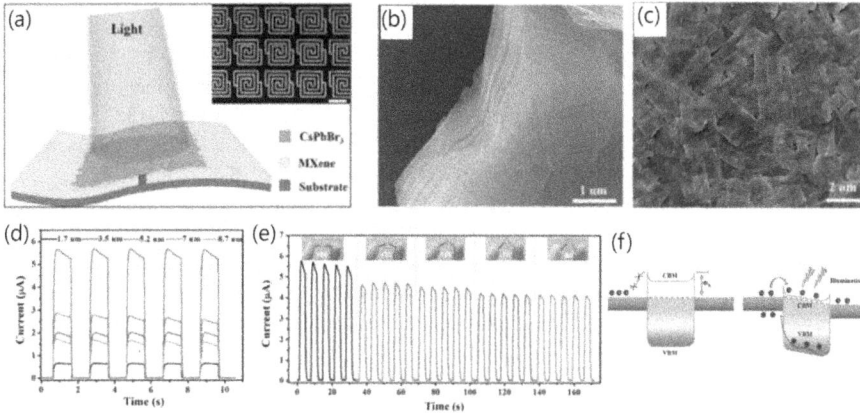

FIGURE 5.15 (a) Diagrammatic representation of a perovskites/MXenes photodetector under illumination. (b) The SEM image of Ti_3C_2Tx. (c) The SEM image of $CsPbBr_3$. (d) Time-resolved photocurrent of the photodetector at a different thickness of MXene electrodes. (e) Time-resolved photocurrent of the photodetector arrays at several bending angles from 180° to 60. (f) Diagrammatic representation of energy band structure of the perovskites/MXenes photodetector at the equilibrium state (left) and under light (right).

Source: Reprinted with the permission of [96], copyrights from the Wiley

the devices still preserve exceptional flexibility and stability. This work offers an economical process of enormously creating flexible large-area photodetectors for extended photo communication and wearable technology.

REFERENCES

1. Bonaccorso F, Sun Z, Hasan T and Ferrari A 2010 Graphene photonics and optoelectronics *Nat. Photonics* **4** 611–622
2. Zhang Y, Liu T, Meng B, Li X, Liang G, Hu X and Wang Q J 2013 Broadband high photoresponse from pure monolayer graphene photodetector *Nat. Comm.* **4** 1–11
3. Bera K P, Haider G, Huang Y-T, Roy P K, Paul Inbaraj C R, Liao Y-M, Lin H-I, Lu C-H, Shen C and Shih W Y 2019 Graphene sandwich stable perovskite quantum-dot light-emissive ultrasensitive and ultrafast broadband vertical phototransistors *ACS Nano* **13** 12540–12552
4. Wang X, Wang P, Wang J, Hu W, Zhou X, Guo N, Huang H, Sun S, Shen H and Lin T 2015 Ultrasensitive and broadband MoS_2 photodetector driven by ferroelectrics *Adv. Mater.* **27** 6575–6581
5. Wu D, Guo J, Wang C, Ren X, Chen Y, Lin P, Zeng L, Shi Z, Li X J and Shan C-X 2021 Ultrabroadband and high-detectivity photodetector based on WS2/Ge heterojunction through defect engineering and interface passivation *ACS Nano.* **15** 10119–10129
6. Khan M F, Rehman S, Akhtar I, Aftab S, Ajmal H M S, Khan W, Kim D-K and Eom J 2019 High mobility ReSe2 field effect transistors: Schottky-barrier-height-dependent photoresponsivity and broadband light detection with Co decoration *2D Mater.* **7** 015010
7. Lai J, Liu X, Ma J, Wang Q, Zhang K, Ren X, Liu Y, Gu Q, Zhuo X and Lu W 2018 Anisotropic broadband photoresponse of layered type-II weyl semimetal MoTe2 *Adv. Mater.* **30** 1707152

8. Price C C, Frey N C, Jariwala D and Shenoy V B 2019 Engineering zero-dimensional quantum confinement in transition-metal dichalcogenide heterostructures *ACS Nano* **13** 8303–8311

9. Kuc A, Zibouche N and Heine T 2011 Influence of quantum confinement on the electronic structure of the transition metal sulfide T S 2 *Phy. Rev. B* **83** 245213

10. Lopez-Sanchez O, Lembke D, Kayci M, Radenovic A and Kis A 2013 Ultrasensitive photodetectors based on monolayer MoS 2 *Nat. Nanotechnol.* **8** 497–501

11. Zeng H, Dai J, Yao W, Xiao D and Cui X 2012 Valley polarization in MoS 2 monolayers by optical pumping *Nat. Nanotechnol.* **7** 490–493

12. Splendiani A, Sun L, Zhang Y, Li T, Kim J, Chim C-Y, Galli G and Wang F 2010 Emerging photoluminescence in monolayer MoS_2 *Nano Lett.* **10** 1271–1275

13. Eda G and Maier S A 2013 Two-dimensional crystals: Managing light for optoelectronics *ACS Nano* **7** 5660–5665

14. Mak K F, He K, Lee C, Lee G H, Hone J, Heinz T F and Shan J 2013 Tightly bound trions in monolayer MoS 2 *Nat. Mat.* **12** 207–211

15. Ross J S, Wu S, Yu H, Ghimire N J, Jones A M, Aivazian G, Yan J, Mandrus D G, Xiao D and Yao W 2013 Electrical control of neutral and charged excitons in a monolayer semiconductor *Nat. Comm.* **4** 1–6

16. Jo S H, Kang D H, Shim J, Jeon J, Jeon M H, Yoo G, Kim J, Lee J, Yeom G Y and Lee S 2016 A high-performance WSe2/h-BN photodetector using a triphenylphosphine (PPh3)-based n-doping technique *Adv. Mater.* **28** 4824–4831

17. Qiao H, Yuan J, Xu Z, Chen C, Lin S, Wang Y, Song J, Liu Y, Khan Q and Hoh H Y 2015 Broadband photodetectors based on graphene-Bi2Te3 heterostructure *ACS Nano* **9** 1886–1894

18. Deng W, Chen Y, You C, Liu B, Yang Y, Shen G, Li S, Sun L, Zhang Y and Yan H 2018 High detectivity from a lateral graphene-MoS_2 schottky photodetector grown by chemical vapor deposition *Adv. Electron. Mater.* **4** 1800069

19. Li A, Chen Q, Wang P, Gan Y, Qi T, Wang P, Tang F, Wu J Z, Chen R and Zhang L 2019 Ultrahigh-sensitive broadband photodetectors based on dielectric shielded MoTe2/Graphene/SnS2 p-g-n junctions *Adv. Mater.* **31** 1805656

20. Huang Z, Zhang T, Liu J, Zhang L, Jin Y, Wang J, Jiang K, Fan S and Li Q 2019 Amorphous MoS_2 photodetector with ultra-broadband response *ACS Appl. Electron. Mater.* **1** 1314–1321

21. Jia C, Huang X, Wu D, Tian Y, Guo J, Zhao Z, Shi Z, Tian Y, Jie J and Li X 2020 An ultrasensitive self-driven broadband photodetector based on a 2D-WS 2/GaAs type-II Zener heterojunction *Nanoscale* **12** 4435–4444

22. Li H, Ye L and Xu J 2017 High-performance broadband floating-base bipolar phototransistor based on WSe2/BP/MoS_2 heterostructure *Acs Photonics* **4** 823–829

23. Ahn J, Kang J-H, Kyhm J, Choi H T, Kim M, Ahn D-H, Kim D-Y, Ahn I-H, Park J B and Park S 2020 Self-powered visible-invisible multiband detection and imaging achieved using high-performance 2D MoTe2/MoS_2 semivertical heterojunction photodiodes *ACS Appl. Mater. Interfaces* **12** 10858–10866

24. Elahi E, Khan M F, Rehman S, Khalil H W, Rehman M A, Kim D-K, Kim H, Khan K, Shahzad M and Iqbal M W 2020 Enhanced electrical and broad spectral (UV-Vis-NIR) photodetection in a Gr/ReSe 2/Gr heterojunction *Dalt. Trans.* **49** 10017–10027

25. Hu G, Albrow-Owen T, Jin X, Ali A, Hu Y, Howe R C, Shehzad K, Yang Z, Zhu X and Woodward R I 2017 Black phosphorus ink formulation for inkjet printing of optoelectronics and photonics *Nat. Comm.* **8** 1–10

26. Polman A, Knight M, Garnett E C, Ehrler B and Sinke W C 2016 Photovoltaic materials: Present efficiencies and future challenges *Science* **352**

27. Jean J, Brown P R, Jaffe R L, Buonassisi T and Bulović V 2015 Pathways for solar photovoltaics *Energy & Environmental Science* **8** 1200–1219

28. Kim S, Van Quy H and Bark C W 2020 Photovoltaic technologies for flexible solar cells: Beyond silicon *Mater. Today Energy* 100583

29. Nathan A, Ahnood A, Cole M T, Lee S, Suzuki Y, Hiralal P, Bonaccorso F, Hasan T, Garcia-Gancedo L and Dyadyusha A 2012 Flexible electronics: The next ubiquitous platform *Proc. IEEE* **100** 1486–1517

30. Wong W S and Salleo A 2009 *Flexible electronics: Materials and applications* vol 11 (Springer Science & Business Media)

31. Das S, Pandey D, Thomas J and Roy T 2019 The role of graphene and other 2D materials in solar photovoltaics *Adv. Mater.* **31** 1802722

32. Wang Q H, Kalantar-Zadeh K, Kis A, Coleman J N and Strano M S 2012 Electronics and optoelectronics of two-dimensional transition metal dichalcogenides *Nat. Nanotechnol.* **7** 699–712

33. Wang L, Huang L, Tan W C, Feng X, Chen L, Huang X and Ang K W 2018 2D photovoltaic devices: Progress and prospects *Small Methods* **2** 1700294

34. Ma J, Bai H, Zhao W, Yuan Y and Zhang K 2018 High efficiency graphene/MoS_2/Si Schottky barrier solar cells using layer-controlled MoS2 films *Solar Energy* **160** 76–84

35. Liu Y, Cai Y, Zhang G, Zhang Y W and Ang K W 2017 Al-doped black phosphorus p-n homojunction diode for high performance photovoltaic *Adv. Funct. Mater.* **27** 1604638

36. Wi S, Kim H, Chen M, Nam H, Guo L J, Meyhofer E and Liang X 2014 Enhancement of photovoltaic response in multilayer MoS_2 induced by plasma doping *ACS Nano* **8** 5270–5281

37. Cheng R, Li D, Zhou H, Wang C, Yin A, Jiang S, Liu Y, Chen Y, Huang Y and Duan X 2014 Electroluminescence and photocurrent generation from atomically sharp WSe2/MoS_2 heterojunction p-n diodes *Nano Lett.* **14** 5590–5597

38. Liu Z, Li J and Yan F 2013 Package-free flexible organic solar cells with graphene top electrodes *Adv. Mater.* **25** 4296–4301

39. Lee J, Pak S, Lee Y-W, Cho Y, Hong J, Giraud P, Shin H S, Morris S M, Sohn J I and Cha S 2017 Monolayer optical memory cells based on artificial trap-mediated charge storage and release *Nat. Comm.* **8** 1–8

40. Wu X, Ge R, Chen P A, Chou H, Zhang Z, Zhang Y, Banerjee S, Chiang M H, Lee J C and Akinwande D 2019 Thinnest nonvolatile memory based on monolayer h-BN *Adv. Mater.* **31** 1806790

41. Lei S, Wen F, Li B, Wang Q, Huang Y, Gong Y, He Y, Dong P, Bellah J and George A 2015 Optoelectronic memory using two-dimensional materials *Nano Lett.* **15** 259–265

42. Lee D, Hwang E, Lee Y, Choi Y, Kim J S, Lee S and Cho J H 2016 Multibit MoS_2 photoelectronic memory with ultrahigh sensitivity *Adv. Mater.* **28** 9196–9202

43. Xiang D, Liu T, Xu J, Tan J Y, Hu Z, Lei B, Zheng Y, Wu J, Neto A C and Liu L 2018 Two-dimensional multibit optoelectronic memory with broadband spectrum distinction *Nat. Comm.* **9** 1–8

44. Huang W, Yin L, Wang F, Cheng R, Wang Z, Sendeku M G, Wang J, Li N, Yao Y and Yang X 2019 Multibit optoelectronic memory in top-floating-gated van der Waals heterostructures *Adv. Funct. Mater.* **29** 1902890

45. Mukherjee B, Zulkefli A, Watanabe K, Taniguchi T, Wakayama Y and Nakaharai S 2020 Laser-assisted multilevel non-volatile memory device based on 2D van-der-Waals few-layer-ReS2/h-BN/graphene heterostructures *Adv. Funct. Mater.* **30** 2001688

46. Bolotsky A, Butler D, Dong C, Gerace K, Glavin N R, Muratore C, Robinson J A and Ebrahimi A 2019 Two-dimensional materials in biosensing and healthcare: From in vitro diagnostics to optogenetics and beyond *ACS Nano* **13** 9781–9810

47. Chen X, Andrés M V and Zhang L 2018 Innovative 2D nanomaterial integrated fiber optic sensors for biochemical applications. In: *Micro-structured and specialty optical fibres V* (International Society for Optics and Photonics) p. 1068107

48. Zhu C, Du D and Lin Y 2015 Graphene and graphene-like 2D materials for optical biosensing and bioimaging: A review *2D Mater.* **2** 032004

49. Borisov S M and Wolfbeis O S 2008 Optical biosensors *Chem. Rev.* **108** 423–461

50. Loan P T K, Zhang W, Lin C T, Wei K H, Li L J and Chen C H 2014 Graphene/MoS₂ heterostructures for ultrasensitive detection of DNA hybridisation *Adv. Mater.* **26** 4838–4844

51. Rout C S, Late D J and Morgan H 2019 *Fundamentals and sensing applications of 2D materials* (Woodhead Publishing).

52. Shavanova K, Bakakina Y, Burkova I, Shtepliuk I, Viter R, Ubelis A, Beni V, Starodub N, Yakimova R and Khranovskyy V 2016 Application of 2D non-graphene materials and 2D oxide nanostructures for biosensing technology *Sensors* **16** 223

53. Liu J, Jalali M, Mahshid S and Wachsmann-Hogiu S 2020 Are plasmonic optical biosensors ready for use in point-of-need applications? *Analyst* **145** 364–384

54. Xing F, Meng G-X, Zhang Q, Pan L-T, Wang P, Liu Z-B, Jiang W-S, Chen Y and Tian J-G 2014 Ultrasensitive flow sensing of a single cell using graphene-based optical sensors *Nano Lett.* **14** 3563–3569

55. Shorie M, Kumar V, Kaur H, Singh K, Tomer V K and Sabherwal P 2018 Plasmonic DNA hotspots made from tungsten disulfide nanosheets and gold nanoparticles for ultrasensitive aptamer-based SERS detection of myoglobin *Microchim. Acta* **185** 1–8

56. Zhai Y, Zhou Y, Yang X, Wang F, Ye W, Zhu X, She D, Lu W D and Han S-T 2020 Near infrared neuromorphic computing via upconversion-mediated optogenetics *Nano Energ.* **67** 104262

57. Mak K F, Ju L, Wang F and Heinz T F 2012 Optical spectroscopy of graphene: From the far infrared to the ultraviolet *Solid State Commun.* **152** 1341–1349

58. Xu X, Yao W, Xiao D and Heinz T F 2014 Spin and pseudospins in layered transition metal dichalcogenides *Nat. Phys.* **10** 343–350

59. Neto A C, Guinea F, Peres N M, Novoselov K S and Geim A K 2009 The electronic properties of graphene *Rev. Mod. Phys.* **81** 109

60. Novoselov K S, Geim A K, Morozov S V, Jiang D-E, Zhang Y, Dubonos S V, Grigorieva I V and Firsov A A 2004 Electric field effect in atomically thin carbon films *Science* **306** 666–669

61. Sun Z, Martinez A and Wang F 2016 Optical modulators with 2D layered materials *Nat. Photonics* **10** 227–238

62. Nichols P L, Liu Z, Yin L, Turkdogan S, Fan F and Ning C-Z 2015 Cd x Pb1-x S alloy nanowires and heterostructures with simultaneous emission in mid-infrared and visible wavelengths *Nano Lett.* **15** 909–916

63. Wen Y, Wang Q, Yin L, Liu Q, Wang F, Wang F, Wang Z, Liu K, Xu K and Huang Y 2016 Epitaxial 2D PbS nanoplates arrays with highly efficient infrared response *Adv. Mater.* **28** 8051–8057

64. Fogler M, Butov L and Novoselov K 2014 High-temperature superfluidity with indirect excitons in van der Waals heterostructures *Nat. Comm.* **5** 1–5

65. Mak K F and Shan J 2016 Photonics and optoelectronics of 2D semiconductor transition metal dichalcogenides *Nat. Photonics* **10** 216

66. Tongay S, Suh J, Ataca C, Fan W, Luce A, Kang J S, Liu J, Ko C, Raghunathanan R and Zhou J 2013 Defects activated photoluminescence in two-dimensional semiconductors: Interplay between bound, charged and free excitons *Sci. Rep.* **3** 1–5

67. Farmer D B, Golizadeh-Mojarad R, Perebeinos V, Lin Y-M, Tulevski G S, Tsang J C and Avouris P 2009 Chemical doping and electron– hole conduction asymmetry in graphene devices *Nano Lett.* **9** 388–392

68. Peters E C, Lee E J, Burghard M and Kern K 2010 Gate dependent photocurrents at a graphene pn junction *Appl. Phys. Lett.* **97** 193102

69. Lemme M C, Koppens F H, Falk A L, Rudner M S, Park H, Levitov L S and Marcus C M 2011 Gate-activated photoresponse in a graphene p-n junction *Nano Lett.* **11** 4134–4137

70. Yang W, Shang J, Wang J, Shen X, Cao B, Peimyoo N, Zou C, Chen Y, Wang Y and Cong C 2016 Electrically tunable valley-light emitting diode (vLED) based on CVD-grown monolayer WS2 *Nano Lett.* **16** 1560–1567

71. He Y-M, Clark G, Schaibley J R, He Y, Chen M-C, Wei Y-J, Ding X, Zhang Q, Yao W and Xu X 2015 Single quantum emitters in monolayer semiconductors *Nat. Nanotechnol.* **10** 497–502

72. Baugher B W, Churchill H O, Yang Y and Jarillo-Herrero P 2014 Optoelectronic devices based on electrically tunable p-n diodes in a monolayer dichalcogenide *Nat. Nanotechnol.* **9** 262–267

73. Pospischil A, Furchi M M and Mueller T 2014 Solar-energy conversion and light emission in an atomic monolayer p-n diode *Nat. Nanotechnol.* **9** 257–261

74. Ross J S, Klement P, Jones A M, Ghimire N J, Yan J, Mandrus D, Taniguchi T, Watanabe K, Kitamura K and Yao W 2014 Electrically tunable excitonic light-emitting diodes based on monolayer WSe 2 p-n junctions *Nat. Nanotechnol.* **9** 268

75. Das S and Appenzeller J 2013 WSe2 field effect transistors with enhanced ambipolar characteristics *Appl. Phys. Lett.* **103** 103501

76. Jo S, Ubrig N, Berger H, Kuzmenko A B and Morpurgo A F 2014 Mono-and bilayer WS2 light-emitting transistors *Nano Lett.* **14** 2019–2025

77. Geim A K and Grigorieva I V 2013 Van der Waals heterostructures *Nature* **499** 419–425

78. Ye Y, Ye Z, Gharghi M, Zhu H, Zhao M, Wang Y, Yin X and Zhang X 2014 Exciton-dominant electroluminescence from a diode of monolayer MoS$_2$ *Appl. Phys. Lett.* **104** 193508

79. Ross J S, Rivera P, Schaibley J, Lee-Wong E, Yu H, Taniguchi T, Watanabe K, Yan J, Mandrus D and Cobden D 2017 Interlayer exciton optoelectronics in a 2D heterostructure p-n junction *Nano Lett.* **17** 638–643

80. Naguib M, Kurtoglu M, Presser V, Lu J, Niu J, Heon M, Hultman L, Gogotsi Y and Barsoum M W 2011 Two-dimensional nanocrystals produced by exfoliation of Ti$_3$AlC$_2$ *Adv. Mater.* **23** 4248–4253

81. Naguib M, Mochalin V N, Barsoum M W and Gogotsi Y 2014 25th anniversary article: MXenes: A new family of two-dimensional materials *Adv. Mater.* **26** 992–1005

82. Hantanasirisakul K and Gogotsi Y 2018 Electronic and optical properties of 2D transition metal carbides and nitrides (MXenes) *Adv. Mater.* **30** 1804779

83. Jiang X, Kuklin A V, Baev A, Ge Y, Ågren H, Zhang H and Prasad P N 2020 Two-dimensional MXenes: From morphological to optical, electric, and magnetic properties and applications *Phys. Rep.* **848** 1–58

84. Li R, Zhang L, Shi L and Wang P 2017 MXene Ti3C2: An effective 2D light-to-heat conversion material *ACS Nano* **11** 3752–3759

85. Fu H C, Ramalingam V, Kim H, Lin C H, Fang X, Alshareef H N and He J H 2019 MXene-contacted silicon solar cells with 11.5% efficiency *Adv. Energy Mater.* **9** 1900180

86. Li Y, Shao H, Lin Z, Lu J, Liu L, Duployer B, Persson P O, Eklund P, Hultman L and Li M 2020 A general Lewis acidic etching route for preparing MXenes with enhanced electrochemical performance in nonaqueous electrolyte *Nat. Mat.* **19** 894–899

87. Hantanasirisakul K, Zhao M Q, Urbankowski P, Halim J, Anasori B, Kota S, Ren C E, Barsoum M W and Gogotsi Y 2016 Fabrication of Ti3C2Tx MXene transparent thin films with tunable optoelectronic properties *Adv. Electron. Mater.* **2** 1600050

88. Zhang C, Anasori B, Seral-Ascaso A, Park S H, McEvoy N, Shmeliov A, Duesberg G S, Coleman J N, Gogotsi Y and Nicolosi V 2017 Transparent, flexible, and conductive 2D titanium carbide (MXene) films with high volumetric capacitance *Adv. Mater.* **29** 1702678

89. Kang Z, Ma Y, Tan X, Zhu M, Zheng Z, Liu N, Li L, Zou Z, Jiang X and Zhai T 2017 MXene-silicon Van Der Waals heterostructures for high-speed self-driven photodetectors *Adv Electron. Mater.* **3** 1700165

90. Zhang X, Shao J, Yan C, Qin R, Lu Z, Geng H, Xu T and Ju L 2021 A review on optoelectronic device applications of 2D transition metal carbides and nitrides *Mater. Des.* 109452

91. Dillon A D, Ghidiu M J, Krick A L, Griggs J, May S J, Gogotsi Y, Barsoum M W and Fafarman A T 2016 Highly conductive optical quality solution-processed films of 2D titanium carbide *Adv. Funct. Mater.* **26** 4162–4168

92. Ahn S, Han T H, Maleski K, Song J, Kim Y H, Park M H, Zhou H, Yoo S, Gogotsi Y and Lee T W 2020 A 2D titanium carbide MXene flexible electrode for high-efficiency light-emitting diodes *Adv. Mater.* **32** 2000919

93. Chaudhari N K, Jin H, Kim B, San Baek D, Joo S H and Lee K 2017 MXene: An emerging two-dimensional material for future energy conversion and storage applications *J. Mater. Chem. A* **5** 24564–24579

94. Velusamy D B, El-Demellawi J K, El-Zohry A M, Giugni A, Lopatin S, Hedhili M N, Mansour A E, Fabrizio E D, Mohammed O F and Alshareef H N 2019 MXenes for plasmonic photodetection *Adv. Mater.* **31** 1807658

95. Zhang X, Shao J, Yan C, Wang X, Wang Y, Lu Z, Qin R, Huang X, Tian J and Zeng L 2021 High performance broadband self-driven photodetector based on MXene (Ti3C2Tx)/ GaAs Schottky junction *Mater. Des.* 109850

96. Deng W, Huang H, Jin H, Li W, Chu X, Xiong D, Yan W, Chun F, Xie M and Luo C 2019 All-sprayed-processable, large-area, and flexible perovskite/MXene-based photodetector arrays for photocommunication *Adv. Opt. Mater.* **7** 1801521

6 Magnetic Properties of 2D Materials

Rajesh Katoch and Shilpee Jain

CONTENTS

6.1   INTRODUCTION

Magnetism is a quantum mechanical phenomenon that arises due to alignment of neighboring spins of individual atoms over a characteristic length scale, driven by exchange interactions between the neighboring spins. The characteristic length scale over which this alignment extends is governed by the strength of the exchange interactions, magnetostatic energy, magnetic anisotropy, etc. At finite temperatures,

DOI: 10.1201/9781003247890-6

reduction in dimensions brings an abrupt end to the long-range order and weakening of magnetic interactions. In thin films, the shape anisotropy tends to align magnetic moments in-plane to minimize the demagnetization field. However, at the surface of magnetic films, surface anisotropy causes the spins alignment perpendicular to the surface while the exchange energy aligns the spins parallel to each other. This causes a large demagnetization field, which destroys the magnetic ordering and increases the coercivity of the films. For bulk samples, magnetostatic energy prevails over surface anisotropy and aligns the spin in-plane of the sample, thereby reducing the demagnetization field. Likewise, magnetic ordering may be lost due to thermal fluctuations if the thermal energy is sufficient to overcome the magnetostatic energy. For instance, in small particles, as the particle diameter decreases, the anisotropy energy decreases. Consequently, below a critical size, the thermal energy can overcome the anisotropy energy and reverse the directions of spontaneous magnetization even without the application of external field, a phenomenon called super-paramagnetism. In fact, for an isotropic two-dimensional Heisenberg system with finite-range interactions, long-range magnetic order does not exist at finite temperature, according to Mermin-Wagner theorem.[1] However, the experimental observation of long-range order even in magnetic monolayers has renewed the dormant interest and opened up avenues of opportunities in 2D magnetic materials.[2,3] It has been suggested that large magnetic anisotropy and/or long-range dipole-dipole interactions can suppress finite temperature fluctuations and stabilize magnetic ordering in 2D materials.[3] Additionally, it was suggested that the essential condition for long-range ferromagnetic ordering in 2D materials is the presence of spin-wave excitation gap, which in turn is a direct consequence of magnetic anisotropy. Due to the immense interest, the number of 2D magnetic materials are increasing by the day. In this chapter, we will focus on various 2D magnetic materials and their properties, methods of synthesis and characterization, and their applications.

6.2 2D MAGNETIC MATERIALS AND PROPERTIES

6.2.1 GRAPHENE

Any discussion on 2D material properties should ideally begin with the mention of graphene,[4] which acted as a catalyst for 2D materials research due to its exceptional electrical,[5] mechanical,[6] optical,[7] thermal properties[8] and unique band structure. Graphene is identified to be an important material for spin transport applications due to long spin diffusion length.[9] In its purest form, graphene, like other graphitic allotropes, is non-magnetic. However, the carbon atoms in the sheet are sp^2 hybridized, and each sheet has an unsaturated bond, making it plausible to functionalize the pristine structure by external means to induce magnetism, which has been confirmed by several theoretical and experimental studies.[9–17] For instance, a magnetic moment of $1\mu_B$ per hydrogen chemisorbed defect and 1.12–1.53 μ_B per vacancy defect was deduced from first principles calculations, for a single carbon atom point defect in the hexagonal lattice.[10]

Further, it was determined that the magnetic coupling was either ferromagnetic or antiferromagnetic depending on the defects belonging to the same or different hexagonal sublattice on graphene sheet, respectively. Another theoretical study on

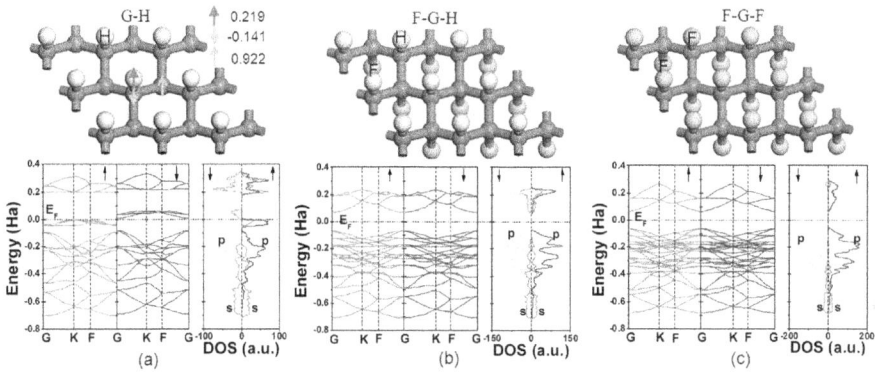

FIGURE 6.1 Relaxed geometric structure and electronic structure of half-hydrogenated graphene (a), half-fluorinated and half-hydrogenated graphene (b), fully fluorinated graphene (c). The arrows and numbers in (a) are for magnetic moments.

Source: Reprinted with permission from Ref.[13]

hydrogenated graphene (graphane)[13] found that the application of electric field converts it into half hydrogenated state, which is a semiconductor with a small indirect bandgap (0.43 eV) along with the appearance of magnetic moment (0.922 μ_B) at the site of unsaturated C atom (Figure 6.1(a)[13]), which polarizes its neighboring C(-0.141 μ_B) and H($+0.219$ μ_B) atoms. The unsaturated C atoms couple ferromagnetically, mediated by the hydrogenated C atom via p-p interactions. Further decorating the half-hydrogenated sheet with F atoms (Figure 6.1 (b)) yields a nonmagnetic semiconductor with direct bandgap (3.7 eV), while a fully fluorinated graphene structure remained nonmagnetic, direct bandgap (3.4 eV) semiconductor (Figure 6.1 (c)). Similar studies reveal that functionalizing graphene with F-atoms creates an antiferromagnetic semiconductor that stabilized to ferromagnetic metallic state on application of magnetic field with a resultant large room temperature magnetoresistance (2,200%).[14] The magnetic properties of graphene quantum dots (GQDs) of various shapes depend on the edge states: the zero-energy states (ZESs) give rise to spin paramagnetism, whereas the dispersed edge states (DESs) are responsible for a temperature independent diamagnetic character. Hexagonal, circular, and randomly shaped GQDs are diamagnetic as they contain mainly DESs. The edge states of the triangular GQDs are of ZES type, giving them a paramagnetic character.[15] In a recent study, simultaneous application of strain, temperature, and magnetic field resulted in inducing magnetism in liquid phase exfoliated graphene flakes and observation of magnetic domains by magnetic force microscopy (MFM).[18]

6.2.2 TRANSITION METAL DICHALCOGENIDES (TMDs)

MoS_2 is the most stable layered TMD in nature. It consists of three covalently bonded hexagonal atomic layers (i.e., S-Mo-S), with middle layer of Mo atoms sandwiched between two layers of S atoms with the interlayer separation ~ 6.2 Å, due to which the van der Waals forces between the adjacent layers are very weak. Consequently, individual layers can be easily exfoliated by chemical or mechanical means. Its tunable bandgap (i.e., E_g ~1.3 eV (indirect) in bulk to E_g ~1.8 eV (direct) in isolated layers),[19] large in-plane mobility (200 cm^2 V^{-1}s^{-1}),[20] and mechanical stability[21] make it one of the most widely investigated TMDs. Magnetism in MoS_2 has been extensively investigated theoretically and experimentally.[22] SQUID magnetometry measurements on single crystals of MoS_2 (~10–100 μm thick) displayed weak ferromagnetism in otherwise diamagnetic system, and the ferromagnetism persisted from 10 K up to RT. The existence of ferromagnetism was attributed to the presence of zigzag edges at the grain boundaries.[22] Due to the sandwich structure of MoS_2, a 2D diluted magnetic semiconductor (DMS) like state has been predicted for MoS_2 from first principles calculations by doping with transition metal ions at Mo site.[23] From binding energy considerations, ferromagnetism was observed to be stable in MoS_2 monolayer for Mn, Zn, Cd, and Hg doping; antiferromagnetic for Fe and Co doping; whereas any magnetic ordering is suppressed by Jahn-Teller distortions in other dopants of these groups. These observations were further supported by a subsequent study on Mn doped monolayer MoS_2[25], which showed that dopant concentrations in the range of 10–15% might be sufficient to provide ferromagnetism in Mn/MoS_2 DMSs at room temperature. To validate these findings, in situ doping of monolayer MoS_2 with Mn via vapor phase deposition techniques was investigated.[26] Photoluminescence (PL) and X-ray photoelectron spectroscopy (XPS) revealed substrate dependence of modifications to the band structure by incorporation of Mn. It was revealed that only inert substrates (i.e., graphene) permit the incorporation of significant amount of Mn (2%) in MoS_2, while with reactive substrates (i.e., SiO_2 and sapphire), incorporation of Mn was not successful. Theoretical studies predict giant spin-splitting induced by spin-orbit coupling in MoS_2 monolayers, which results in an out-of-plane spin polarization.[27] Based on this, atomically thin MoS_2 based spin-valves (i.e., NiFe/ MoS_2 /NiFe) were fabricated.[24] The junctions show spin-valve signals up to 240 K with moderate MR of ~0.4% at 10 K (Figure 6.2[24]). The MR dramatically improved to 0.73% when the NiFe(Py) layer was capped with a thin gold layer. A recent study has demonstrated room temperature ferromagnetism in in-situ Fe-doped MoS_2 monolayers grown by chemical vapor deposition (CVD) technique.[28] Fe-doped MoS_2 related emission in magnetic circular dichroism (MCD) and PL studies showed circular dichroism (CD) ($\rho \approx 40\%$) at both 4 K and RT. Since luminescence in transition metal loses its CD above T_C, the presence of it at 300 K is a signature of ferromagnetic state, which is further verified by the presence of magnetic hysteresis loops at 4 K and 300 K.

6.2.3 TRANSITION METAL HALIDES (TMHS)

CrI₃: Bulk magnetic susceptibility and magnetization measurements show that CrI_3 is a soft ferromagnet below the Curie temperature (T_c = 61 K), with its easy axis pointing perpendicular to the layers, and a saturation magnetization consistent with a spin S = 3/2 state of the Cr atoms.[29] Another magnetic phase transition at T ≈ 50 K is found in low-field magnetization measurements with B applied parallel to the layers, but there is a lack of understanding about its nature and details of the magnetic state currently.[30] Low temperature magneto-optical Kerr effect (MOKE) measurements on atomically thin CrI_3 crystals showed hysteretic Kerr rotation at low fields (B ~ 0.1T),

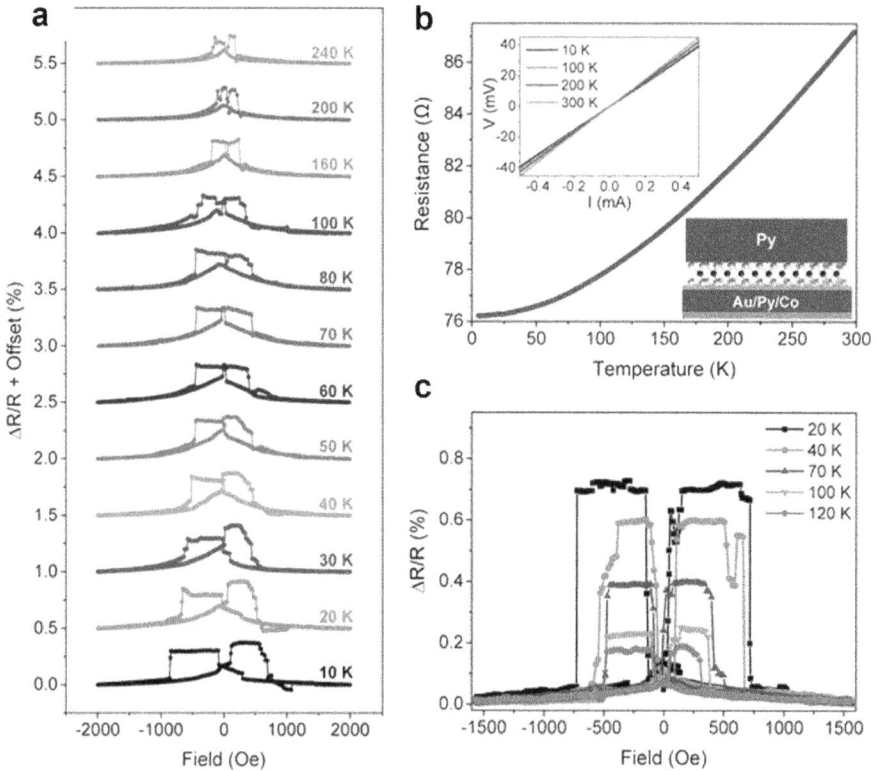

FIGURE 6.2 (a) Magnetoresistance (MR) curves at different temperatures. The spin-valve effect is found up to 240 K, and the TMR ratio decreases as temperature rises up. (b) Characterizations of the devices with $Py/MoS_2/Au/Py/Co$ structure. The junction resistance versus temperature, showing a metallic behavior. The linear I-V curve (inset) exhibits an Ohmic contact between MoS_2 and electrodes. (c) MR curves at different temperatures. The spin valve effect is found up to 120 K, and the MR ratio increases as the temperature is reduced.

Source: Reprinted with permission from Ref.[24]

applied perpendicular to the plane of the layer, which saturated at high fields in mono and trilayer samples only.[2] In the case of bilayers, the Kerr angle vanishes at low fields and only shows non-hysteretic Kerr rotation at high fields (B ~ 0.65T). This field dependence of Kerr angle suggests that each monolayer of CrI_3 possesses finite magnetization perpendicular to the layer that corresponds to the easy axes of magnetization in bulk crystals. Likewise, the absence of Kerr rotation in bilayers suggests that the magnetization of subsequent monolayers is directed opposite to each other so that the magnetization ceases to exist at B = 0. Only at fields sufficiently large enough to reorient and point magnetic moments in the field direction, finite magnetization, and Kerr rotation is observed in bilayer samples. This implies that interlayer exchange coupling in CrI_3 monolayers is antiferromagnetic in nature. This

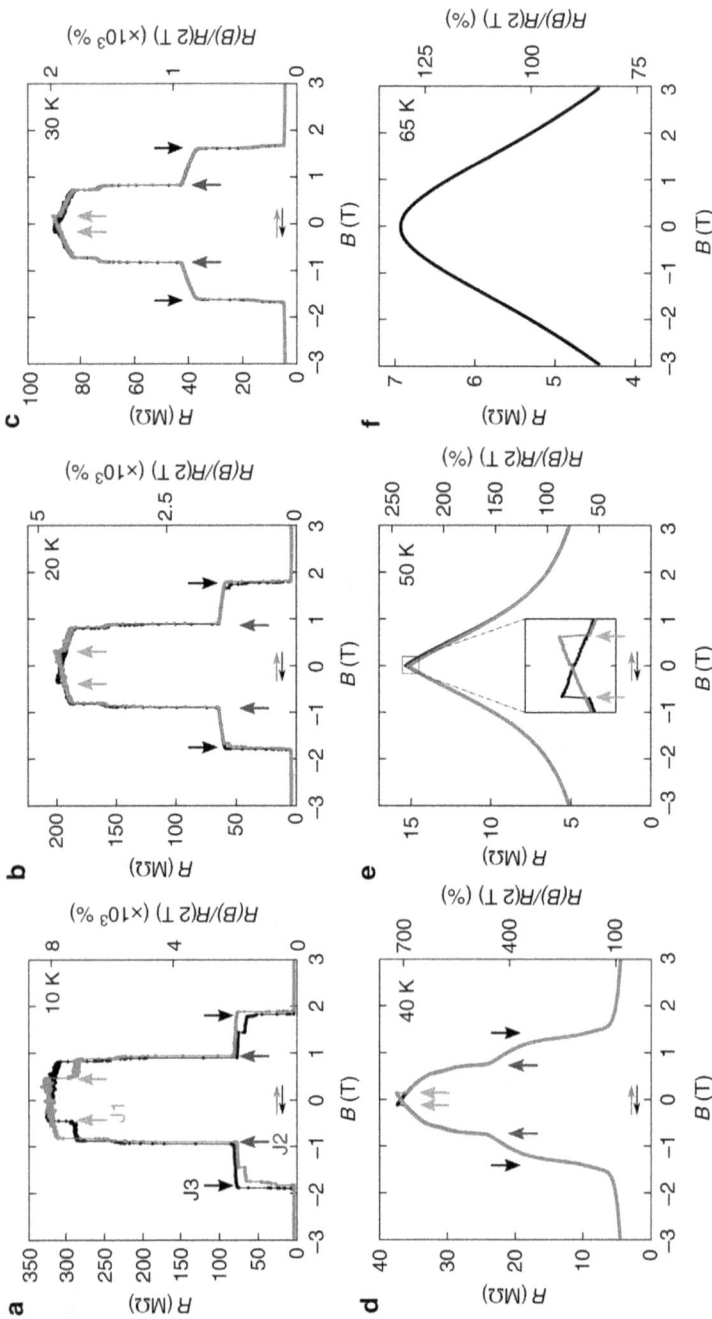

FIGURE 6.3 Large tunneling magnetoresistance in vertical junctions. (a–f) Tunneling resistance (left axis) and resistance ratio $R(B)/R(2\,T)$ (right axis) of the device measured at the temperature indicated in each panel (with $V = 0.5\,V$ and B applied perpendicular to the CrI_3 layers). The red and black dots correspond to data measured upon sweeping the field in opposite directions as indicated by the horizontal arrows of the corresponding color. The resistance ratio increases upon lowering temperature and reaches 8,000% at 10 K. The arrows of different color point to the magnetoresistance jumps that are seen in all devices, irrespective of the thickness of the CrI_3 crystal. Jump J1 is always accompanied by hysteresis; at low temperature, jumps J2 and J3 occur in all devices at the same value of the applied magnetic field, irrespective of sweeping direction. All jumps shift to lower field values upon increasing temperature and disappear above 50 K.

Source: Reprinted with permission from Ref.[30]

is in stark contrast with the ferromagnetic exchange interaction in bulk CrI_3 crystals at low temperature.[29] However, ab initio calculations for the high temperature phase (T > 200 K)[29] suggest the presence of antiferromagnetic interlayer exchange interactions.[31–33] Very large tunnel magnetoresistance (TMR) ~ 10,000% depending on the magnetic state of layered CrI_3 was obtained in a recent study,[30] demonstrating strong coupling between transport and magnetism in magnetic van der Waals semiconductors (Figure 6.3). A recent study has shown that in bilayer CrI_3, pressure induces a transition from layered antiferromagnetic to ferromagnetic phase.[32] Further, using scanning magnetic circular dichroism (MCD) microscopy, it was shown that in trilayer CrI_3, pressure can create coexisting domains multiple phases, i.e., ferromagnetic (one) and antiferromagnetic (two). The observed changes in magnetic order were explained by changes in the stacking arrangement.

6.2.4 TERNARY TRANSITION METAL CHALCOGENIDES AND HALIDES (TMCHS)

$Cr_2Ge_2Te_6$: Another 2D vdW material system for which persistent magnetism down to monolayer is demonstrated is $Cr_2Ge_2Te_6$.[3] Bulk $Cr_2Ge_2Te_6$ is a semiconducting ferromagnetic below 68 K with an out-of-plane easy axis of magnetization and small coercivity (3.4 mT).[34,35] To obtain 2D vdW layers, $Cr_2Ge_2Te_6$ atomic layers were mechanically exfoliated with adhesive tape on 260 nm SiO_2/Si substrates.[3] Based on the temperature and magnetic-field-dependent magneto-optic Kerr microscopy, long-range ferromagnetic order in pristine $Cr_2Ge_2Te_6$ bilayers was established. The Kerr rotation intensity under 0.075T fell to zero at 30 K for bilayer samples, indicating the destruction of magnetic ordering. The temperature-dependent magnetic measurements revealed strong dimensionality effect as the T_c monotonically decreased for 68 K for bulk samples to about 30 K in the bilayer samples, indicating that interlayer magnetic interactions are essential for establishing the ferromagnetic order in $Cr_2Ge_2Te_6$ samples. Likewise, field dependence of T_c in different thickness samples displayed a rapid increase in T_c with increase in field intensity at low fields (B ≤ 0.3T), providing an effective way of engineering the magnetic anisotropy with magnetic fields. The ability to tune magnetism in 2D materials with electric field opens up new avenues to develop compact spintronic devices.[36–39] For instance, electrostatic gating of thin (~22 nm) $Cr_2Ge_2Te_6$ crystals was shown to modulate the magnetic phase transition and magnetic anisotropy and increase its Curie temperature.[39] The exfoliated $Cr_2Ge_2Te_6$ in contact with a polymer gel forms electric double-layer transistor device with high carrier density (~4×10^{14} cm^{-2} for V_G = 3.9V). Magnetoresistance (MR) curves exhibited hysteresis at 60 K, 120 K, and 180 K, indicating that ferromagnetic order could be stabilized well beyond T_c at high doping levels. DFT calculations of magnetic anisotropy energy (MAE) reveal that the magnetic easy-axis rotates from out-of-plane in pristine samples to in-plane in heavily doped $Cr_2Ge_2Te_6$. To explain the rotation of magnetic easy axis, it was suggested that heavy doping promotes a double-exchange mechanism mediated by free carriers, which dominates over the superexchange mechanism of the pristine samples. While interesting in its own right, the effective charge layer and enhanced ferromagnetic state is limited to top layer only (~ 1 nm). In a recent work, it was shown that the magnetic easy-axis of few-layered $Cr_2Ge_2Te_6$ can be gradually tuned from an out-of-plane direction to in-plane direction in a solid ion conductor (SIC) gated FET device by electric field stimulation, with tuning depth ~10 nm.[40] Additionally, using anomalous Hall effect and magnetoresistance studies, the authors reported significant enhancement in T_c from 65 K to 180 K.

Fe₃GeTe₃: Unlike CrI_3 and $Cr_2Ge_2Te_6$, bulk Fe_3GeTe_2 is an itinerant vdW ferromagnet with relative high transition temperature (T_c ~220–230 K) and strong perpendicular anisotropy, whose tunable magnetic properties by changing thickness, chemical composition, external gate doping, makes it very attractive for spintronic applications.[38,41–43] The presence of strong magnetocrystalline anisotropy supports the sustenance of long-range magnetic order in 2D limit.[42,44] Micromechanical cleavage of this material is difficult with conventional methods, and to increase the chances of obtaining monolayer flakes, exfoliation is performed on freshly evaporated gold[41] or on Al_2O_3.[38] For instance, exfoliated Fe_3GeTe_2 flakes (on SiO_2/Si and Au coated SiO_2/Si substrates) were shown to exhibit robust 2D ferromagnetism with strong out-of-plane anisotropy down to monolayer.[41] Using magneto-transport (anomalous Hall effect), MOKE and RMCD measurements, a crossover from 3D to 2D Ising ferromagnetism for thickness less than 4 nm was evidenced. The transitions temperature (T_c) sharply declined from 220 K in five layers, down to 180 K for bilayer and further down to 130 K for a monolayer, clearly showing the effect of dimensionality on ferromagnetism in thin structures Likewise, the magnetic hysteresis loops in polar RMCD shrunk as the temperature was increased and completely vanished at 130 K for a monolayer sample, suggesting the destruction of magnetic order. Another similar study on exfoliated Fe_3GeTe_3 layers on Al_2O_3 coated SiO_2/Si substrates showed that as the number of layers in exfoliated Fe_3GeTe_3 flakes are reduced, the anomalous Hall resistance increases,[38] while the ferromagnetism persists down to a monolayer. Further temperature-dependent magnetic studies revealed that at T = 100 K, ferromagnetism disappears in mono and bilayer samples, whereas it survives in thicker samples. In wafer-scale 2D ferromagnetic Fe_3GeTe_2 thin films grown via MBE, the Curie temperature can be enhanced by raising the Fe composition. Additionally, upon interface coupling with antiferromagnetic MnTe, the coercive field dramatically increases 50% from 0.65 to 0.94 Tesla upon cooling. The large-scale layer-by-layer growth and controllable magnetic properties make Fe_3GeTe_2 a promising candidate for spintronic applications (Figure 6.4[42]). To further verify this strong dimensionality effect on ferromagnetism in Fe_3GeTe_3 system, analysis of thickness dependence of T_C revealed a monotonic decrease from 180 K in bulk to 20 K in monolayer samples. From ab initio calculations, it was argued that magnetocrystalline anisotropy gives rise to an energy gap in the magnon dispersion, which suppresses low-frequency, long-wavelength spin waves (magnon excitations), and protects the magnetic order below a finite T_c. A very interesting finding of this study is the elevation of T_c much beyond the bulk value by using an ionic gate. In an ionic field-effect transistor setup, $LiClO_4$ dissolved in polyethylene oxide matrix covers both the Fe_3GeTe_3 flake and the side gate. A positive gate voltage, Vg, intercalates lithium ions into the Fe_3GeTe_3 flake. Large variations in coercivity were observed by modifying the gate voltage. Additionally, the ferromagnetic T_c was raised to ~ 300 K as a function of doping. It was shown that the gate-induced electrons sequentially fill the sub-bands originating from the Fe d_{3z^2}, d_{xz}, and d_{yz} orbitals, resulting in sharp peaks in the DOS that

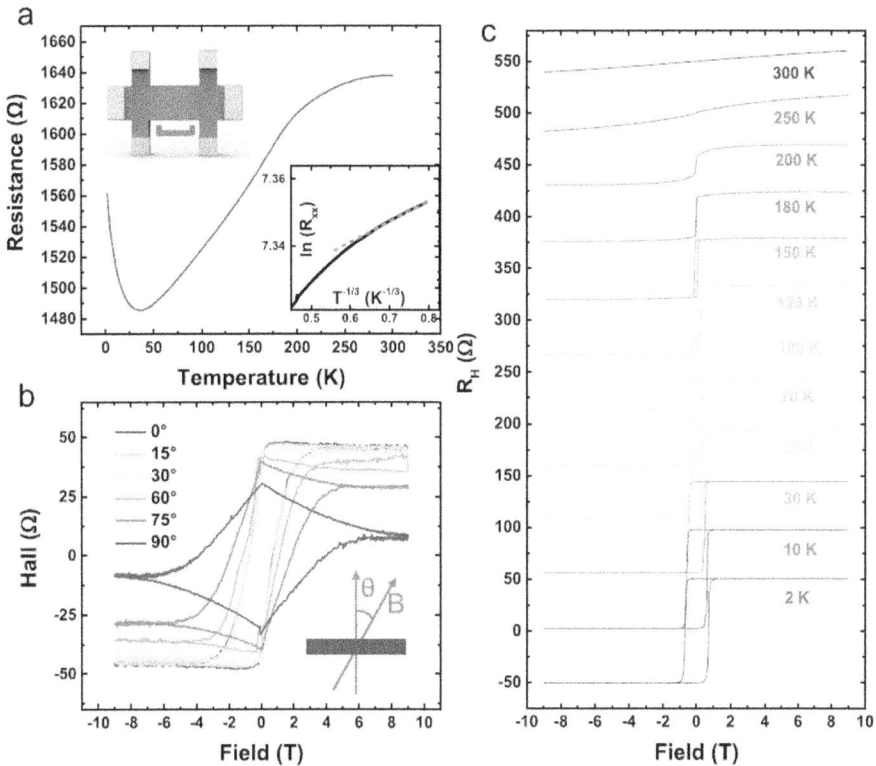

FIGURE 6.4 Transport properties of 8 nm Fe_3GeTe_2 thin film. (a) R-T curve, showing a metallic characteristic. The bottom inset is a fit to the variable range hopping model, $\ln(Rxx) \sim T\text{-}1/3$. The scale bar of the top inset Hall-bar structure is 1 mm. (b) Angle-dependent anomalous Hall data. The measurement geometry is displayed in the inset. With the magnetic field changing from perpendicular ($\theta = 0°$) to parallel ($\theta = 90°$) to the sample surface, the coercive field increases largely, from which the easy axis can be determined to be out-of-plane. (c) Anomalous Hall data at different temperatures with the offset of 50 Ω. With rising temperature, the coercive field decreases successively.

Source: Reprinted with permission from Ref.[42]

correspond to the Fermi level passing through the flat band edges of the sub-bands, leading to sharp increase in T_c upon doping.

6.2.5 TRANSITION METAL PHOSPHOROUS TRICHALCOGENIDES (TMPCs)

With a general formula (MPX_3; M = Mn, Fe, Ni, Co; X = S) is a family of quasi-2D antiferromagnetic materials that has drawn immense attention in recent years.[45–50] All these compounds are isostructural (i.e., monoclinic with C 2/m space group symmetry). In this family, transition metal ions are arranged in a honeycomb lattice structure in *ab* plane of individual layers, which are weakly held by van der Waals forces. The spins in each layer are aligned in the zigzag direction along the a axis,[45,46] while the adjacent spin chains are anti-aligned (Figure 6.5 (a)), forming a zigzag antiferromagnetic order. Amongst them, $MnPS_3$ is described by an isotropic Heisenberg Hamiltonian,[51,52] $FePS_3$ constitutes an Ising system,[51,53] $NiPS_3$ by the anisotropic Heisenberg Hamiltonian,[51] whereas $CoPS_3$ appears to have an XY-like anisotropy.[54] Thus, this family of 2D vdW materials form an excellent model system for studying 2D magnetism with spin S = 5/2, 2, 3/2, and 1 for $MnPS_3$, $FePS_3$, $CoPS_3$, and $NiPS_3$, respectively. Below Neel temperature (T_N ~155 K), Ni ions in bulk $NiPS_3$ are arranged in a honeycomb lattice structure in *ab* plane of individual layers, which are weakly held by van der Waals forces. The two-dimensional magnetic structure of $NiPS_3$ comprises of ferromagnetic chains that are antiferromagnetically coupled in a zigzag manner. The moments in $NiPS_3$ lie in the *ab* planes with spins in each layer aligned in ferromagnetic chains in a zigzag manner along the a axis,[45,46] while the adjacent spin chains are anti-aligned (Figure 6.5), forming a zigzag antiferromagnetic order. Further inelastic neutron scattering studies and fitting of the data with linear spin wave theory using Heisenberg Hamiltonian with single ion anisotropy demonstrates the presence of small easy axis anisotropy and a magnon excitation gap of ~ 7meV.[46] To study the thickness dependence of anti-ferromagnetic ordering, a later study investigated the temperature dependence of phonon spectra using Raman spectroscopy and quasi-electron scattering (QES) as a function of thickness down to the monolayer limit.[48] It was observed that antiferromagnetic ordering persists down to two layers but is drastically suppressed in the monolayer. The Néel temperature, however, shows weak dependence on number of layers as long as it is greater than or equal to two. In comparison, in Ising-type antiferromagnet $FePS_3$, antiferromagnetic ordering persists down to the monolayer limit.[53] It is also in contrast to the case of ferromagnetic CrI_3 or $Cr_2Ge_2Te_6$, where the Curie temperature decreases as the thickness is decreased but remains finite in the monolayer limit.[2,3] These investigations indicate that the intra-layer exchange interactions are much stronger than the interlayer ones. In the monolayer samples, the monotonic increase in intensity of QES signal with decreasing temperature, even down to zero kelvin, suggests that spin fluctuations keep growing, destroying the antiferromagnetic ordering in accordance with Kosterlitz-Thouless (KT) transition. Recent studies have dedicated efforts in understanding the electronic and optical properties of this compound.[49,50] From first principles calculations, the zigzag AFM order was determined to be the lowest energy state for all values of the thickness in $NiPS_3$ thin films, whereas for single layer, the zigzag and Néel orders were found to be virtually degenerate ($\delta E \approx 0.2$ meV/Ni).[49] This facilitates large spin fluctuations and suppression of magnetic order in single layer. Additionally, it was shown that (Figure 6.5 (b)[49]) application of electric field E ~ 0.7 eV/Å along z-axis induces a metal-insulator transition, with simultaneous vanishing of magnetic moments and electronic bandgap. A further photoluminescence (PL) study[50] demonstrated the presence of highly anisotropic excitons (Wannier like) with optical anisotropic axis controlled by zigzag antiferromagnetic order. The strong exciton-magnetic coupling was revealed from the fact that the exciton emission spectra diminished and vanished near the Néel temperature (T_N~155 K). The excitons also exhibit a strong dependence on thickness of the samples as the PL intensity fell off and

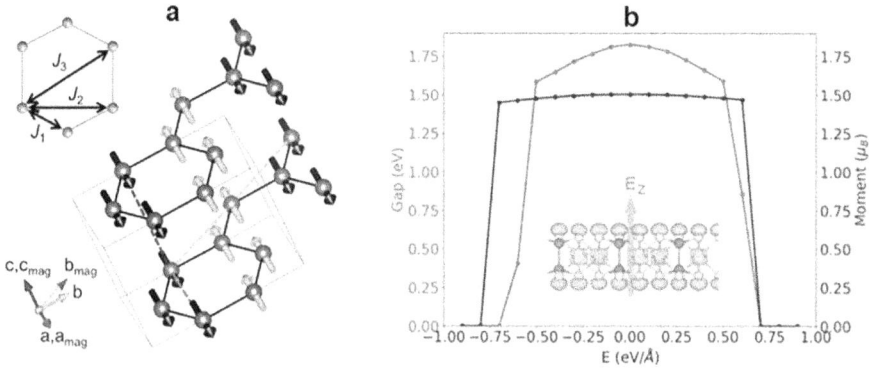

FIGURE 6.5 (a) The magnetic structure of NiPS$_3$ with the crystallographic unit cell and the unit cell used in the calculation of the magnetic dynamic structure factor. The insert shows the exchange interactions between the first, second, and third nearest intraplanar neighbors. (b) Evolution of the electronic bandgap (crimson line) and the nickel magnetic moment (blue line) for monolayer NiPS$_3$ with external applied electric field. E is taken to be positive along the z direction and is shown schematically with the crystal structure of monolayer NiPS$_3$.

Source: Reprinted with permission from Refs[46,49]

peak blue shifts with decreasing layer number. Besides these, the observation of strong exciton-phonon coupling from linear dichroism (LD) studies pave the path for future investigations into exciton-magnon coupling in 2D materials.

6.2.6 TOPOLOGICAL INSULATORS

Topological insulators (TIs) form a special class of materials that are insulating in the bulk but conducting at the surface. The surface states are bound by time-reversal symmetry (TRS), and the carriers are spin-momentum locked, which gives rise to intrinsic spin-polarized currents. TIs in two dimensions (2D) are described as a quantum spin Hall (QSH) phase, in which the spin-orbit interaction takes the role of an effective magnetic field that acts in opposite directions for opposite spins, and TRS is preserved. QSH was first investigated in graphene[55] following the measurement of QH effect.[56] However, it has since eluded detection experimentally due to its weak spin-orbit coupling and small bandgap (~ 10^{-3} eV).[57] For stable 2D TI phases, strong spin-orbit coupling and a reasonably large bandgap are necessary. Some well-known topological insulators are Bi$_2$Se$_3$, Bi$_2$Te$_3$, Sb$_2$Te$_3$ and their solid solutions. Magnetic topological insulators (TIs) provide an important platform to explore emergent phenomena such as quantum anomalous Hall (QAH) effect,[59,60] Majorana modes,[61] magnetoelectric effect,[58] etc. Magnetism arises in TIs by doping. For example, in Cr-doped (Bi, Sb)$_2$Te$_3$ thin film, QAH was demonstrated experimentally.[59] Linear magnetoresistance (MR), persisting up to RT, induced by 2D magneto-transport was reported in Bi$_2$Se$_3$ nanoribbons.[58] For a magnetic field parallel to the surface of a nanoribbon (a-b plane), the observed MR effect is negligible compared to the MR in perpendicular fields (Figure 6.6). The MR showed an |cos(θ)| angular dependence

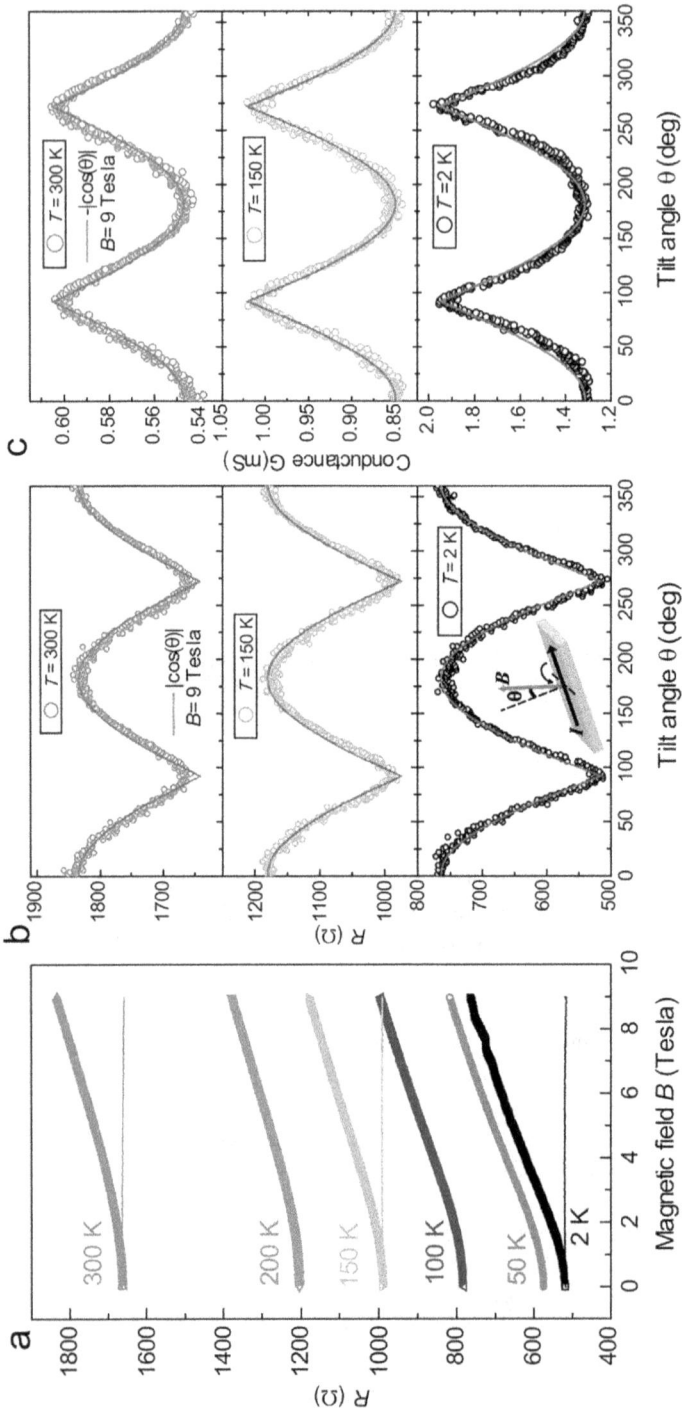

FIGURE 6.6 2D linear magneto-resistance persisting to room temperature. (a) Resistance vs. magnetic field of sample #2 from T = 2 to 300 K. The data in the perpendicular field are shown as dots, while the data in the parallel field are shown as solid lines for T = 2, 150, and 300 K. (b) R vs. rotation angle at T = 2, 150, and 300 K. The function |cos(θ)| is shown as solid red lines. (c) Conductance of Bi_2Se_3 nanoribbon sample #2 vs. rotation angle at T = 2, 150, and 300 K. Red solid lines represent the function |cos(θ)|, which is seen to describe the conductance data well for all temperatures.

Source: Reprinted with permission from Ref.[58]

at arbitrary angle, illustrating the 2D origin of linear MR. Likewise, Mn doping of Bi_2Se_3, Bi_2Te_3 and Sb_2Te_3 thin films grown by MBE gives rise to a large magnetic gap of ~90 meV at 1 K for Mn-doped Bi_2Te_3, which displays ferromagnetic hysteresis and out-of-plane spin texture.[62] This is due to enhanced wavefunction overlap in a self-organized heterostructure of alternate Mn-Bi_2Te_4 and Bi_2Te_3 multilayers. On the other hand, Bi_2Se_3 with 6% Mn doping reveal only a temperature-independent non-magnetic gap due to relatively weaker spin-orbit interaction in-plane magnetization and a T_C of 6 K. Likewise, Cr doping (~ up to 12%) in epitaxial thin films of Bi_2Se_3 grown by molecular beam epitaxy (MBE) show ferromagnetic hysteresis loops at 2 K.[63] SQUID magnetometry and polarized neutron reflectometry (PNR) measurements reveal a uniform magnetic moment ~ $2\mu_B$ per Cr ion across the sample, with no enhanced magnetism near the surface or interface and a ferromagnetic $T_C \sim 8.5$ K

6.3 SYNTHESIS AND CHARACTERIZATION OF 2D MATERIALS

Although magnetism can be induced in 2D materials by doping, defect engineering, interfacial effects, or other modifications, it may not be stable for a long time and susceptible to thermal fluctuations in the absence of strong exchange interactions. For better control and stability, it is therefore necessary that 2D magnetic crystals with magnetic ions, which possess intrinsic magnetism, are grown with precise synthesis routes. Secondly, several theoretically predicted 2D magnetic systems have to be realized experimentally, especially those having T_C around room temperature. Several well-established techniques such as chemical vapor deposition (CVD), chemical vapor transport (CVD), molecular beam epitaxy (MBE), pulsed laser deposition (PLD), wet chemical methods, exfoliation, etc. are used. Keeping in mind specific characteristics desired and intended application, generally two approaches are employed to obtain 2D magnetic materials: top-down and bottom-up approach.

6.3.1 TOP-DOWN APPROACH

In 2D materials, different atomic layers are bound together by weak van der Waals (vdW) forces. Consequently, sheets with monolayer thickness can be easily separated from the bulk material by exfoliation either mechanically or by liquid phase exfoliation (LPE) without damaging the in-plane bonds of the layer. This forms the basis of top-down approach. Mechanical exfoliation uses mechanical forces to separate atomic layers/sheet, which can be easily achieved by using an adhesive substance such as a scotch tape. In fact, this was used to isolate graphene from graphite in the pathbreaking work of Novoselov and Geim.[4] The bulk material like graphite, which is in the form of hexagonal sheets, is attached to the adhesive on the scotch tape and then peeled into a thin flake by using another adhesive surface or substrate. This process is repeated several times to get the desired thickness and further transfer the obtained sheets onto an appropriate substrate. Strong vdW interactions between Au and the top-most layer of bulk crystal/film are often exploited for mechanical exfoliation of large area monolayers of TMDs.[64] Several other 2D layered materials, such as h-BN,[65] CrI_3,[2] $Cr_2Ge_2Te_6$,[40] Sb_2Te_3,[66] $NiPS_3$,[67] etc. are also obtained through this method. Despite its nondestructive nature, absence of any chemical reaction, ability to obtain high quality crystals, there are some disadvantages, such as low production yield, labor intensive, low scalability, little control over microstructure, etc., which limits its use in various practical applications. To overcome these disadvantages, chemical exfoliation or liquid phase exfoliation (LPE) is used. In this technique, solvents are responsible for exfoliating the layered materials by intercalation, ion exchange, or shear force assisted exfoliation.[68] The common intercalants utilized are acids/bases,[69,70] oxidizing/reducing agents,[71,72] etching agents,[73,74] and some functional molecules.[75–77] For effective exfoliation, the choice of solvent used depends on the strength of vdW forces between layers, surface energy, surface tension of the solvents, etc. All classes of 2D materials, including graphene,[78] MoS_2,[79] h-BN,[80] $Cr_2Ge_2Te_6$,[81] Bi_2Te_3,[82] Bi_2Se_3,[83] $NiPS_3$,[84] etc. can be easily obtained with this technique. LPE is a low-cost process and suitable for providing high yield of nanosheets of 2D layered materials. However, the produced sheets are of relatively poor crystal quality due to the damage caused by solvents and the shear forces used in this process.

6.3.2 BOTTOM-UP APPROACH

6.3.2.1 Chemical Vapor Deposition (CVD)

CVD or CVT is a vacuum deposition technique to synthesize uniform thin films over large areas. It is the most often used technique in semiconductor industry, specially in the process of atomic layer deposition. A typical process consists of controlled deposition of volatile precursors of desired material, which react and decompose on to a substrate. It facilitates layer by layer deposition of material on to a substrate. The deposition rate and the quality of films is dependent on temperature, gas flow rate, and the distance between the precursors and the substrates. CVD method is preferable for making 2D ultrathin films with relatively high crystallinity, conformal coverage, and controlled microstructure, shape, size, and thickness. It has been found very useful in synthesis and fabrication of graphene.[85] Several other 2D materials, like h-BN,[86,87] TMDs,[88,89] $NiPS_3$,[90] and topological insulators,[91] etc. with scalable size, precise thickness, and excellent properties were synthesized by this technique. CVD is a low-cost technique for producing thin films with high growth rates, good reproducibility, and varied morphology. However, it is susceptible to contamination due to low vacuum and lacking precise control over thickness of films. To overcome these limitations, molecular beam epitaxy (MBE) can be used.

6.3.2.2 Molecular Beam Epitaxy (MBE)

It is a technique for epitaxial growth of thin films and heterostructures, where one or more molecular or atomic beams of desired materials (e.g., gallium, arsenic) are generated by placing the source materials in heated evaporation cells. The substrate is heated to the necessary temperature and rotated to promote 2D and 3D growth mechanisms. The deposition is carried at ultrahigh vacuum (UHV) (10^{-8}–10^{-12} Torr) and very small deposition rates (~ML per second) to ensure homogeneity and epitaxial growth. To monitor the deposition rate and thickness of the films, reflection high-energy electron diffraction (RHEED) is often used. Apart from growing single crystals, it has been used to grow 2D layers of high purity such as graphene on alumina and h-BN,[92,93] $MoSe_2$ on GaAs,[94] Bi_2Te_3, Bi_2Se_3[62,63] as well. High cost, requirements of UHV, and very slow deposition rates are drawbacks from the point of view of high throughput and scalability compared to CVD.

6.3.2.3 Wet Chemical Methods

In these methods, the precursors react with each other in a solution or liquid phase under controlled temperature, pressure, concentration, etc. For instance, hydrothermal/solvothermal synthesis is a common technique used for preparing thin films and nanoparticles at typically lower temperature (100°C to 250°C). The precursors are dissolved in a solution, sealed off, and heated above boiling point in Teflon coated vessel. The high temperature and pressure stimulate chemical reaction and synthesis of high quality 2D materials.[95–98] Likewise, 2D materials have been synthesized with template-assisted synthesis,[99–101] microwave-assisted synthesis,[102–104] etc. Since device applications need high level uniformity and crystallinity, to achieve better crystallinity and uniformity, post-deposition heat treatment is often required to obtain high-quality crystals. It is difficult to monitor the growth of materials and control the dimensions of the end products limiting their use in 2D materials synthesis where high degree of control over the crystallinity and thickness is desirable.

6.3.3 CHARACTERIZATION TECHNIQUES

In 2D materials, due to quantum confinement, the magnetism may be destroyed or modulated by small electric or magnetic stimulus or simply due to thermal fluctuations. Thus, the characterization tools have to be sensitive yet subtle so as to not transform the virgin state of the material. Some of prevalent characterization techniques being utilized for probing the magnetic properties of 2D materials are vibrating sample magnetometry (VSM), superconducting quantum interference device (SQUID), magnetic force microscopy (MFM), magneto optic Kerr-effect (MOKE), magneto-circular dichroism (MCD), Raman spectroscopy, angle-resolved photoemission spectroscopy (ARPES), electron spin resonance (ESR), etc. Here we discuss working principles of a few of them.

6.3.3.1 Vibrating-Sample Magnetometer (VSM)

VSM is often used to investigate magnetism in 2D vdW magnets.[105–108] It is based on Faraday's law of electromagnetic induction, where an electromagnetic force (emf) is generated in a pickup coil when there is a change in magnetic flux due to the vibration (up and down movement) of a magnetized sample near it. The induced emf gives rise to an electric current whose magnitude is proportional to the magnetization of the sample. For measurement in VSM, a sample is vibrated between a pair of pickup coils and a dc magnetic field is applied to the sample. A lock-in amplifier can be used to detect the signal from the coils. The lock-in amplifier also can be used to give a dc voltage output, which is proportional to the magnetization of the material. The magnetization of the sample depends upon the strength of the magnetic field and its magnetic susceptibility.

6.3.3.2   Superconducting Quantum Interference Device Magnetometer (SQUID)

It is a versatile tool for detecting and measuring extremely small magnetic fields (10^{-15} T) and currents (10^{-12} A) often associated with 2D vdW magnets.[16,22,63,109] Its working principle is based on the tunneling of charge carriers (Cooper pairs) through a Josephson junction and quantization of flux in a superconductor. In the most widely used dc SQUID, two Josephson junctions are connected to each other in a dc biased superconducting loop. The dc current flows equally in both the arms. When an external magnetic field is applied to the superconducting loop, a screening current opposing the change in magnetic flux circulates in the loop. The screening current is in-phase and out-of-phase in the two arms where the current is limited by the Josephson junction. When the current in any arm exceeds a critical value (i.e., Ic), a voltage develops across the junction. This voltage is read out as function of applied magnetic field or magnetic flux Φa. Since the magnetic flux in a superconducting loop is quantized (i.e., $\varphi = h/2e$), therefore, the critical current in the loop is modulated by the flux with a period equivalent to the flux quantum. Hence, the SQUID has a sensitivity of one flux quantum. In addition to the magnetic flux, it can measure electrical current ($\sim 10^{-12}$ A) and electrical resistance ($\sim 10^{-12}$ Ω).

6.3.3.3   Magneto-Optic Kerr Effect (MOKE)

Magneto-optic Kerr effect is the phenomenon of rotation of the plane of polarization of a light beam on reflection from the surface of a magnetized sample. MOKE can readily imagine domains in magnetic films and 2D vdW sheets/layers at the relatively low resolution of $\sim$1 μm.[2,3,41,109] The basic principle of this phenomenon is the circular birefringence due to differential reflection (phase difference or rotation) of the incident radiation. For magnetization along the vertical z-axis (polar MOKE), an additional phase difference will arise between left and right circularly polarized photons traveling along the z-axis upon reflection from the surface.[110] For linearly polarized light, polarization direction rotates upon reflection by an angle θ_K, known as the Kerr rotation angle. For most materials, the amount of rotation is small ($\sim$0.10 degrees) and depends on both the direction and the magnitude of the magnetization of the sample.[111]

6.3.3.4   Magnetic Force Microscopy

Magnetic force microscopy (MFM) is a scanning-probe-based technique, in which a sharp metallic tip covered with a magnetic layer, such as Co-Cr, is utilized to probe local magnetic forces with high spatial resolution required in the case of 2D materials.[18,112,113] It does not require any special sample preparation and can be done at ambient or temperature/environment-controlled conditions. This is a two-pass technique wherein the surface topography is measured in tapping mode during the first pass, and magnetic forces are mapped on the second pass with respect to the topography data. The magnetic interactions between the super paramagnetic particles of the magnetic tip (Co-Cr) and the stray field, a field originating due to the ferromagnetic materials on the surface, will lead to deflection of the tip as it experiences magnetic forces on a sample surface while scanning. The deflection of the tip depends upon the strength of the local magnetostatic interaction between the tip and the magnetic features in the sample. The mapping of magnetic forces to topography AFM scans produces a 2D image with lateral resolution ~ 50 nm and can be used to characterize the domain structure of magnetic materials.

6.4 APPLICATIONS OF 2D MAGNETIC MATERIALS

To meet the ever-growing demands of high sensitivity, high packing density, low power consumption along with compatibility to existing CMOS architecture, 2D materials are considered as ideal candidates due to their unique physical properties. 2D vdW magnetic materials can be used in fabrication of spin valves, spin field-effect transistors, magnetic tunnel junctions, Hall sensors, etc., to name a few.

6.4.1 SPIN VALVE

A spin valve is the fundamental building block of spintronic devices. It consists of two or more layers of conducting magnetic materials, whose electrical resistance can change depending on the relative alignment of the magnetization in the layers. A logical "off," or the high-resistance state, is represented by the antiparallel configuration of two ferromagnetic layers, while logical "on," or the low-resistance state, arises due to parallel alignment of ferromagnetic layers. Spin transport in graphene[11] has been successfully demonstrated at room temperature and significant improvement in spin injection efficiency in graphene through monolayer h-BN has been achieved in a later study.[114] Atomically thin MoS_2 based spin-valves (i.e., NiFe/ MoS_2/NiFe) show moderate MR of ~0.4% at 10 K, which dramatically improved to 0.73% when the NiFe layer was capped with a thin gold layer.[24] Further efforts are required to improve the spin-injection efficiency, increasing the spin diffusion length and understanding spin filtering effects at the interface to improve the MR ratio.

6.4.2 HALL EFFECT SENSORS

These devices detect the presence and magnitude of small magnetic fields using the Hall effect. Graphene-based Hall sensors have been extensively investigated and performance evaluated due to its low carrier densities and high mobility, which promises extremely high magnetic sensitivity and low electronic noise, respectively. They show ultrahigh sensitivity of 2093V/AT with resolution of $1mG/Hz^{0.5}$ @ $3kHz^{115}$ and low noise (i.e., Hooge's parameter $\alpha_H < 2 \times 10^{-3}$) with sensitivity of 87 V/AT.[116] Likewise, Hall sensors based on h-BN/graphene/h-BN van der Waals heterostructures show sensitivities up to 97 V/AT room temperature.[117] Further, Hall sensor based on monolayer MoS_2 on a hexagonal boron nitride (h-BN) thin film under optimum bias conditions revealed a maximum sensitivity ~3000 V/AT.[118]

6.4.3 SPIN FIELD-EFFECT TRANSISTORS

A spin field-effect transistor exploits the electrical control of spin current through the Rashba spin-orbit interaction, i.e., generation of an effective magnetic field by an electric field applied to the gate terminal to control the spin transport (via spin precession) from the source FM electrode to the drain FM electrode. At zero gate voltage, the transistor is in ON state since no spin precession results in spin direction parallel to the drain FM. Varying the gate voltage can alter the spin precession angle and result in a modulation of current across the spin transport channel. In spin-FET heterostructure of graphene and MoS_2, spin transport in the graphene channel could be switched ON and OFF by tuning the spin absorption into the MoS_2 layer, by virtue of strong spin-orbit coupling of MoS_2, with a gate electrode.[119] The spin transport can also be modulated by using spin lifetime anisotropy effect (SLA), as demonstrated in $MoSe_2$/graphene[120] and WS_2/graphene heterostructures.[121]

6.4.4 MAGNETIC TUNNEL JUNCTION (MTJ)

An MTJ consists of two FM electrodes with a thin insulating (dielectric) barrier sandwiched between them, where the electrons can tunnel through insulating barrier and cause charge flow. However, issues related to controlling the thickness of insulating barrier, its interfacial roughness, thermal and mechanical stability have limited the efforts for scaling up. In this context, 2D materials as tunnel barrier have huge potential due to their unique properties. Several 2D materials such as graphene,[123,124] h-BN,[125] MoS_2,[126,127] WS_2[128] have been investigated as tunnel barrier for MTJ devices (Figure 6.7). However, the magnetoresistance ratio (MR%) obtained from these have been relatively low ≤10% due to issues such as contamination during exfoliation, limited understanding of substate interactions, or due to reduced spin polarization in the buffer layer.

6.5 CONCLUSIONS AND OUTLOOK

Magnetism in the 2D vdW materials is a fascinating subject where the role of spin-orbit coupling resulting in magnetic anisotropy is the key for long-range magnetic order. The weak interlayer coupling and flexible lateral configurations of vdW materials, along with the tunability of magnetic properties such as magnetic anisotropy, transition temperature (T_C), relative orientation of magnetic easy-axis through strain, electric field, magnetic field, doping, etc. is of intense scientific appeal. Due to these unique physical properties and potential applications, 2D magnetic materials have received enormous attention in recent years. Exploring intrinsic magnetism in such materials is crucial not only from the point of view of understanding intricate low-dimensional magnetic interactions but also in the context of design and development of futuristic devices exploiting spin and charge degrees of freedom. Notwithstanding the current difficulties in experimental synthesis and characterization of 2D magnetic materials, significant research activity, emergence of novel materials heterostructures, and intriguing physical phenomenon await discovery in this nascent yet very stimulating field in the foreseeable future.

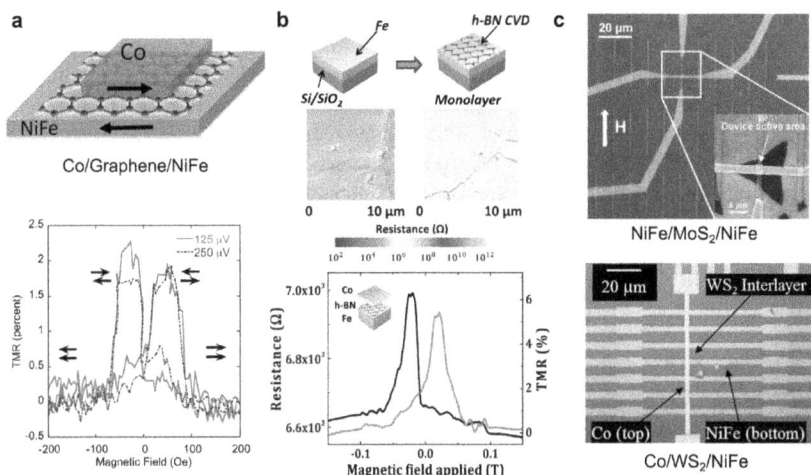

FIGURE 6.7 (a) The graphene-inserted magnetic tunnel junction (MTJ) device (top) and measured TMR characteristics (bottom) at temperature of 4 K and with low bias voltages (125 and 250 μV) applied. The CVD-grown graphene is transferred onto the 20 nm NiFe FM electrode. The arrows in the TMR graph indicate the relative orientation of the NiFe (lower) and Co (higher) FM contacts. (b) The Co/h-BN/Fe MTJ device structure is fabricated by annealing the Fe substrate, then directly growing h-BN on the Fe catalytic substrate via the low-pressure CVD technique (top), and finally depositing the 15 nm Co top electrode by evaporation. The conductive AFM (atomic force microscope) measurement yields the resistance values of 105 Ω for the Fe substrate and 107 Ω for the monolayer h-BN/Fe stack. The relatively high TMR ratio of about 6% is reported in this work (bottom), arising from suppression of the FM electrode oxidation due to direct CVD growth of h-BN on top of the FM electrode. (c) Monolayer TMD is used as the spacer layer in the ferromagnetic sandwich structure to result in the spin valve effect. The NiFe/MoS$_2$/NiFe device with monolayer MoS$_2$ grown by CVD and transferred by the wet process (top) and the Co/WS$_2$/NiFe device with the exfoliated single-layer WS$_2$ (bottom) exhibit the magnetoresistance ratios of up to 0.73% and 0.47% at low temperatures, respectively.

Source: Reprinted with permission from Ref.[122]

REFERENCES

1. Mermin, N. D.; Wagner, H. Absence of Ferromagnetism or Antiferromagnetism in One- or Two-Dimensional Isotropic Heisenberg Models. *Phys. Rev. Lett.* **1966**, vol. 17, pp. 1133–1136. https://doi.org/10.1103/PhysRevLett.17.1133.

2. Huang, B.; Clark, G.; Navarro-Moratalla, E.; Klein, D. R.; Cheng, R.; Seyler, K. L.; Zhong, Di.; Schmidgall, E.; McGuire, M. A.; Cobden, D. H.; Yao, W.; Xiao, D.; Jarillo-Herrero, P.; Xu, X. Layer-Dependent Ferromagnetism in a van Der Waals Crystal Down to the Monolayer Limit. *Nature* **2017**, vol. 546, pp. 270–273. https://doi.org/10.1038/nature22391.

3. Gong, C.; Li, L.; Li, Z.; Ji, H.; Stern, A.; Xia, Y.; Cao, T.; Bao, W.; Wang, C.; Wang, Y.; Qiu, Z. Q.; Cava, R. J.; Louie, S. G.; Xia, J.; Zhang, X. Discovery of Intrinsic Ferromagnetism in Two-Dimensional van Der Waals Crystals. *Nature* **2017**, vol. 546, pp. 265–269. https://doi.org/10.1038/nature22060.

4. Novoselov, K. S.; Geim, A. K.; Morozov, S. V.; Jiang, D.; Zhang, Y.; Dubonos, S. V.; Grigorieva, I. V.; Firsov, A. A. Electric Field in Atomically Thin Carbon Films. *Science (80-.).* **2004**, vol. 306, pp. 666–669. https://doi.org/10.1126/science.1102896.

5. Bolotin, K. I.; Sikes, K. J.; Jiang, Z.; Klima, M.; Fudenberg, G.; Hone, J.; Kim, P.; Stormer, H. L. Ultrahigh Electron Mobility in Suspended Graphene. *Solid State Commun.* **2008**, vol. 146, pp. 351–355. https://doi.org/10.1016/j.ssc.2008.02.024.

6. Lee, C.; Wei, X.; Kysar, J. W.; Hone, J. Measurement of the Elastic Properties and Intrinsic Strength of Monolayer Graphene. *Science (80-.).* **2008**, vol. 321, pp. 385–388. https://doi.org/10.1126/science.1157996.

7. Blake, P.; Hill, E. W.; Castro Neto, A. H.; Novoselov, K. S.; Jiang, D.; Yang, R.; Booth, T. J.; Geim, A. K. Making Graphene Visible. *Appl. Phys. Lett.* **2007**, vol. 91, pp. 063124. https://doi.org/10.1063/1.2768624.

8. Balandin, A. A. Thermal Properties of Graphene and Nanostructured Carbon Materials. *Nat. Mater.* **2011**, vol. 10, pp. 569–581. https://doi.org/10.1038/nmat3064.

9. Han, W.; Kawakami, R. K.; Gmitra, M.; Fabian, J. Graphene Spintronics. *Nat. Nanotechnol.* **2014**, vol. 9, pp. 794–807. https://doi.org/10.1038/nnano.2014.214.

10. Yazyev, O. V.; Helm, L. Defect-Induced Magnetism in Graphene. *Phys. Rev. B—Condens. Matter Mater. Phys.* **2007**, vol. 75, pp. 125408. https://doi.org/10.1103/PhysRevB.75.125408.

11. Tombros, N.; Jozsa, C.; Popinciuc, M.; Jonkman, H. T.; Van Wees, B. J. Electronic Spin Transport and Spin Precession in Single Graphene Layers at Room Temperature. *Nature* **2007**, vol. 448, pp. 571–574. https://doi.org/10.1038/nature06037.

12. Gui, G.; Li, J.; Zhong, J. Band Structure Engineering of Graphene by Strain: First-Principles Calculations. *Phys. Rev. B—Condens. Matter Mater. Phys.* **2008**, vol. 78, pp. 075435. https://doi.org/10.1103/PhysRevB.78.075435.

13. Zhou, J.; Wu, M. M.; Zhou, X.; Sun, Q. Tuning Electronic and Magnetic Properties of Graphene by Surface Modification. *Appl. Phys. Lett.* **2009**, vol. 95, pp. 103108. https://doi.org/10.1063/1.3225154.

14. Li, L.; Qin, R.; Li, H.; Yu, L.; Liu, Q.; Luo, G.; Gao, Z.; Lu, J. Functionalized Graphene for High-Performance Two-Dimensional Spintronics Devices. *ACS Nano* **2011**, vol. 5, pp. 2601–2610. https://doi.org/10.1021/nn102492g.

15. Espinosa-Ortega, T.; Luk'Yanchuk, I. A.; Rubo, Y. G. Magnetic Properties of Graphene Quantum Dots. *Phys. Rev. B—Condens. Matter Mater. Phys.* **2013**, vol. 87, pp. 205434. https://doi.org/10.1103/PhysRevB.87.205434.

16. Tang, T.; Liu, F.; Liu, Y.; Li, X.; Xu, Q.; Feng, Q.; Tang, N.; Du, Y. Identifying the Magnetic Properties of Graphene Oxide. *Appl. Phys. Lett.* **2014**, vol. 104, pp. 123104. https://doi.org/10.1063/1.4869827.

17. Varykhalov, A.; Sánchez-Barriga, J.; Shikin, A. M.; Biswas, C.; Vescovo, E.; Rybkin, A.; Marchenko, D.; Rader, O. Electronic and Magnetic Properties of Quasifreestanding Graphene on Ni. *Phys. Rev. Lett.* **2008**, vol. 101, pp. 157601. https://doi.org/10.1103/PhysRevLett.101.157601.

18. Alimohammadian, M.; Sohrabi, B. Observation of Magnetic Domains in Graphene Magnetized by Controlling Temperature, Strain and Magnetic Field. *Sci. Rep.* **2020**, vol. 10, pp. 21325 https://doi.org/10.1038/s41598-020-78262-w.

19. Mak, K. F.; Lee, C.; Hone, J.; Shan, J.; Heinz, T. F. Atomically Thin MoS$_2$: A New Direct-Gap Semiconductor. *Phys. Rev. Lett.* **2010**, vol. 105, pp. 136805. https://doi.org/10.1103/PhysRevLett.105.136805.

20. Radisavljevic, B.; Radenovic, A.; Brivio, J.; Giacometti, V.; Kis, A. Single-Layer MoS$_2$ Transistors. *Nat. Nanotechnol.* **2011**, vol. 6, pp. 147–150. https://doi.org/10.1038/nnano.2010.279.

21. Ataca, C.; Şahin, H.; Aktuörk, E.; Ciraci, S. Mechanical and Electronic Properties of MoS₂ Nanoribbons and Their Defects. *J. Phys. Chem. C* **2011**, vol. 115, pp. 3934–3941. https://doi.org/10.1021/jp1115146.

22. Tongay, S.; Varnoosfaderani, S. S.; Appleton, B. R.; Wu, J.; Hebard, A. F. Magnetic Properties of MoS 2: Existence of Ferromagnetism. *Appl. Phys. Lett.* **2012**, vol. 101, pp. 123105. https://doi.org/10.1063/1.4753797.

23. Cheng, Y. C.; Zhu, Z. Y.; Mi, W. B.; Guo, Z. B.; Schwingenschlögl, U. Prediction of Two-Dimensional Diluted Magnetic Semiconductors: Doped Monolayer MoS₂ Systems. *Phys. Rev. B—Condens. Matter Mater. Phys.* **2013**, vol. 87, pp. 100401. https://doi.org/10.1103/PhysRevB.87.100401.

24. Wang, W.; Narayan, A.; Tang, L.; Dolui, K.; Liu, Y.; Yuan, X.; Jin, Y.; Wu, Y.; Rungger, I.; Sanvito, S.; Xiu, F. Spin-Valve Effect in NiFe/MoS₂/NiFe Junctions. *Nano Lett.* **2015**, vol. 15, pp. 5261–5267. https://doi.org/10.1021/acs.nanolett.5b01553.

25. Ramasubramaniam, A.; Naveh, D. Mn-Doped Monolayer MoS₂: An Atomically Thin Dilute Magnetic Semiconductor. *Phys. Rev. B—Condens. Matter Mater. Phys.* **2013**, vol. 87, pp. 195201. https://doi.org/10.1103/PhysRevB.87.195201.

26. Zhang, K.; Feng, S.; Wang, J.; Azcatl, A.; Lu, N.; Addou, R.; Wang, N.; Zhou, C.; Lerach, J.; Bojan, V.; Kim, M. J.; Chen, L.-Q.; Wallace, R. M.; Terrones, M.; Zhu, J.; Robinson, J. A. Manganese Doping of Monolayer MoS₂: The Substrate Is Critical. *Nano Lett.* **2015**, vol. 15, pp. 6586–6591. https://doi.org/10.1021/acs.nanolett.5b02315.

27. Zhu, Z. Y.; Cheng, Y. C.; Schwingenschlögl, U. Giant Spin-Orbit-Induced Spin Splitting in Two-Dimensional Transition-Metal Dichalcogenide Semiconductors. *Phys. Rev. B—Condens. Matter Mater. Phys.* **2011**, vol. 84., pp. 153402. https://doi.org/10.1103/PhysRevB.84.153402.

28. Fu, S.; Kang, K.; Shayan, K.; Yoshimura, A.; Dadras, S.; Wang, X.; Zhang, L.; Chen, S.; Liu, N.; Jindal, A.; Li, X.; Pasupathy, A. N.; Vamivakas, A. N.; Meunier, V.; Strauf, S.; Yang, E. H. Enabling Room Temperature Ferromagnetism in Monolayer MoS₂ via in Situ Iron-Doping. *Nat. Commun.* **2020**, vol. 11, pp. 2034. https://doi.org/10.1038/s41467-020-15877-7.

29. McGuire, M. A.; Dixit, H.; Cooper, V. R.; Sales, B. C. Coupling of Crystal Structure and Magnetism in the Layered, Ferromagnetic Insulator Cri3. *Chem. Mater.* **2015**, vol. 27, pp. 612–620. https://doi.org/10.1021/cm504242t.

30. Wang, Z.; Gutiérrez-Lezama, I.; Ubrig, N.; Kroner, M.; Gibertini, M.; Taniguchi, T.; Watanabe, K.; Imamoğlu, A.; Giannini, E.; Morpurgo, A. F. Very Large Tunneling Magnetoresistance in Layered Magnetic Semiconductor CrI3. *Nat. Commun.* **2018**, vol. 9, pp. 2516. https://doi.org/10.1038/s41467-018-04953-8.

31. Soriano, D.; Cardoso, C.; Fernández-Rossier, J. Interplay between Interlayer Exchange and Stacking in CrI3 Bilayers. *Solid State Commun.* **2019**, vol. 299, pp. 113662. https://doi.org/10.1016/j.ssc.2019.113662.

32. Song, T.; Fei, Z.; Yankowitz, M.; Lin, Z.; Jiang, Q.; Hwangbo, K.; Zhang, Q.; Sun, B.; Taniguchi, T.; Watanabe, K.; McGuire, M. A.; Graf, D.; Cao, T.; Chu, J. H.; Cobden, D. H.; Dean, C. R.; Xiao, D.; Xu, X. Switching 2D Magnetic States via Pressure Tuning of Layer Stacking. *Nat. Mater.* **2019**, vol. 18, pp. 1298–1302. https://doi.org/10.1038/s41563-019-0505-2.

33. Sivadas, N.; Okamoto, S.; Xu, X.; Fennie, C. J.; Xiao, D. Stacking-Dependent Magnetism in Bilayer CrI3. *Nano Lett.* **2018**, vol. 18, pp. 7658–7664. https://doi.org/10.1021/acs.nanolett.8b03321.

34. Carteaux, V.; Brunet, D.; Ouvrard, G.; Andre, G. Crystallographic, Magnetic and Electronic Structures of a New Layered Ferromagnetic Compound Cr2Ge2Te6. *J. Phys. Condens. Matter* **1995**, vol. 7, pp. 69–87. https://doi.org/10.1088/0953-8984/7/1/008.

35. Ji, H.; Stokes, R. A.; Alegria, L. D.; Blomberg, E. C.; Tanatar, M. A.; Reijnders, A.; Schoop, L. M.; Liang, T.; Prozorov, R.; Burch, K. S.; Ong, N. P.; Petta, J. R.; Cava, R. J. A Ferromagnetic Insulating Substrate for the Epitaxial Growth of Topological Insulators. *J. Appl. Phys.* **2013**, vol. 114, pp. 114907. https://doi.org/10.1063/1.4822092.

36. Jiang, S.; Li, L.; Wang, Z.; Mak, K. F.; Shan, J. Controlling Magnetism in 2D CrI3 by Electrostatic Doping. *Nat. Nanotechnol.* **2018**, vol. 13, pp. 549–553. https://doi.org/10.1038/s41565-018-0135-x.

37. Jiang, S.; Shan, J.; Mak, K. F. Electric-Field Switching of Two-Dimensional van Der Waals Magnets. *Nat. Mater.* **2018**, vol. 17, pp. 406–410. https://doi.org/10.1038/s41563-018-0040-6.

38. Deng, Y.; Yu, Y.; Song, Y.; Zhang, J.; Wang, N. Z.; Sun, Z.; Yi, Y.; Wu, Y. Z.; Wu, S.; Zhu, J.; Wang, J.; Chen, X. H.; Zhang, Y. Gate-Tunable Room-Temperature Ferromagnetism in Two-Dimensional Fe3GeTe2. *Nature* **2018**, vol. 563, pp. 94–99. https://doi.org/10.1038/s41586-018-0626-9.

39. Verzhbitskiy, I. A.; Kurebayashi, H.; Cheng, H.; Zhou, J.; Khan, S.; Feng, Y. P.; Eda, G. Controlling the Magnetic Anisotropy in Cr2Ge2Te6 by Electrostatic Gating. *Nat. Electron.* **2020**, vol. 3, pp. 460-465. https://doi.org/10.1038/s41928-020-0427-7.

40. Zhuo, W.; Lei, B.; Wu, S.; Yu, F.; Zhu, C.; Cui, J.; Sun, Z.; Ma, D.; Shi, M.; Wang, H.; Wang, W.; Wu, T.; Ying, J.; Wu, S.; Wang, Z.; Chen, X. Manipulating Ferromagnetism in Few-Layered Cr2Ge2Te6. *Adv. Mater.* **2021**, vol. 33, pp. 2008586. https://doi.org/10.1002/adma.202008586.

41. Fei, Z.; Huang, B.; Malinowski, P.; Wang, W.; Song, T.; Sanchez, J.; Yao, W.; Xiao, D.; Zhu, X.; May, A. F.; Wu, W.; Cobden, D. H.; Chu, J. H.; Xu, X. Two-Dimensional Itinerant Ferromagnetism in Atomically Thin Fe3GeTe2. *Nat. Mater.* 2018, vol. 17, pp. 778–782 https://doi.org/10.1038/s41563-018-0149-7.

42. Liu, S.; Yuan, X.; Zou, Y.; Sheng, Y.; Huang, C.; Zhang, E.; Ling, J.; Liu, Y.; Wang, W.; Zhang, C.; Zou, J.; Wang, K.; Xiu, F. Wafer-Scale Two-Dimensional Ferromagnetic Fe3GeTe2 Thin Films Grown by Molecular Beam Epitaxy. *NPJ 2D Mater. Appl.* **2017**, vol. 1, pp. 30. https://doi.org/10.1038/s41699-017-0033-3.

43. Park, S. Y.; Kim, D. S.; Liu, Y.; Hwang, J.; Kim, Y.; Kim, W.; Kim, J. Y.; Petrovic, C.; Hwang, C.; Mo, S. K.; Kim, H. J.; Min, B. C.; Koo, H. C.; Chang, J.; Jang, C.; Choi, J. W.; Ryu, H. Controlling the Magnetic Anisotropy of the van Der Waals Ferromagnet Fe3GeTe2 through Hole Doping. *Nano Lett.* **2020**, vol. 20, pp. 95–100. https://doi.org/10.1021/acs.nanolett.9b03316.

44. Zhuang, H. L.; Kent, P. R. C.; Hennig, R. G. Strong Anisotropy and Magnetostriction in the Two-Dimensional Stoner Ferromagnet Fe3GeTe2. *Phys. Rev. B* **2016**, vol. 93, pp. 134407 https://doi.org/10.1103/PhysRevB.93.134407.

45. Wildes, A. R.; Simonet, V.; Ressouche, E.; McIntyre, G. J.; Avdeev, M.; Suard, E.; Kimber, S. A. J.; Lançon, D.; Pepe, G.; Moubaraki, B.; Hicks, T. J. Magnetic Structure of the Quasi-Two-Dimensional Antiferromagnet NiPS3. *Phys. Rev. B—Condens. Matter Mater. Phys.* **2015**, vol. 92, pp. 224408. https://doi.org/10.1103/PhysRevB.92.224408.

46. Lançon, D.; Ewings, R. A.; Guidi, T.; Formisano, F.; Wildes, A. R. Magnetic Exchange Parameters and Anisotropy of the Quasi-Two-Dimensional Antiferromagnet NiPS3. *Phys. Rev. B* **2018**, vol. 98, pp. 134414. https://doi.org/10.1103/PhysRevB.98.134414.

47. Kim, S. Y.; Kim, T. Y.; Sandilands, L. J.; Sinn, S.; Lee, M. C.; Son, J.; Lee, S.; Choi, K. Y.; Kim, W.; Park, B. G.; Jeon, C.; Kim, H. D.; Park, C. H.; Park, J. G.; Moon, S. J.; Noh, T. W. Charge-Spin Correlation in van Der Waals Antiferromagnet NiPS3. *Phys. Rev. Lett.* **2018**, vol. 120, pp. 136402. https://doi.org/10.1103/PhysRevLett.120.136402.

48. Kim, K.; Lim, S. Y.; Lee, J. U.; Lee, S.; Kim, T. Y.; Park, K.; Jeon, G. S.; Park, C. H.; Park, J. G.; Cheong, H. Suppression of Magnetic Ordering in XXZ-Type Antiferromagnetic Monolayer NiPS 3. *Nat. Commun.* **2019**, vol. 10, pp. 345. https://doi.org/10.1038/s41467-018-08284-6.

49. Lane, C.; Zhu, J. X. Thickness Dependence of Electronic Structure and Optical Properties of a Correlated van Der Waals Antiferromagnetic NiPS3 Thin Film. *Phys. Rev. B* **2020**, vol. 102, pp. 075124. https://doi.org/10.1103/PhysRevB.102.075124.

50. Hwangbo, K.; Zhang, Q.; Jiang, Q.; Wang, Y.; Fonseca, J.; Wang, C.; Diederich, G. M.; Gamelin, D. R.; Xiao, D.; Chu, J. H.; Yao, W.; Xu, X. Highly Anisotropic Excitons and Multiple Phonon Bound States in a van Der Waals Antiferromagnetic Insulator. *Nat. Nanotechnol.* **2021**, vol. 16, pp. 655–660. https://doi.org/10.1038/s41565-021-00873-9.

51. Joy, P. A.; Vasudevan, S. Magnetism in the Layered Transition-Metal Thiophosphates MPS3 (M=Mn, Fe, and Ni). *Phys. Rev. B* **1992**, vol. 46, pp. 5425–5433. https://doi.org/10.1103/PhysRevB.46.5425.

52. Wildes, A. R.; Rønnow, H. M.; Roessli, B.; Harris, M. J.; Godfrey, K. W. Static and Dynamic Critical Properties of the Quasi-Two-Dimensional Antiferromagnet MnPS3. *Phys. Rev. B—Condens. Matter Mater. Phys.* **2006**, vol. 74, pp. 094422. https://doi.org/10.1103/PhysRevB.74.094422.

53. Lançon, D.; Walker, H. C.; Ressouche, E.; Ouladdiaf, B.; Rule, K. C.; McIntyre, G. J.; Hicks, T. J.; Rønnow, H. M.; Wildes, A. R. Magnetic Structure and Magnon Dynamics of the Quasi-Two-Dimensional Antiferromagnet FePS3. *Phys. Rev. B* **2016**, vol. 94, pp. 214407. https://doi.org/10.1103/PhysRevB.94.214407.

54. Wildes, A. R.; Simonet, V.; Ressouche, E.; Ballou, R.; McIntyre, G. J. The Magnetic Properties and Structure of the Quasi-Two-Dimensional Antiferromagnet CoPS3. *J. Phys. Condens. Matter* **2017**, vol. 29, pp. 455801. https://doi.org/10.1088/1361-648X/aa8a43.

55. Kane, C. L.; Mele, E. J. Z2 Topological Order and the Quantum Spin Hall Effect. *Phys. Rev. Lett.* **2005**, vol. 95, pp. 146802. https://doi.org/10.1103/PhysRevLett.95.146802.

56. Geim, A. K.; Novoselov, K. S. The Rise of Graphene. *Nat. Mater.* **2007**, vol. 6, pp. 183–191. https://doi.org/10.1038/nmat1849.

57. Yao, Y.; Ye, F.; Qi, X. L.; Zhang, S. C.; Fang, Z. Spin-Orbit Gap of Graphene: First-Principles Calculations. *Phys. Rev. B—Condens. Matter Mater. Phys.* **2007**, vol. 75, pp. 041401. https://doi.org/10.1103/PhysRevB.75.041401.

58. Tang, H.; Liang, D.; Qiu, R. L. J.; Gao, X. P. A. Two-Dimensional Transport-Induced Linear Magneto-Resistance in Topological Insulator Bi2Se3 Nanoribbons. *ACS Nano* **2011**, vol. 5, pp. 7510–7516. https://doi.org/10.1021/nn2024607.

59. Chang, C. Z.; Zhang, J.; Feng, X.; Shen, J.; Zhang, Z.; Guo, M.; Li, K.; Ou, Y.; Wei, P.; Wang, L. L.; Ji, Z. Q.; Feng, Y.; Ji, S.; Chen, X.; Jia, J.; Dai, X.; Fang, Z.; Zhang, S. C.; He, K.; Wang, Y.; Lu, L.; Ma, X. C.; Xue, Q. K. Experimental Observation of the Quantum Anomalous Hall Effect in a Magnetic Topological Insulator. *Science (80-.).* **2013**, vol. 340, pp. 167–170. https://doi.org/10.1126/science.1234414.

60. Xie, H.; Wang, D.; Cai, Z.; Chen, B.; Guo, J.; Naveed, M.; Zhang, S.; Zhang, M.; Wang, X.; Fei, F.; Zhang, H.; Song, F. The Mechanism Exploration for Zero-Field Ferromagnetism in Intrinsic Topological Insulator MnBi2Te4 by Bi2Te3 Intercalations. *Appl. Phys. Lett.* **2020**, vol. 116, pp. 221902. https://doi.org/10.1063/5.0009085.

61. Liu, T.; He, J. J.; Nori, F. Majorana Corner States in a Two-Dimensional Magnetic Topological Insulator on a High-Temperature Superconductor. *Phys. Rev. B* **2018**, vol. 98, pp. 245413. https://doi.org/10.1103/PhysRevB.98.245413.

62. Rienks, E. D. L.; Wimmer, S.; Sánchez-Barriga, J.; Caha, O.; Mandal, P. S.; Růžička, J.; Ney, A.; Steiner, H.; Volobuev, V. V.; Groiss, H.; Albu, M.; Kothleitner, G.; Michalička, J.; Khan, S. A.; Minár, J.; Ebert, H.; Bauer, G.; Freyse, F.; Varykhalov, A.; Rader, O.; Springholz, G. Large Magnetic Gap at the Dirac Point in Bi2Te3/ MnBi2Te4 Heterostructures. *Nature* **2019**, vol. 576, pp. 423–428. https://doi.org/10.1038/ s41586-019-1826-7.

63. Collins-Mcintyre, L. J.; Harrison, S. E.; Schönherr, P.; Steinke, N. J.; Kinane, C. J.; Charlton, T. R.; Alba-Veneroa, D.; Pushp, A.; Kellock, A. J.; Parkin, S. S. P.; Harris, J. S.; Langridge, S.; Van Der Laan, G.; Hesjedal, T. Magnetic Ordering in Cr-Doped Bi2Se3 Thin Films. *EPL* **2014**, vol.107, pp. 57009. https://doi.org/10.1209/0295-5075/107/57009.

64. Velický, M.; Donnelly, G. E.; Hendren, W. R.; McFarland, S.; Scullion, D.; DeBenedetti, W. J. I.; Correa, G. C.; Han, Y.; Wain, A. J.; Hines, M. A.; Muller, D. A.; Novoselov, K. S.; Abruńa, H. D.; Bowman, R. M.; Santos, E. J. G.; Huang, F. Mechanism of Gold-Assisted Exfoliation of Centimeter-Sized Transition-Metal Dichalcogenide Monolayers. *ACS Nano* **2018**, vol. 12, pp. 10463–10472. https://doi.org/10.1021/acsnano.8b06101.

65. Pierret, A.; Loayza, J.; Berini, B.; Betz, A.; Plaçais, B.; Ducastelle, F.; Barjon, J.; Loiseau, A. Excitonic Recombinations in h—BN: From Bulk to Exfoliated Layers. *Phys. Rev. B—Condens. Matter Mater. Phys.* **2014**, vol. 89, pp. 035414. https://doi. org/10.1103/PhysRevB.89.035414.

66. Shahil, K. M. F.; Hossain, M. Z.; Goyal, V.; Balandin, A. A. Micro-Raman Spectroscopy of Mechanically Exfoliated Few-Quintuple Layers of Bi2Te3, Bi2Se3, and Sb2Te3 Materials. *J. Appl. Phys.* **2012**, vol. 111, pp. 54305. https://doi.org/10.1063/1.3690913.

67. Kuo, C. T.; Balamurugan, K.; Shiu, H. W.; Park, H. J.; Sinn, S.; Neumann, M.; Han, M.; Chang, Y. J.; Chen, C. H.; Kim, H. D.; Park, J. G.; Noh, T. W. The Energy Band Alignment at the Interface between Mechanically Exfoliated Few-Layer NiPS3 Nanosheets and ZnO. *Curr. Appl. Phys.* **2016**, vol. 16, pp. 404–408. https://doi. org/10.1016/j.cap.2016.01.001.

68. Paton, K. R.; Varrla, E.; Backes, C.; Smith, R. J.; Khan, U.; O'Neill, A.; Boland, C.; Lotya, M.; Istrate, O. M.; King, P.; Higgins, T.; Barwich, S.; May, P.; Puczkarski, P.; Ahmed, I.; Moebius, M.; Pettersson, H.; Long, E.; Coelho, J.; O'Brien, S. E.; McGuire, E. K.; Sanchez, B. M.; Duesberg, G. S.; McEvoy, N.; Pennycook, T. J.; Downing, C.; Crossley, A.; Nicolosi, V.; Coleman, J. N. Scalable Production of Large Quantities of Defect-Free Few-Layer Graphene by Shear Exfoliation in Liquids. *Nat. Mater.* **2014**, vol. 13, pp. 624–630. https://doi.org/10.1038/nmat3944.

69. Zhu, J.; Zhuang, X.; Yang, J.; Feng, X.; Hirano, S. I. Graphene-Coupled Nitrogen-Enriched Porous Carbon Nanosheets for Energy Storage. *J. Mater. Chem. A* **2017**, vol. 5, pp. 16732–16739. https://doi.org/10.1039/c7ta04752e.

70. Zhuang, X.; Gehrig, D.; Forler, N.; Liang, H.; Wagner, M.; Hansen, M. R.; Laquai, F.; Zhang, F.; Feng, X. Conjugated Microporous Polymers with Dimensionality-Controlled Heterostructures for Green Energy Devices. *Adv. Mater.* **2015**, vol. 27, pp. 3789-3796. https://doi.org/10.1002/adma.201501786.

71. Li, D. O.; Gilliam, M. S.; Chu, X. S.; Yousaf, A.; Guo, Y.; Green, A. A.; Wang, Q. H. Covalent Chemical Functionalization of Semiconducting Layered Chalcogenide Nanosheets. *Mol. Syst. Des. Eng.* **2019**, vol. 4, pp. 962–973. https://doi.org/10.1039/c9me00045c.

72. Guo, S.; Nishina, Y.; Bianco, A.; Ménard-Moyon, C. A Flexible Method for Covalent Double Functionalization of Graphene Oxide. *Angew. Chemie—Int. Ed.* **2020**, vol. 59, pp. 1542–1547. https://doi.org/10.1002/anie.201913461.

73. Er, E.; Hou, H. L.; Criado, A.; Langer, J.; Möller, M.; Erk, N.; Liz-Marzán, L. M.; Prato, M. High-Yield Preparation of Exfoliated 1T-MoS$_2$ with SERS Activity. *Chem. Mater.* **2019**, vol. 31, pp. 5725–5734. https://doi.org/10.1021/acs.chemmater.9b01698.

74. Alzakia, F. I.; Tan, S. C. Liquid-Exfoliated 2D Materials for Optoelectronic Applications. *Adv. Sci.* **2021**, vol. 8, pp. 2003864. https://doi.org/10.1002/advs.202003864.

75. Van Druenen, M.; Davitt, F.; Collins, T.; Glynn, C.; O'Dwyer, C.; Holmes, J. D.; Collins, G. Covalent Functionalization of Few-Layer Black Phosphorus Using Iodonium Salts and Comparison to Diazonium Modified Black Phosphorus. *Chem. Mater.* **2018**, vol. 30, pp. 4667–4674. https://doi.org/10.1021/acs.chemmater.8b01306.

76. Ohashi, M.; Shirai, S.; Nakano, H. Direct Chemical Synthesis of Benzyl-Modified Silicane from Calcium Disilicide. *Chem. Mater.* **2019**, vol. 31, pp. 4720–4725. https://doi.org/10.1021/acs.chemmater.9b00715.

77. Sturala, J.; Luxa, J.; Matějková, S.; Plutnar, J.; Hartman, T.; Pumera, M.; Sofer, Z. Exfoliation of Calcium Germanide by Alkyl Halides. *Chem. Mater.* **2019**, vol. 31, pp. 10126–10134. https://doi.org/10.1021/acs.chemmater.9b03391.

78. Gayathri, S.; Jayabal, P.; Kottaisamy, M.; Ramakrishnan, V. Synthesis of Few Layer Graphene by Direct Exfoliation of Graphite and a Raman Spectroscopic Study. *AIP Adv.* **2014**, vol. 4, pp. 027116. https://doi.org/10.1063/1.4866595.

79. Joshi, R. K.; Shukla, S.; Saxena, S.; Lee, G. H.; Sahajwalla, V.; Alwarappan, S. Hydrogen Generation via Photoelectrochemical Water Splitting Using Chemically Exfoliated MoS$_2$ Layers. *AIP Adv.* **2016**, vol. 6, pp. 015315. https://doi.org/10.1063/1.4941062.

80. Zhang, K.; Feng, Y.; Wang, F.; Yang, Z.; Wang, J. Two Dimensional Hexagonal Boron Nitride (2D-HBN): Synthesis, Properties and Applications. *J. Mater. Chem. C.* **2017**, vol. 5, pp. 11992–12022. https://doi.org/10.1039/c7tc04300g.

81. Wang, N.; Tang, H.; Shi, M.; Zhang, H.; Zhuo, W.; Liu, D.; Meng, F.; Ma, L.; Ying, J.; Zou, L.; Sun, Z.; Chen, X. Transition from Ferromagnetic Semiconductor to Ferromagnetic Metal with Enhanced Curie Temperature in Cr2Ge2Te6 via Organic Ion Intercalation. *J. Am. Chem. Soc.* **2019**, vol. 141, pp. 17166–17173. https://doi.org/10.1021/jacs.9b06929.

82. Teweldebrhan, D.; Goyal, V.; Balandin, A. A. Exfoliation and Characterization of Bismuth Telluride Atomic Quintuples and Quasi-Two-Dimensional Crystals. *Nano Lett.* **2010**, vol. 10, pp. 1209–1218. https://doi.org/10.1021/nl903590b.

83. Ambrosi, A.; Sofer, Z.; Luxa, J.; Pumera, M. Exfoliation of Layered Topological Insulators Bi2Se3 and Bi2Te3 via Electrochemistry. *ACS Nano* **2016**, vol. 10, pp. 11442–11448. https://doi.org/10.1021/acsnano.6b07096.

84. Liu, J.; Wang, Y.; Fang, Y.; Ge, Y.; Li, X.; Fan, D.; Zhang, H. A Robust 2D Photo-Electrochemical Detector Based on NiPS3 Flakes. *Adv. Electron. Mater.* **2019**, vol. 5, pp. 1900726. https://doi.org/10.1002/aelm.201900726.

85. Habib, M. R.; Liang, T.; Yu, X.; Pi, X.; Liu, Y.; Xu, M. A Review of Theoretical Study of Graphene Chemical Vapor Deposition Synthesis on Metals: Nucleation, Growth, and the Role of Hydrogen and Oxygen. *Rep. Prog. Phys.* **2018**, vol. 81, pp. 036501. https://doi.org/10.1088/1361-6633/aa9bbf.

86. Khan, M. H.; Liu, H. K.; Sun, X.; Yamauchi, Y.; Bando, Y.; Golberg, D.; Huang, Z. Few-Atomic-Layered Hexagonal Boron Nitride: CVD Growth, Characterization, and Applications. *Mater. Today.* **2017**, vol. 20, pp. 611–628. https://doi.org/10.1016/j.mattod.2017.04.027.

87. Cartamil-Bueno, S. J.; Cavalieri, M.; Wang, R.; Houri, S.; Hofmann, S.; van der Zant, H. S. J. Mechanical Characterization and Cleaning of CVD Single-Layer h-BN Resonators. *NPJ 2D Mater. Appl.* **2017**, vol. 1, pp. 16. https://doi.org/10.1038/s41699-017-0020-8.

88. Zhang, Y.; Yao, Y.; Sendeku, M. G.; Yin, L.; Zhan, X.; Wang, F.; Wang, Z.; He, J. Recent Progress in CVD Growth of 2D Transition Metal Dichalcogenides and Related Heterostructures. *Adv. Mater.* **2019**, vol.31, pp. 1901694. https://doi.org/10.1002/adma.201901694.

89. Pawbake, A. S.; Pawar, M. S.; Jadkar, S. R.; Late, D. J. Large Area Chemical Vapor Deposition of Monolayer Transition Metal Dichalcogenides and Their Temperature Dependent Raman Spectroscopy Studies. *Nanoscale* **2016**, vol. 8, pp. 3008–3018. https://doi.org/10.1039/c5nr07401k.

90. Samal, R.; Sanyal, G.; Chakraborty, B.; Rout, C. S. Two-Dimensional Transition Metal Phosphorous Trichalcogenides (MPX3): A Review on Emerging Trends, Current State and Future Perspectives. *J. Mater. Chem. A.* **2021**, vol. 9, pp. 2560–2591. https://doi.org/10.1039/d0ta09752g.

91. Naveed, M.; Cai, Z.; Bu, H.; Fei, F.; Shah, S. A.; Chen, B.; Rahman, A.; Zhang, K.; Xie, F.; Song, F. Temperature-Dependent Growth of Topological Insulator Bi2Se3for Nanoscale Fabrication. *AIP Adv.* **2020**, vol. 10, pp. 115202. https://doi.org/10.1063/5.0021125.

92. Zhan, N.; Olmedo, M.; Wang, G.; Liu, J. Layer-by-Layer Synthesis of Large-Area Graphene Films by Thermal Cracker Enhanced Gas Source Molecular Beam Epitaxy. *Carbon N. Y.* **2011**, vol. 49, pp. 2046–2052. https://doi.org/10.1016/j.carbon.2011.01.033.

93. Garcia, J. M.; Wurstbauer, U.; Levy, A.; Pfeiffer, L. N.; Pinczuk, A.; Plaut, A. S.; Wang, L.; Dean, C. R.; Buizza, R.; Van Der Zande, A. M.; Hone, J.; Watanabe, K.; Taniguchi, T. Graphene Growth on H-BN by Molecular Beam Epitaxy. *Solid State Commun.* **2012**, vol. 152, pp. 975–978. https://doi.org/10.1016/j.ssc.2012.04.005.

94. Chen, M. W.; Ovchinnikov, D.; Lazar, S.; Pizzochero, M.; Whitwick, M. B.; Surrente, A.; Baranowski, M.; Sanchez, O. L.; Gillet, P.; Plochocka, P.; Yazyev, O. V.; Kis, A. Highly Oriented Atomically Thin Ambipolar MoSe2 Grown by Molecular Beam Epitaxy. *ACS Nano* **2017**, vol. 11, pp. 6355–6361. https://doi.org/10.1021/acsnano.7b02726.

95. Xie, B.; Chen, Y.; Yu, M.; Sun, T.; Lu, L.; Xie, T.; Zhang, Y.; Wu, Y. Hydrothermal Synthesis of Layered Molybdenum Sulfide/N-Doped Graphene Hybrid with Enhanced Supercapacitor Performance. *Carbon N. Y.* **2016**, vol. 99, pp. 35-42. https://doi.org/10.1016/j.carbon.2015.11.077.

96. Han, S.; Hu, L.; Liang, Z.; Wageh, S.; Al-Ghamdi, A. A.; Chen, Y.; Fang, X. One-Step Hydrothermal Synthesis of 2D Hexagonal Nanoplates of α-Fe2O3/Graphene Composites with Enhanced Photocatalytic Activity. *Adv. Funct. Mater.* **2014**, vol. 24, pp. 5719–5727. https://doi.org/10.1002/adfm.201401279.

97. Shen, J.; Yan, B.; Shi, M.; Ma, H.; Li, N.; Ye, M. One Step Hydrothermal Synthesis of TiO2-Reduced Graphene Oxide Sheets. *J. Mater. Chem.* **2011**, vol. 21, pp. 3415–3421. https://doi.org/10.1039/c0jm03542d.

98. Xu, F.; Xu, C.; Chen, H.; Wu, D.; Gao, Z.; Ma, X.; Zhang, Q.; Jiang, K. The Synthesis of Bi2S3/2D-Bi2WO6 Composite Materials with Enhanced Photocatalytic Activities. *J. Alloys Compd.* **2019**, vol. 780, pp. 634–642. https://doi.org/10.1016/j.jallcom.2018.11.397.

99. Mayyas, M.; Li, H.; Kumar, P.; Ghasemian, M. B.; Yang, J.; Wang, Y.; Lawes, D. J.; Han, J.; Saborio, M. G.; Tang, J.; Jalili, R.; Lee, S. H.; Seong, W. K.; Russo, S. P.; Esrafilzadeh, D.; Daeneke, T.; Kaner, R. B.; Ruoff, R. S.; Kalantar-Zadeh, K. Liquid-Metal-Templated Synthesis of 2D Graphitic Materials at Room Temperature. *Adv. Mater.* **2020**, vol. 32, pp. 2001997. https://doi.org/10.1002/adma.202001997.

100. Xiao, X.; Wang, H.; Urbankowski, P.; Gogotsi, Y. Topochemical Synthesis of 2D Materials. *Chem. Soc. Rev.* **2018**, vol. 47, pp. 8744–8765. https://doi.org/10.1039/c8cs00649k.

101. Chen, K. H. M.; Lin, H. Y.; Yang, S. R.; Cheng, C. K.; Zhang, X. Q.; Cheng, C. M.; Lee, S. F.; Hsu, C. H.; Lee, Y. H.; Hong, M.; Kwo, J. Van Der Waals Epitaxy of Topological Insulator Bi2Se3 on Single Layer Transition Metal Dichalcogenide MoS2. *Appl. Phys. Lett.* **2017**, vol. 111, pp. 083106. https://doi.org/10.1063/1.4989805.

102. Bera, S.; Behera, P.; Mishra, A. K.; Krishnan, M.; Patidar, M. M.; Venkatesh, R.; Ganesan, V. Weak Antilocalization in Sb2Te3 Nano-Crystalline Topological Insulator. *Appl. Surf. Sci.* **2019**, vol. 496, pp. 143654. https://doi.org/10.1016/j.apsusc.2019.143654.

103. Zhang, X.; Wen, F.; Xiang, J.; Wang, X.; Wang, L.; Hu, W.; Liu, Z. Wearable Non-Volatile Memory Devices Based on Topological Insulator Bi2Se3/Pt Fibers. *Appl. Phys. Lett.* **2015**, vol. 107, pp. 103109. https://doi.org/10.1063/1.4930822.

104. Singh, S.; Hong, S.; Jeon, W.; Lee, D.; Hwang, J. Y.; Lim, S.; Kwon, G. D.; Pribat, D.; Shin, H.; Kim, S. W.; Baik, S. Graphene-Templated Synthesis of C -Axis Oriented Sb2Te3 Nanoplates by the Microwave-Assisted Solvothermal Method. *Chem. Mater.* **2015**, vol. 27, pp. 2315–2321. https://doi.org/10.1021/cm502749y.

105. Lukins, R. E. Vibrating Sample Magnetometer 2D and 3D Magnetization Effects Associated with Different Initial Magnetization States. *AIP Adv.* **2017**, vol. 7, pp. 056801. https://doi.org/10.1063/1.4973750.

106. Chen, M.; Hu, C.; Luo, X.; Hong, A.; Yu, T.; Yuan, C. Ferromagnetic Behaviors in Monolayer MoS_2 Introduced by Nitrogen-Doping. *Appl. Phys. Lett.* **2020**, vol. 116, pp. 073102. https://doi.org/10.1063/5.0001572.

107. Mendes, J. B. S.; Rezende, S. M.; Holanda, J. Rashba-Edelstein Magnetoresistance in Two-Dimensional Materials at Room Temperature. *Phys. Rev. B* **2021**, vol. 104, pp. 014408. https://doi.org/10.1103/PhysRevB.104.014408.

108. Wan, Y.; Xue, M.; Cheng, X.; Peng, Y.; Li, P.; Yang, S.; Liu, M.; Kan, E.; Yang, J.; Dai, L. Surface-Sensitive Magnetic Characterization Technique for Ultrathin Ferromagnetic Film with Perpendicular Magnetic Anisotropy. *AIP Adv.* **2020**, vol. 10, pp. 065019. https://doi.org/10.1063/5.0012321.

109. Mak, K. F.; Shan, J.; Ralph, D. C. Probing and Controlling Magnetic States in 2D Layered Magnetic Materials. *Nat. Rev. Phys.* **2019**, vol. 1, pp. 646–661. https://doi.org/10.1038/s42254-019-0110-y.

110. Gibertini, M.; Koperski, M.; Morpurgo, A. F.; Novoselov, K. S. Magnetic 2D Materials and Heterostructures. *Nat. Nanotechnol.* **2019**, vol. 14, pp. 408–419 https://doi.org/10.1038/s41565-019-0438-6.

111. Cullity, B. D.; Graham, C. D. *Introduction to Magnetic Materials*, 2nd ed. John Wiley & Sons, Inc., 2009, pp. 289.

112. Li, H.; Qi, X.; Wu, J.; Zeng, Z.; Wei, J.; Zhang, H. Investigation of MoS_2 and Graphene Nanosheets by Magnetic Force Microscopy. *ACS Nano* **2013**, vol. 7, pp. 2842–2849. https://doi.org/10.1021/nn400443u.

113. Li, L. H.; Chen, Y. Electric Contributions to Magnetic Force Microscopy Response from Graphene and MoS_2 Nanosheets. *J. Appl. Phys.* **2014**, vol. 116, pp. 213904. https://doi.org/10.1063/1.4903040.

114. Yamaguchi, T.; Inoue, Y.; Masubuchi, S.; Morikawa, S.; Onuki, M.; Watanabe, K.; Taniguchi, T.; Moriya, R.; Machida, T. Electrical Spin Injection into Graphene through Monolayer Hexagonal Boron Nitride. *Appl. Phys. Express* **2013**, vol.6, pp. 073001. https://doi.org/10.7567/APEX.6.073001.

115. Huang, L.; Zhang, Z.; Chen, B.; Ma, X.; Zhong, H.; Peng, L. M. Ultra-Sensitive Graphene Hall Elements. *Appl. Phys. Lett.* **2014**, vol. 104, pp. 183106. https://doi.org/10.1063/1.4875597.

116. Ciuk, T.; Petruk, O.; Kowalik, A.; Jozwik, I.; Rychter, A.; Szmidt, J.; Strupinski, W. Low-Noise Epitaxial Graphene on SiC Hall Effect Element for Commercial Applications. *Appl. Phys. Lett.* **2016**, vol. 108, pp. 223504. https://doi.org/10.1063/1.4953258.

117. Dankert, A.; Karpiak, B.; Dash, S. P. Hall Sensors Batch-Fabricated on All-CVD h-BN/Graphene/h-BN Heterostructures. *Sci. Rep.* **2017**, vol. 7, pp. 15231. https://doi.org/10.1038/s41598-017-12277-8.

118. Joo, M. K.; Kim, J.; Lee, G.; Kim, H.; Lee, Y. H.; Suh, D. Feasibility of Ultra-Sensitive 2D Layered Hall Elements. *2D Mater.* **2017**, vol. 4, pp. 021029. https://doi.org/10.1088/2053-1583/aa735d.

119. Yan, W.; Txoperena, O.; Llopis, R.; Dery, H.; Hueso, L. E.; Casanova, F. A Two-Dimensional Spin Field-Effect Switch. *Nat. Commun.* **2016**, vol. 7, pp. 13372. https://doi.org/10.1038/ncomms13372.
120. Ghiasi, T. S.; Ingla-Aynés, J.; Kaverzin, A. A.; Van Wees, B. J. Large Proximity-Induced Spin Lifetime Anisotropy in Transition-Metal Dichalcogenide/Graphene Heterostructures. *Nano Lett.* **2017**, vol. 17, pp. 7528–7532. https://doi.org/10.1021/acs.nanolett.7b03460.
121. Omar, S.; Madhushankar, B. N.; Van Wees, B. J. Large Spin-Relaxation Anisotropy in Bilayer-Graphene/WS2 Heterostructures. *Phys. Rev. B* **2019**, vol. 100, pp. 155415. https://doi.org/10.1103/PhysRevB.100.155415.
122. Ahn, E. C. 2D Materials for Spintronic Devices. *NPJ 2D Mater. Appl.* **2020**, vol. 4, pp. 17. https://doi.org/10.1038/s41699-020-0152-0.
123. Mohiuddin, T. M. G.; Hill, E.; Elias, D.; Zhukov, A.; Novoselov, K.; Geim, A. Graphene in Multilayered CPP Spin Valves. *IEEE Trans. Magn.* **2008**, vol. 44, pp. 2624–2627. https://doi.org/10.1109/TMAG.2008.2003065.
124. Park, J. H.; Lee, H. J. Out-of-Plane Magnetoresistance in Ferromagnet/Graphene/Ferromagnet Spin-Valve Junctions. *Phys. Rev. B—Condens. Matter Mater. Phys.* **2014**, vol. 89, pp. 165417. https://doi.org/10.1103/PhysRevB.89.165417.
125. Piquemal-Banci, M.; Galceran, R.; Caneva, S.; Martin, M. B.; Weatherup, R. S.; Kidambi, P. R.; Bouzehouane, K.; Xavier, S.; Anane, A.; Petroff, F.; Fert, A.; Robertson, J.; Hofmann, S.; Dlubak, B.; Seneor, P. Magnetic Tunnel Junctions with Monolayer Hexagonal Boron Nitride Tunnel Barriers. *Appl. Phys. Lett.* **2016**, vol. 108, pp. 102404. https://doi.org/10.1063/1.4943516.
126. Dankert, A.; Venkata Kamalakar, M.; Wajid, A.; Patel, R. S.; Dash, S. P. Tunnel Magnetoresistance with Atomically Thin Two-Dimensional Hexagonal Boron Nitride Barriers. *Nano Res.* **2015**, vol. 8, pp. 1357–1364. https://doi.org/10.1007/s12274-014-0627-4.
127. Tonkikh, A. A.; Voloshina, E. N.; Werner, P.; Blumtritt, H.; Senkovskiy, B.; Güntherodt, G.; Parkin, S. S. P.; Dedkov, Y. S. Structural and Electronic Properties of Epitaxial Multilayer H-BN on Ni(111) for Spintronics Applications. *Sci. Rep.* **2016**, vol. 6, pp. 23547. https://doi.org/10.1038/srep23547.
128. Iqbal, M. Z.; Iqbal, M. W.; Siddique, S.; Khan, M. F.; Ramay, S. M. Room Temperature Spin Valve Effect in NiFe/WS2/Co Junctions. *Sci. Rep.* **2016**, vol. 6, pp. 21038. https://doi.org/10.1038/srep21038.

7 2D Materials
Mechanical Properties and Applications

*Deeksha Nagpal, Astakala Anil Kumar,
Jashandeep Singh, Ajay Vasishth, Shashank Priya,
Ashok Kumar, and Shyam Sundar Pattnaik*

CONTENTS

DOI: 10.1201/9781003247890-7

7.1 INTRODUCTION

In recent years, 2D materials attracted researchers' attention due to interesting mechanical properties such as flexibility, mechanical strength, elastic modulus, and the friction transfer between the different layers. The 2D materials can be analyzed for their myriad flaws and functionalization in numerous ways using the methodologies of synthesis, isolation, and propagation. Exfoliation and chemical vapor deposition are two common techniques used in the synthesis of layered 2D materials. The mechanical characteristics such as brittleness and flexibility of 2D materials are responsible for the distortion out of the plane in terms of angstroms. However, they have a big impact on the effective characteristics. Ripples or corrugations are the terms used to describe these little out-of-plane deformations. These are distinguishable from wider buckling deformations, such as wrinkles or crumples, that can occur from 100 nm to several microns as a result of contact with another substrate [1–2]. Addressing the functionality of 2D materials in actual devices, their durability during ambience and varying environments, is required for practical application. The investigation on the dependability of 2D materials based systems on electrical or thermal stress assessment methods along with device stability requires attention of the researchers. The device safety, performance and the impact of the device is an important aspect of fabricating a device. The characteristic material used in the device fabrication plays a key role in device performance. Currently, the 2D materials that are used in several mechanical devices will have no adverse impact on the environment and are expected to enhance the device performance. In the mechanical viewpoint, 2D material structure attributes to its unique layered structure with good flexibility and high mechanical strength, making it an attractive active material in the fabrication of several mechanical devices such as sensors, piezoelectric devices, etc. It is possible to alter the mechanical behavior of the material with the incorporation of defects/functional groups [3].

According to Hooke's law, the strain developed under the application of stress will have the linear relationship, which may be regarded as a mechanical comparable to Ohm's law and is among the most underlying mechanical characteristics of mechanical components. If we introduce a uniaxial tensile or compressive stress to the substance, then the equation is followed as $\sigma = E\,\varepsilon$, where σ, E, ε represents stress, modulus of elasticity, and strain, respectively. Hooke's law may be used to describe both bulk quantities and 2D nanosheets. To analyze the in-plane characteristic of the elastic constants of a monolayer, Hooke's law is extensively applied. The ratio of applied stress in two different orientations is termed as the Poisson's ratio "m" [4, 5].

Surface adhesion is a fascinating field of research and an important mechanical characteristic of 2D materials. Layers with an atomic thickness and unequalled flexibility can adapt to contacting surfaces far better in comparison to bulk materials, which leads to significantly improved contact area and stickiness [6].

7.2 GRAPHENE AS A 2D MATERIAL

7.2.1 MECHANICAL PROPERTIES OF GRAPHENE

The interest in nanoscience has grown among academia and industry since the discovery of monolayer graphene. Graphene possesses remarkable mechanical and functional properties and has significant potential to unveil new devices in the market.

The development of improved devices is becoming a priority among researchers all around the world. However, creating graphene-based products, particularly high-quality and well-structured graphene in large quantities, becomes a significant difficulty. As a result, researchers continue to be concerned about manufacturing high-quality graphene, and many production methods have been used to manufacture high-quality graphene. Graphene is the origin of all graphitic forms. As a result, all carbon compounds and graphene-related materials have the same fundamental chemical structure. The sp^2 hybrid bonds cluster the carbon atoms symmetrically in a hexagonal lattice to form a thick carbon layer. Sensors, actuators, electronics, and biomedical devices are just a few of the sectors where graphene-based materials and composites provide incentives.

In the natural world, the thinnest feasible 2D membrane is monoatomic graphene. As a result of its great strength and ductility, it is ideal for a wide range of applications. Studies of graphene in various matrices have shown its high stiffness and strength under tension while also revealing the complete stress-strain response under compressive loads. Denser graphene has a nonuniform strain distribution in the out-of-plane direction, which reduces its elastic properties. It's worth to note that boundary dislocations operating on graphene in air cause wrinkling because the buckling instabilities threshold for such a thin plate is so low.

Graphene and its related materials have outstanding mechanical properties, making them a suitable material for engineering components. Graphene's thickness has no effect on its in-plane elastic modulus. As a result, single-layer graphene was predicted to have the same in-plane elastic modulus as bulk graphite. Significant differences owing to ripples and other variables, on the other hand, need more sophisticated theoretical models for improved consistency [7]. The hexagonal lattice is developed by their stable sp^2 bonds, which restrict various in-plane deformations. In suspended monolayers of graphene, 1 nm ripples height with less than 25 nm periodicity have been observed as a result of temperature variations and elastic stresses [8]. The graphene has a negative thermal coefficient with decreased in-plane stiffness due to its non-flat arrangement [9, 10]. The out-of-plane wrinkling in graphene may result in reduction of the elastic modulus of a single-layer graphene while increasing its toughness or chemical activity.

The width of a monolayer of a 2D material is usually considered as the interlayer thickness. The single sheet of graphene is often regarded as 0.334 nanometers thick, while atomic force microscopy measurements have shown that the thickness varies between 0.4–1.7 nm. [11]. The 2D membranes are also evaluated for their mechanical behavior and linearity using bend stiffness, which is related to ripple production and electron scattering. Flexible and transparent electronics, as well as corrosion-resistant coatings, might benefit from chemical vapor deposition (CVD) graphene [12–15]. When graphene purity and quality are not a major concern, exfoliation and reduction of GO seem to be the best methods for producing vast amounts of graphene at minimal cost. Exfoliated graphene might be utilized in conductive coatings, inks, energy storage, and lithium-ion batteries [16–18].

The heat generated in graphene-based systems raises the temperature, which has a negative impact on graphene's mechanical characteristics. Figure 7.1 (a) explains that the increase in temperature results in the linear reduction in Young's modulus [19]. The stress-strain relationship is linear at smaller strains, but then stress grows

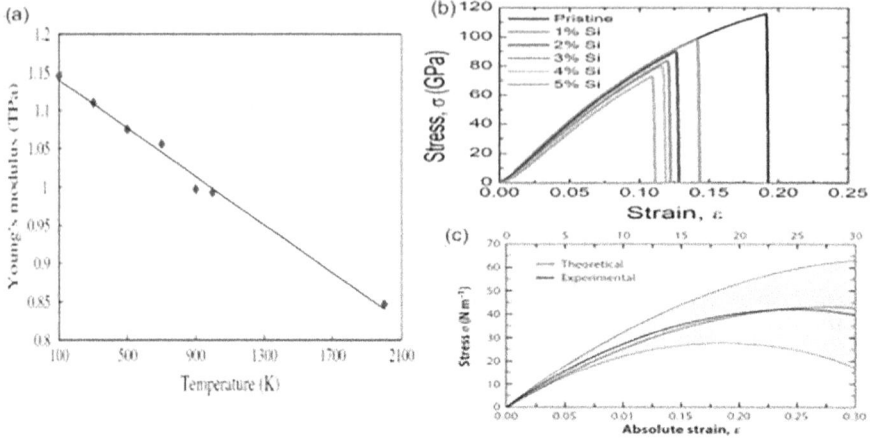

FIGURE 7.1 Mechanical characteristics of graphene. (a) Temperature-dependent change in Young's modulus [19]. (b) For varied percentages of silicon doping, the stress-strain curve of silicon-doped graphene with zigzag loading [20]. (c) Data comparison of stress-strain curve [5, 21].

nonlinearly with strain until it reaches its peak, known as maximum tensile stress, at which time the material severely breaks and undergoes brittle breakdown. It is clear from Figure 7.1 (b) that the stress-strain curve exhibits catastrophic fracture even after doping graphene by silicon atoms [20]. In addition, Figure 7.1 (c) analyzes the failure stress associated with an expected failure stress of 42.4 Nm^{-1}.

7.2.2 DEFECTS AND THEIR INFLUENCE ON MECHANICAL CHARACTERISTICS OF GRAPHENE

The density of the flaws, as well as the layout of the faults, impact mechanical properties. In the ideal crystal structure, deformations and disclinations affect a crystal's mechanical properties. Defects can impact the shear strength, fracture toughness, durability, and mechanical characteristics of crystals in general, and it's important to figure out how to recover, characterize, and utilize these impacts. Atomic force microscope (AFM) nanoindentation can be used to detect the effect of defects on mechanical properties [22]. It is found that carbon tends to establish an interlayer, connecting around flaws, in irradiated multilayer graphene, which partially restores the damaged modulus. Multilayer graphene has a greater elasticity than monolayer graphene as an outcome of this interlayer bonding, which makes it more resistant to radiation.

The CVD graphene is an excellent technique for investigating the impact of flaws and surface defects. The Young's modulus and compressive strength of CVD graphene are lower than those of softly exfoliated graphene [23–25]. Graphene produced through CVD process is prone to defects and grain boundaries. Point defects reduce graphene's elastic flexibility and compressive strength, while grain boundaries defects can increase or decrease graphene's strength according to their tilt angles [26–29]. The harmonic assumption was utilized, with coefficients indicating force constants, to construct discrete deformations in graphene when generalized for eigen deformations [30]. An analytical system can be more easily addressed with the help of discrete Fourier transformation. Core values and structures were assessed in accordance with published data. The fracture toughness simply declined with the defect density, in contrast to the behavior of the modulus of elasticity, and this may be handled by typical fracture continuum models.

7.2.3 APPLICATIONS OF GRAPHENE

In spite of possessing the dimension in atomic scale, the 2D materials show better mechanical strength and endurance, allowing them to be used in a variety of disciplines. Consequently, 2D materials are not only used as additives to strengthen them but also used in extremely sensitive resonators and versatile electronic devices. There are a few 2D materials that possess pressure-dependent functionalities, such as variations in electrical and chemical characteristics, that can be used to produce chemical monitors and piezoelectric sensors.

Graphene resonators are the most extreme 2D nanoscale resonators since the thickness is of atomic order and the addition of few atoms increases the significant fraction of the volume of graphene [31]. The graphene exhibits a higher surface to volume ratio with higher value of Young's modulus than existing microelectromechanical system (MEMS) resonators. As a result, graphene may be used to achieve high-frequency and high-Q resonators. The MEMS voltage-controlled oscillators as well as quantum information systems are two more uses of graphene nano-mechanical resonators [32–33]. Furthermore, graphene resonators show dynamic intermodal coupling, making them attractive in quantum information technology.

Graphene has cleared the direction for future usage in engines, actuators, detectors, energy harvesters, and resonators, among other things. Graphene is chemically inert, has elevated thermal stability, and is resistant to gas diffusion. This makes graphene superior to any other material in protecting metals from surface oxidation. The copper (Cu) and copper/nickel (Cu/Ni) alloys can be protected from air oxidation by using techniques such as CVD-grown graphene sheets [34]. In addition to corrosion resistance, graphene might be used to create a variety of hybrid coatings for electronics, including self-cleaning or anti-fouling coatings and fire-resistant coatings.

The 2D materials are suitable for adaptable electrical devices because of their ultra-thinness, great elasticity, and robustness. There are no performance degradation issues with graphene field-effect transistors on adaptable substrates at high loads of up to 5%. Graphene was chosen for medication delivery, gene therapy, genetic analysis, and tissue engineering because of its porous structure and stable chemical characteristics.

7.3 TRANSITION METAL DICHALCOGENIDES (TMDs)

7.3.1 MECHANICAL PROPERTIES OF TMDs

The TMDs such as tungsten disulfide (WS_2), molybdenum disulfide (MoS_2), and tungsten di-selenide (WSe_2) have been in attention due to their favorable energy gap and also due to their transition near the monolayer from indirect bandgap in bulk to direct bandgap [35]. The mechanical characteristics of TMD coatings are aimed toward less-defective, exfoliated MoS_2 flakes. The MoS_2 has an overall breaking strength of 15±3 N/m (23 GPa), which has been several times lower than that of monolayer of graphene but in contrast to steel has been higher. There seems to be no defect-related Raman spikes in 2D TMDs [36]. Using first-principles density functional computations, the mechanical characteristics of many additional TMDs were also studied.

Monolayer of MoS_2 has been associated with 9.61 eV bending modulus, whereas graphene has a bending modulus of 1.4 eV, which is very less in comparison to the monolayer of MoS_2. This difference in the value of bending modulus has been due to the fact that monolayer of MoS_2 has been associated with three atomic planes in comparison to the graphene [37]. The first-principle technique was used to determine the bending rigidities of different TMDs that would not require empirical factors such as thickness [38].

The elastic mechanical characteristics of the MoS_2 nanosheets were measured by applying a force in the middle of the suspended area with an atomic force microscope tip. When the tip and material come into contact, the deformation of the nanosheets, the deflection of the atomic force microscope cantilever, as well as the movement of the atomic force microscope instrument's monitoring piezo tube are all linked. When the tip and material come into contact, the deformation of the nanosheets, the deflection of the atomic force microscope cantilever, as well as the movement of the atomic force microscope instrument's monitoring piezo tube are all linked. Elastic deformations are calculated by the equation:

$$\delta = \Delta Zpiezo- \Delta Zc \qquad (1)$$

Where $\Delta Zpiezo$ is the displacement of piezo-tube, and ΔZc is the fluctuation of cantilever. MoS_2 nanosheets have a very significant value of Young's modulus, E = 0.33 ± 0.07 TPa [39]. Figure 7.2 depicts the elastic characteristics of two-dimensional monolayers and heterostructures [40].

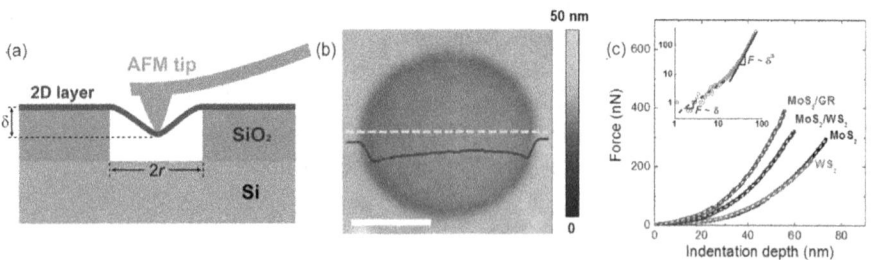

FIGURE 7.2 Mechanical properties of TMDs: (a) Representation of the indentation analysis. (b) Illustration of an AFM topography of a monolayer of MoS_2 by taking scale bar 500 nm. (c) Curves of force-displacement for distinct chemical vapor deposited monolayers and heterostructures [40].

TMDs materials must be grown on a huge scale before they can be used in actual devices. The most efficient technique to accomplish massive growth is by chemical vapor deposition (CVD). The sulfurization of metallic thin films and the vapor phase reaction of metal oxides containing chalcogen precursors are the two types of this technique.

7.3.2 Defects and Their Influence on Mechanical Characteristics of TMDs

MoS_2, one of the most prevalent 2D semiconductors, has a grain boundary arrangement that may be exploited to change its mechanical properties. According to DFT calculations, the defect production energy of grain boundaries in polycrystalline MoS_2 is less than that of the grain core. This grain boundary (GB) model's estimated formation energy is about 2.50 eV/unit, which is fairly comparable to that of its crystallographic counterpart (2.53 eV/unit), indicating that grain boundary creation is reasonably straightforward [41–42].

The TMDs are made by directly sulfurizing a metal oxide thin layer, but there are certain drawbacks, such as the complications in controlling the width of the predeposited metal oxide, which impacts wafer-scale homogeneity.

7.3.3 TMD Applications

Before utilizing TMDs materials in real devices, they must be produced on a large scale. Chemical vapor deposition (CVD) is the most effective method for achieving enormous growth. The single-layer MoS_2 can be used in a wide range of applications, including composite reinforcement and the manufacturing of flexible electronic equipment. Similar to graphene field-effect transistors, devices constructed of TMDs on adaptable substrates demonstrate great stability and better performance. The development of field-effect transistor with the MoS_2 as an active material exhibited a better on/off ratio of 107 and a carrier mobility of 30 cm^2/Vs at a moderate bending radius of 0.75 to 1 mm without significant degradation [43–44]. The MoS_2 has a strain-dependent electronic structure that may be utilized to control excitons by localized strain. It has been demonstrated that, under localized strain, the restricted point of MoS_2 may focus and gather excitons from adjacent MoS_2, acting similarly to a funnel, which is utilized to enhance the efficiency of solar power output and light emitters [45]. Aside from that, the MoS_2 drum resonators had a resonance frequency of up to 60 MHz and a Q factor of approximately 3,000 [46].

The sensor based devices and nanogenerators have been developed by using the piezoelectricity of 2D materials. The MoS_2 was used to fabricate strain/force monitors with gauge factors in excess of 1,000 [47]. An extremely high level gauge factor has been obtained on changing the Schottky barrier width with the help of piezoelectric polarization by doing the strain variation. The MoS_2 piezoelectricity has been proposed as a method of enhancing the theoretical efficiency limit of MoS_2 solar panels [48]. The application of the piezoelectric photocatalyst such as zinc oxide, zinc sulfide, and molybdenum disulfide nanostructures in the degradation of methylene blue dye is successfully reported by Fu et al. [49]. It is observed that the degradation efficiency is almost 2.5 times more as compared to the material combination without piezoelectricity. Interestingly, the efficiency remains around 80% after the five exhaustive cycles. Han et al. [50] fabricated the piezoelectric nanogenerator with the layered MoS_2, which is passivated with the activated sulfur vacancies. The sulfur vacancies reduce the carrier concentration on the surface of the MOS_2 resulting in the prevention of piezoelectric polarization by free carrier charges. The piezoelectric output voltage is found to increase significantly by a magnitude of 3–, 2– times for sulfur treated and nanosheet piezoelectric nanogenerators, respectively. Similarly, the MOS_2 nanoflakes were studied as an active material in piezoelectric nanogenerators by Wu et al. [51]. Interestingly, it is observed that the single monolayer of MoS_2 strained by ~ 0.53% provides an output voltage of 15 mV and the current of 20 pA with efficiency of 5.08%. Further research will have to be performed on the application of various other transition metal dichalcogenides as theoretical results predict better performance of these materials

7.4   H-BN (HEXAGONAL BORON NITRIDE) AS A 2D MATERIAL

7.4.1   MECHANICAL PROPERTIES OF H-BN

The h-BN is a white powder that resembles the hexagonal configuration and has a layered structure coating that can be used to alleviate the in graphene problem by opening up the graphene bandgap and therefore increasing the switching functioning of graphene devices. Because graphene's layered heterostructures are advantageous for flexible and transparent devices, its mechanical characteristics are significant. According to nano-indentation studies on h-BN over the micrometer scale implicated the mechanical behavior of h-BN as an equivalent to those of graphene [52].

According to atomic force microscopy bending observations and classical plate theory, the folding modulus of h-BN improves with lowering thickness, approaching the expected value of C_{33}, which reflects elasticity underneath a longitudinal load with transverse displacement [53]. The thickness dependency of the bending modulus is caused by the layer pattern of stacking flaws. The mechanical properties of hexagonal boron nitride nanosheets (h-BNNS) persist reasonably constant and are related with the ratio of radius of circular-region and nano-indenter. The mechanical properties of (h-BNNS) progressively deteriorated with enhanced temperature.

The loading force versus deflection graphs in six and eight layers of h-BNNS reveal the phenomenon with sudden reductions, which might be related to the multilayered structure's transitional deformation. Despite the fact that the higher load force increases because the count of layers increases, the enhanced gradient of higher load force decreases or remains generally steady. Young's modulus for h-BNNS grows from single to two layers, then drops after three layers, and ultimately reaches about 0.7 TPa, which is less as compared to the single-layer h-BNNS Young's modulus. Similarly, the shift in fracture toughness shows the similar pattern, with higher

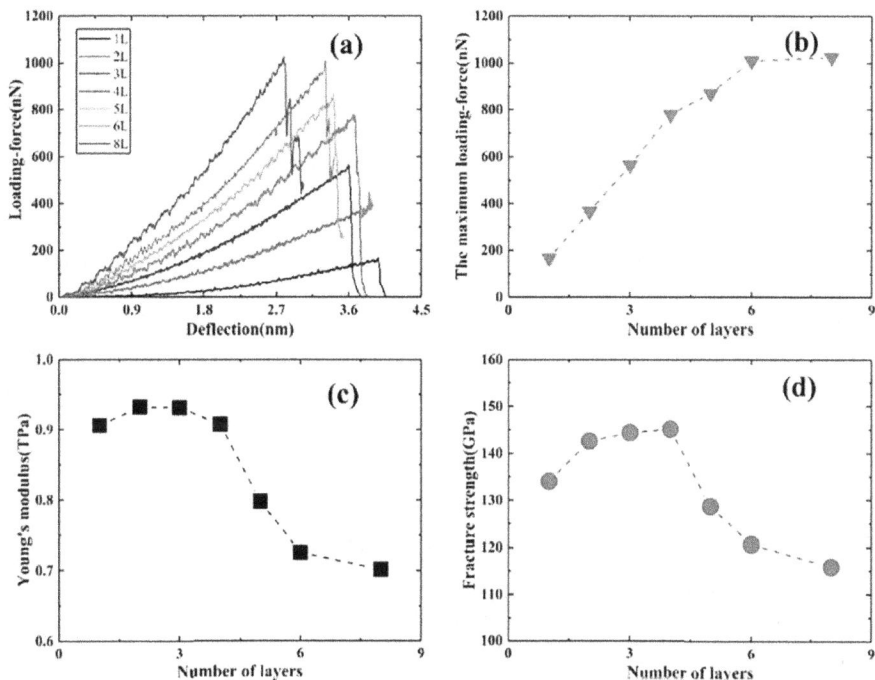

FIGURE 7.3   Effect on h-BNNS (a) loading-force vs. deflection curve. (b) Maximum loading-force. (c) Young's modulus fluctuates with successive layers. (d) The fracture strength varies under various layers [54].

and lower fracture strengths of 145 GPa and 115 GPa, respectively, as shown in Figure 7.3 [54].

The inherent hardness of h-BN sheet with grain boundaries on the surface displayed a clear linear relationship with the surface's inflection angles, indicating that the intrinsic strength dropped linearly as the inflection angles rose.

7.4.2   DEFECT AND THEIR INFLUENCE ON MECHANICAL CHARACTERISTICS OF H-BN

The h-BN sheets shattered at or around the contact, suggesting that the interface was weaker for that material. Due to lattice distortion, faults are invariably formed at the interface during the epitaxial growth of h-BN heterojunctions using the epitaxial CVD technique [55–56]. Furthermore, flaws may be utilized to carefully regulate the device's electrical structure, which influences device's band structure, carrier density, and spin properties. The stress on a particle progressively increases during tensile crack growth, and the stress concentration is present at the graphene-h-BN interface. The first bond deformation of the C-B bond at the surface occurs when the tensile distortion achieves the fracture strain. Afterward the bond fracture expands across the interface direction and induces breakage to the opposite side of the interface unless entire fracture occurs [57].

7.4.3 Applications of h-BN

Versatile "All-2D" devices fabricated by using only 2D materials such as h-BN, MoS_2, and graphene have shown the high value of carrier mobility and on-off rate of 45 cm²/Vs 107, respectively, and opacity of 88% to 95% while tolerating 1.5% to 2% strain [58]. The h-BN nanosheets have already been extensively used as nanofillers in polymers to improve their ability. The thermal conductivity of polymethyl methacrylate (PMMA) lowers from 184 to 160 ppm⁰C⁻¹ at 0.3 wt% h-BN nanosheet concentration, and the modulus of elasticity improves by 22%. Due to the nonaligned character of the nanocomposites, it has been found that the nanoparticle strength is susceptible to the dispersion of nanofillers and that a relatively modest amount of reinforcement is obtained. Other uses for 2D h-BN nanosheets include sensing devices for methane, ethanol, and NH_3 sensors; separation of multiple impurities in oil, solvents, dyes, and trihalomethanes to clean and remove impurities from water; etc.

7.5 BLACK PHOSPHORUS AS 2D MATERIAL

7.5.1 Mechanical Properties of Black Phosphorus

The growing interest in new developing single-element structured material is due to black phosphorus's layer reliant 0.3–2.0 eV bandgap modulation region, which is similar to that of transition metal dichalcogenides and single- and few-layered graphene [59]. Because of its application in various sectors, ongoing improvements are being made to enhance the productivity of phosphorene thin films. The phosphorene has unusual mechanical characteristics, making it a potential new material for nanomechanical devices and systems [60]. It can withstand tensile stresses of about 30%, which is greater than the values recorded for graphene or single-layer MoS_2. Because of its distinctive puckered structure, phosphorus exhibits a negative Poisson's ratio [61]. The Young's modulus of this material is less than that of graphene as well as single-layer MoS_2 [62]. Phosphorene has the capability to be an effective filler for reinforcing polymers and composites due to its exceptional mechanical characteristics. Polyvinylchloride following phosphorene reinforcement has a significant increase in all mechanical measures like Young modulus, composite stiffness, and tensile hardness, making it a strong contender for graphene reinforcement [63]. The black phosphorus experiences crystal structural changes under high pressures, firstly from an orthorhombic to a rhombohedral structure at approximately 5 GPa and subsequently to a basic cubic structure around 10 GPa [64]. An investigation of black phosphorene stress-strain correlations under uniaxial tension was conducted. Increasing the temperature of black phosphorene causes its Young's modulus to decrease along with its fracture toughness and fracture strain [65–66].

7.5.2 Defect and Their Influence on Mechanical Characteristics of Black Phosphorus

During the development of 2D phosphorus sheets, interstitial faults are common. Self-interstitial phosphorus has a substantially lesser formation energy than graphene. The main dislocation of phosphorene is 5/7 pairs with such a buckled configuration, identical to hexagonal 2D materials. The tilt angle controls the grain boundary (GB) energy in phosphorene, which is made up of an assembly of dislocations. Because the production energy of the deformations that make up GBs in phosphorene are vastly shorter than those in graphene, the GBs in phosphorene seem to be more reliable than that in graphene [67]. The vacancies in phosphorene did not remain static; they migrated and aggregated, producing line defects that were much more energetically advantageous. The mobility of flaws was strongly influenced by temperature, with single vacancies moving 16 orders quicker at ambient temperature than those at 70 K [68]. The Poisson's ratio of black phosphorus is relatively low, which restricts its use in equipment due to poor fracture toughness. The fracture process is investigated using molecular dynamics, which leads to the conclusion that stretchable fracture developed across the X axis is produced by the breakdown of interfacial bonds across the Y axis [69].

7.5.3 APPLICATIONS OF BLACK PHOSPHORUS

This unusual mechanical anisotropy may be used to make nano-mechanical resonators, thermoelectric appliances, and motion detectors with adjustable functionalities that aren't possible with isotropic materials. The 2D black phosphorus has been utilized in a nanoelectromechanical resonator, exhibiting sustained resonance vibrations at extremely high frequencies, whereas the fundamental benefit of mechanical anisotropy may be harnessed to enable novel phenomena that are not possible in isotropic 2D materials [70]. A potential use for two-dimensional black phosphorus resonators is mass monitoring or environmental condition assessment. As the environment deteriorates, structural and mechanical characteristics might be affected, resulting in a readily visible change in the resonant frequency of the unique modes.

7.6 MXene AS A 2D MATERIAL

7.6.1 MECHANICAL PROPERTIES OF MXENE

Through selective peeling of the aluminium layer in Ti_3AlC_2 phase using hydrofluoric acid, a novel 2D transition metal carbide and nitride (MXene) was created, which sparked a lot of attention [71–74]. Exfoliation of the MAX phase (abbreviated later) that is firmly bound inside a 2D layer and kept together through weak van der Waals contact may be used to make MXenes. The MAX phase formula is $M_{n+1}AX_n$ (n = 1, 2, 3), where M is a transition metal, A is a group IIIA or IVA element, and X would either be a carbon or nitrogen atom [75–76]. With increase in the value of n, the production of MXene and MAX may be reduced. The $Ti_{n+1}N_n$ has a greater production rate than $Ti_{n+1}C_n$, implying that nitrogen-containing MXene has a lower durability than carbon-containing MXene [77]. The MXene material possesses metal and ceramic properties, with great stiffness and high strength. Corrosion resistance, strong electrical conductivity, and good thermal conductivity are all features of this material. MXene also exhibits high hydrophilicity, adaptability, and plasticity and has increasingly become a 2D materials research focus. According to experimental data, the modulus of elasticity of monolayer $Ti_3C_2(OH)_2$ is around 300 GPa, which is

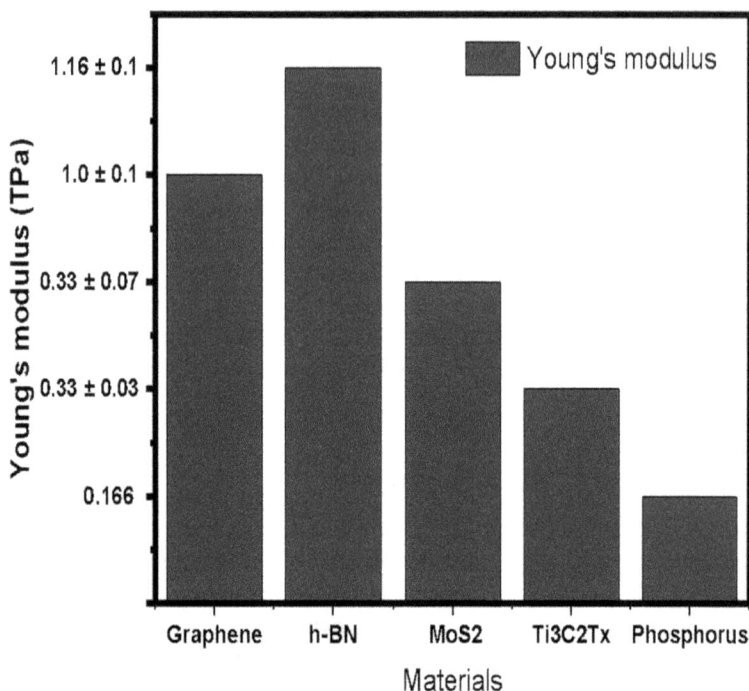

FIGURE 7.4 MXene ($Ti_3C_2T_x$) Young's modulus compared to other 2D materials [81–85].

lower than graphene, equivalent to several transition metal carbides, but larger than most oxides and multilayer clays [77]. The bonding here between M_2X and the terminating group improves the mechanical characteristics. Single-layer M_2X, in-plane elasticity is estimated to be between 42 and 199 N m^{-1}, that is less stiff as compared to graphene (341 N m^{-1}) and monolayer h-BN (276 N m^{-1}) [78–80].

Young's modulus of MXene ($Ti_3C_2T_x$) is observed as 0.33 ± 0.03 TPa [84], which is more than the phosphorus (0.166 TPa) [85], as demonstrated in Figure 7.4. However, h-BN has a higher value of Young's modulus in the comparison of other 2D materials.

7.6.2 DEFECT AND THEIR INFLUENCE ON MECHANICAL CHARACTERISTICS OF MXENE

In contrast to the deformation process, in sensor application, primarily the resistance between the $Ti_3C_2T_x$ lamellae changes due to hydrogel distortion caused by an external force, converting the mechanical signal to an electrical signal [86]. The T_x refers to the surface terminations produced during the etching process, such as hydroxyl (-OH), oxygen (-O), and fluoride (-F) [87]. Substantial flake size is preferred for mechanical reinforcing operations as a film will have even less sheet-to-sheet contact defects, and the considerable in-plane stiffness of the nano-components will be fully used. The bending strength is a key parameter that describes the mechanical

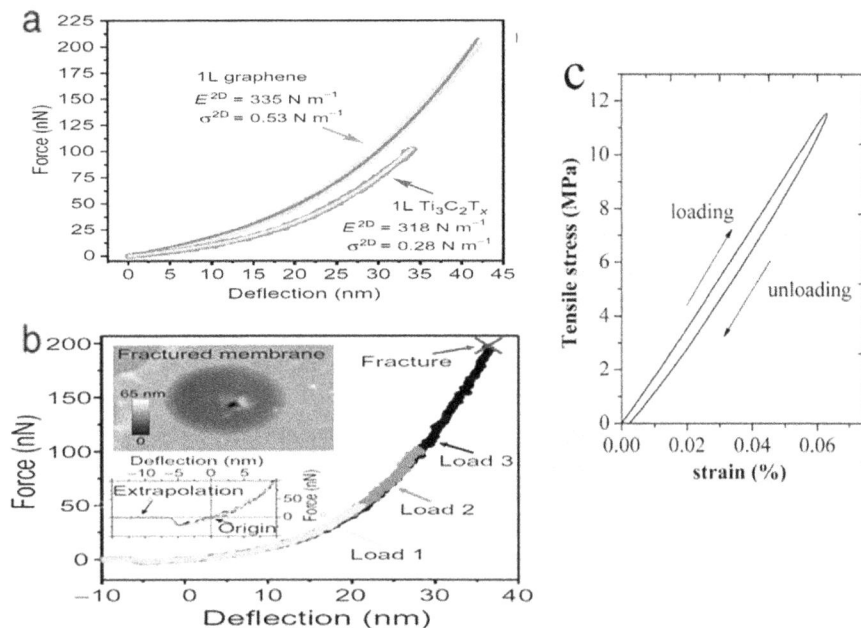

FIGURE 7.5 (a) Force-deflection (F-δ) curves comparison for monolayer graphene and Ti₃C₂Tx substrates [81]. (b) The F-δ curves of a bilayer Ti₃C₂Tₓ flake at various loads. The AFM picture of the broken membrane is shown in the top corner [84]. (c) Inside the linear elastic regime, the loading and unloading curve [90].

behavior of 2D materials undergoing bending deformations [88]. Some factors, such as film thickness and the functionalization component, have an impact on bending rigidity. Bending capability of Ti_2C is lesser in comparison to MoS_2 because of differing atomic configurations and a higher thickness in MoS_2 than Ti_2C [89]. The comparison of force versus deflection curves (F-δ curves) for suspended Ti₃C₂Tx with graphene monolayers is made as graphene is a superior 2D material (Figure 7.5 (a)). The figure depicts the membrane's behavior at the start of the indentation procedure. Before snapping, linear force versus deflection dependency extends till it crosses the arc and assumes this location to be the origin, which would be required to achieve the right F-δ correlation. The trajectory of extension and retraction in each loading cycle, as well as the curves for load changes, are retraced. The MXene flakes have a great flexibility, and no flake breakage takes place during the assessments. This bilayer's fracturing ruptured at a load force of around 200 nN and a deflection of about 38 mm as shown in Figure 7.5 (b) [84].

The linear regimes are not fully elastic, shown by the stress-strain curves. The loading and unloading lines do not coincide when the $Ti_3C_2T_x$ sheet is compressed within its linear elastic range and subsequently unloaded, suggesting that the linear distortion is viscoelastic instead of simply elastic, as shown in Figure 7.5 (c). Unless and until a film is stretched beyond the linear elastic regime, material undergoes plastic deformation, under which the strain increases faster than the stress, implying that sliding between neighboring sub-layers is likely to play a dominating role in the irreversible process of deformation [90].

7.6.3 Applications of MXene

These have been utilized in a variety of applications, including power storage, catalyst supports, methanol oxidation and hydrogen transformation, supercapacitors, and more [91–94]. Around 2013, MXene seems to have been a major research area in the battery sector. Due to its controlled production technique and remarkable characteristics, MXene has also shown potential in wearable sensing applications. The MXene-derived photodynamic and photothermal compounds have a high hydrophilic nature, which facilitates excellent dispersion and sustainability in physiological fluids. With the polar terminating groups, anticancer medicines may be readily grafted upon MXene surfaces. Optimizing the swelling ability of MXene hydrogels can greatly improve the quality of anticancer medicines with loading capabilities as greater as 84% and significant release percent [95].

Because pure MXenes, like all the other 2D materials, are not stretchy, introducing polymers to them can improve their mechanical robustness and sensing range [96]. The $Ti_3C_2T_x$ nanosheets created by MXenes and poly(diallyldimethylammonium chloride) (PDAC) from the original $Ti_3C_2T_x$ MAX phase are used in layer-by-layer assembly to make composites. The film's conductance can exceed up to 2,000 S/m. Composite membrane strain sensor on polydimethylsiloxane (PDMS) could bent up to 40%, whereas the bending sensor on polyethylene terephthalate (PET) can be stretched to 35% [97].

Furthermore, MXenes are immediately added to the produced elastic substrate, and the high conductivity of MXenes, as well as the superior mechanical qualities of the stretchy substrate, are exploited to meet the resistance impact and geometric features of the adjustable piezoresistive sensor. The MXenes can be useful in detecting the strain with the help of crack mechanism. When the MXenes-based monitor is fractured by strain, the characteristics of the monitor changes as the crack size rises.

7.7 OTHER MATERIALS

Silicene, like graphene, is a hexagonal 2D material with composition of silicon [98]. Despite the better characteristics of silicene and the need for high similarity with the silicon-based industries, silicene's high reactivity quickly promotes the spontaneous dissociation of oxygen on the interface and the creation of the Si-O bonding. Because silicene is so unstable in its natural environment, determining its mechanical characteristics empirically is exceedingly challenging [99–100]. The Poisson ratio and in-plane stiffness of silicene are estimated to be 0.3 and $C = 62$ J/m², respectively, by applying the calculated changes in elastic energy to a quadratic deformation model [101]. The mica is a silica-based material that has a complex composition and is a two-dimensional substance. Multilayered mica has a 190 GPa elastic modulus as determined by AFM [102]. Mica also has a minimal pre-strain of 0.25 N/m and excessive breaking force of 4 to 9 GPa. Mica's mechanical characteristics suggest it might be used as an adjustable extremely thin insulating/dielectric substrate. In recent years, research has focused on the formation of moiré patterns on the various 2D materials and their effect on the chemical/ physical properties. The van der Waals defects, or interlayer deformations, can be used to quantify Moiré patterns in actual materials systems [103]. Throughout the lattice stackings, these patterns provide areas of commensurability and incommensurability. Moiré textures on h-BN, SiC, and faces of face centered cubic (FCC) metals have been seen experimentally.

7.8 CONCLUSIONS AND FUTURE PROSPECTUS

Many alternatives are emerging at the convergence of 2D materials mechanics. Recent advances in the study of mechanical characteristics of 2D materials, as well as their applicability as a specialized type of mechanical system, have been summarized. Nature's strongest 2D substance has been confirmed to be graphene. Other 2D materials, like MoS_2, offer comparable mechanical properties to traditional materials. The anisotropic mechanical properties of 2D materials with asymmetric in-plane lattices, such as black phosphorus, would be considerably more difficult to study but will be highly intriguing. The mechanical characteristics of 2D materials have been observed to decline as the number of layers increases, and defects can exacerbate this decline. When it comes to the impact of flaws, the effect of grains and their related properties on the strength/durability of 2D materials is currently being debated. It is necessary to develop synthesis/deposition methodologies for 2D materials with well-controlled grain sizes and orientations in order to overcome these challenges. Defects can strengthen or diminish interlayer interactions, affecting the mechanical characteristics of the system. Both in terms of measurements and usage, there have been significant advancements. Nevertheless, there are still unanswered issues and problems in this area of study. Except for graphene, research corresponding to the mechanical behavior of the other 2D materials is to be investigated.

REFERENCES

1. Shikai Deng and Vikas Berry, "Wrinkled, rippled and crumpled graphene: an overview of formation mechanism, electronic properties, and applications," *Mater. Today*, 19, 4, 197–212, 2016. Doi:10.1016/j.mattod.2015.10.002.
2. Ch. Androulidakis, E. N. Koukaras, M. G. Pastore Carbone, M. Hadjinicolaou, and C. Galiotis, "Wrinkling formation in simply-supported graphenes under tension and compression loadings," *Nanoscale*, 9, 18180–18188, 2017. Doi:10.1039/C7NR06463B.
3. Oleg V. Yazyev and Steven G. Louie, "Topological defects in graphene: dislocations and grain boundaries," *Phys. Rev. B, Condensed Matter*, 81, 19, 2010. Doi:10.1103/PhysRevB.81.195420.
4. Joel I. Gersten and Frederick W. Smith, *The Physics and Chemistry of Materials*, New York, Wiley, 1–856, 2001, ISBN: 978-0-471-05794-9.
5. Changgu Lee, Xiaoding Wei, Jeffrey W. Kysar, and James Hone, "Measurement of the elastic properties and intrinsic strength of monolayer graphene," *Science*, 321, 5887, 385–388, 2008. Doi:10.1126/science.1157996.
6. Ankur Gupta, Tamilselvan Sakthivel, and Sudipta Seal, "Recent development in 2D materials beyond graphene," *Prog. Mater. Sci.*, 73, 44–126, 2015. Doi:10.1016/j.pmatsci.2015.02.002.

7. Lee Changgu, Wei Xiaoding, Li Qunyang, Carpick Robert, Kysar Jeffrey W., and Hone James, "Elastic and frictional properties of graphene," *Phys. Status Solidi B-Basic Solid State Phys.*, 246, 11–12, 2562–2567, 2009. Doi:10.1002/pssb.200982329.

8. Jannik C. Meyer, A. K. Geim, M. I. Katsnelson, K. S. Novoselov, T. J. Booth, and S. Roth, "The structure of suspended graphene sheets," *Nature*, 446, 60, 2007. Doi:10.1038/nature05545.

9. Duhee Yoon, Young-Woo Son, and Hyeonsik Cheong, "Negative thermal expansion coefficient of graphene measured by Raman Spectroscopy," *Nano Lett.*, 11, 8, 3227–3231, 2011. Doi:10.1021/nl201488g.

10. Ryan J. T. Nicholl, Hiram J. Conley, Nickolay V. Lavrik, Ivan Vlassiouk, Yevgeniy S. Puzyrev, Vijayashree Parsi Sreenivas, Sokrates T. Pantelides, and Kirill I. Bolotin, "The effect of intrinsic crumpling on the mechanics of free-standing Graphene." *Nat. Commun.*, 6, 8789, 2015. Doi:10.1038/ncomms9789.

11. Cameron J. Shearer, Ashley D. Slattery, Andrew J. Stapleton, Joseph G. Shapterand, and Christopher T. Gibson, "Accurate thickness measurement of graphene," *Nanotechnology*, 27, 12, 125704, 2016. Doi:10.1088/0957-4484/27/12/125704.

12. Houk Jang, Yong Ju Park, Xiang Chen, Tanmoy Das, Min-Seok Kim, and Jong-Hyun Ahn, "Graphene-based flexible and stretchable electronics," *Adv. Mater.*, 28, 4184–4202, 2016. Doi:10.1002/adma.201504245.

13. A. J. M. Giesbers, P. C. P. Bouten, J. F. M. Cillessen, L.van der Tempel, J. H. Klootwijk, A. Pesquera, A. Centeno, A. Zurutuza, and A. R. Balkenende, "Defects, a challenge for graphene in flexible electronics," *Solid State Commun.*, 229, 49–52, 2016. Doi:10.1016/j.ssc.2016.01.002.

14. Maido Merisalu, Tauno Kahro, Jekaterina Kozlova, Ahti Niilisk, Aleksandr Nikolajev, Margus Marandi, Aare Floren, Harry Alles, and Väino Sammelselg, "Graphene—polypyrrole thin hybrid corrosion resistant coatings for copper," *Synth. Met.*, 200, 16–23, 2015. Doi:10.1016/j.synthmet.2014.12.024.

15. N. T. Kirkland, T. Schiller, N. Medhekar, and N. Birbilis, "Exploring graphene as a corrosion protection barrier," *Corros. Sci.*, 56, 1–4, 2012. Doi:10.1016/j.corsci.2011.12.003.

16. Andrea Capasso, Antonio Esau Del Rio Castillo, Haiyan Sun, Alberto Ansaldo, V. Pellegrin, and Francesco Bonaccorso, "Ink-jet printing of graphene for flexible electronics: an environmentally-friendly approach," *Solid State Commun.*, 224, 53–63, 2015. Doi:10.1016/j.ssc.2015.08.011.

17. Wen Qian, Rui Hao, Jian Zhou, Micah Eastman, Beth A. Manhat, Qiang Sun, Andrea M. Goforth, and Jun Jiao, "Exfoliated graphene supported Pt and Pt-based alloys as electrocatalysts for direct methanol fuel cells," *Carbon*, 52, 595–604, 2013. Doi:10.1016/j.carbon.2012.10.031.

18. Haiyan Sun, Antonio Esau Del Rio Castillo, Simone Monaco, Andrea Capasso, Alberto Ansaldo, Mirko Prato, Duc Anh Dinh, Vittorio Pellegrini, Bruno Scrosati, Liberato Manna, and Francesco Bonaccorso, "2016 Binder-free graphene as an advanced anode for lithium batteries," *J. Mater. Chem. A*, 4, 18, 6886–6895, 2016. Doi:10.1039/C5TA08553E.

19. Yingyan Zhang and Y. T. Gu, "Mechanical properties of graphene: effects of layer number, temperature and isotope," *Comput. Mater. Sci.*, 71, 197–200, 2013. Doi:10.1016/j.commatsci.2013.01.032.

20. Md. Habibur Rahman, Shailee Mitra, Mohammad Motalab, and Pritom Bose, "Investigation on the mechanical properties and fracture phenomenon of silicon doped graphene by molecular dynamics simulation," *RSC Adv.*, 10, 31318–31332, 2020. Doi:10.1039/D0RA06085B.

21. Emiliano Cadelano, Pier Luca Palla, Stefano Giordano, and Luciano Colombo, "Nonlinear elasticity of monolayer graphene," *Phys. Rev. Lett.*, 102, 23, 235502, 2009. Doi:10.1103/PhysRevLett.102.235502.

22. Ardavan Zandiatashbar, Gwan-Hyoung Lee, Sung Joo An, Sunwoo Lee, Nithin Mathew, Mauricio Terrones, Takuya Hayashi, Catalin R. Picu, James Hone, and Nikhil Koratkar, "Effect of defects on the intrinsic strength and stiffness of graphene," *Nat. Commun.*, 5, 3186, 2014. Doi:10.1038/ncomms4186.

23. Alfonso Reina, Xiaoting Jia, John Ho, Daniel Nezich, Hyungbin Son, Vladimir Bulovic, Mildred S. Dresselhaus, and Jing Kong, "Large area, few-layer graphene films on arbitrary substrates by chemical vapor deposition," *Nano Lett.*, 9, 1, 30–35, 2009. Doi:10.1021/nl801827v.

24. Carlos S. Ruiz-Vargas, Houlong L. Zhuang, Pinshane Y. Huang, Arend M. van der Zande, Shivank Garg, Paul L. McEuen, David A. Muller, Richard G. Hennig, and Jiwoong Park, "Softened elastic response and unzipping in chemical vapor deposition graphene membranes," *Nano Lett.*, 11, 6, 2259, 2011. Doi:10.1021/nl200429f.

25. Qing-Yuan Lin, Guangyin Jing, Yangbo Zhou, Yifan Wang, Jie Meng, Yaqing Bie, Dapeng Yu, and Zhi-Min Liao, "Stretch-induced stiffness enhancement of graphene grown by chemical vapor deposition," *ACS Nano*, 7, 2, 1171, 2013. Doi:10.1021/nn3053999.

26. Riccardo Dettori, Emiliano Cadelano, and Luciano Colombo, "Elastic fields and moduli in defected graphene," *J. Phys.: Condens. Matter*, 24, 10, 2012. Doi:10.1088/0953-8984/24/10/104020.

27. Nuannuan Jing, Qingzhong Xue, Cuicui Ling, Meixia Shan, Teng Zhang, Xiaoyan Zhou, and Zhiyong Jiao, "Effect of defects on Young's modulus of graphene sheets: a molecular dynamics simulation," *RSC Adv.*, 2, 24, 9124–9129, 2012. Doi:10.1039/C2RA21228E.

28. Rassin Grantab, Vivek B. Shenoy, and Rodney S. Ruoff, "Anomalous strength characteristics of tilt grain boundaries in graphene," *Science*, 330, 6006, 946–948, 2010. Doi:10.1126/science.1196893.

29. Yujie Wei, Jiangtao Wu, Hanqing Yin, Xinghua Shi, Ronggui Yang, and Mildred Dresselhaus, "The nature of strength enhancement and weakening by pentagon-heptagon defects in graphene," *Nat. Mater.*, 11, 9, 759–763, 2012. Doi:10.1038/nmat3370.

30. Pilar Ariza and M. Ortiz, "Discrete dislocations in graphene," *J. Mech. Phys. Sol.*, 58, 5, 710–734, 2010. Doi:10.1016/j.jmps.2010.02.008.

31. J. Scott Bunch, Arend M. van der Zande, Scott S. Verbridge, Ian W. Frank, David M. Tanenbaum, Jeevak M. Parpia, Harold G. Craighead, and Paul L. McEuen, "Electromechanical resonators from graphene sheets," *Science*, 315, 5811, 490–493, 2007. Doi:10.1126/science.1136836.

32. Changyao Chen, Sunwoo Lee, Vikram V. Deshpande, Gwan-Hyoung Lee, Michael Lekas, Kenneth Shepard, and James Hone, "Graphene mechanicaloscillators with tunable frequency," *Nat. Nanotechnol.*, 8, 12, 923–927, 2013. Doi:10.1038/nnano.2013.232.

33. John P. Mathew, Raj N. Patel, Abhinandan Borah, R. Vijay, and Mandar M. Deshmukh, "Dynamical strong coupling and parametric amplification of mechanical modes of graphene Drums," *Nat. Nanotechnol.*, 11, 747–751, 2016. Doi:10.1038/nnano.2016.94.

34. Shanshan Chen, Lola Brown, Mark Levendorf, Weiwei Cai, Sang Yong Ju, Jonathan Edgeworth, Xuesong Li, Carl W. Magnuson, Aruna Velamakanni, Richard D. Piner, Junyong Kang, Jiwoong Park, and Rodney S. Ruoff, "Oxidation resistance of graphene-coated Cu and Cu/Ni alloy," *ACS Nano*, 5, 2, 1321–1327, 2011. Doi:10.1021/nn103028d.

35. Sheneve Z. Butler, Shawna M. Hollen, Linyou Cao, Yi Cui, Jay A. Gupta, Humberto R. Gutiérrez, Tony F. Heinz, Seung Sae Hong, Jiaxing Huang, Ariel F. Ismach, Ezekiel

Johnston-Halperin, Masaru Kuno, Vladimir V. Plashnitsa, Richard D. Robinson, Rodney S. Ruoff, Sayeef Salahuddin, Jie Shan, Li Shi, Michael G. Spencer, Mauricio Terrones, Wolfgang Windl, and Joshua E. Goldberger, "Progress, challenges, and opportunities in two-dimensional materials beyond graphene," *ACS Nano*, 7, 4, 2898–2926, 2013. Doi:10.1021/nn400280c.

36. Kai Liu and Junqiao Wu, "Mechanical properties of two-dimensional materials and heterostructures," *J. Mater. Res.*, 31, 832–844, 2016. Doi:10.1557/jmr.2015.324.

37. Jin-Wu Jiang, Zenan Qi, Harold S. Park, and Timon Rabczuk, "Elastic bending modulus of single-layer molybdenum disulfide (MoS_2): finite thickness effect," *Nanotechnology*, 24, 435705, 2013. Doi:10.1088/0957-4484/24/43/435705.

38. Kang Lai, Wei-Bing Zhang, Fa Zhou, Fan Zeng, and Bi-Yu Tang, "Bending rigidity of transition metal dichalcogenide monolayers from first-principles," *J. Phys. D Appl. Phys.*, 49, 185301, 2016. Doi:10.1088/0022-3727/49/18/185301.

39. A. Castellanos Gomez, M. Poot, G. A. Steele, H. S. J. van der Zant, N. Agrait, and G. Rubio-Bollinger, "Elastic properties of freely suspended MoS_2 nanosheets," *Adv. Mater.*, 24, 6, 772–775, 2012. Doi:10.1002/adma.201103965.

40. Kai Liu, Qimin Yan, Michelle Chen, Wen Fan, Yinghui Sun, Joonki Suh, Deyi Fu, Sangwook Lee, Jian Zhou, Sefaattin Tongay, Jie Ji, Jeffrey B. Neaton, and Junqiao Wu, "Elastic properties of chemical-vapor-deposited monolayer MoS_2, WS2, and their bilayer heterostructures," *Nano Lett.*, 14, 5097–5103, 2014. Doi:10.1021/nl501793a.

41. Jianyang Wu, Pinqiang Cao, Zhisen Zhang, Fulong Ning, Song-Sheng Zheng, Jianying He, and Zhiliang Zhang, "Grain-size-controlled mechanical properties of polycrystalline monolayer MoS_2," *Nano Lett.*, 18, 1543–1552, 2018. Doi:10.1021/acs.nanolett.7b05433.

42. Zhi Gen Yu, Yong-Wei Zhang, and Boris Yakobson, "An anomalous formation pathway for dislocation-sulfur vacancy complexes in polycrystalline monolayer MoS_2," *Nano Lett.*, 15, 10, 6855–6861, 2015. Doi:10.1021/acs.nanolett.5b02769.

43. Hsiao-Yu Chang, Shixuan Yang, Jongho Lee, Li Tao, Wan-Sik Hwang, Debdeep Jena, Nanshu Lu, and Deji Akinwande, "High-performance, highly bendable MoS_2 transistors with high-k dielectrics for flexible low-power systems," *ACS Nano*, 7, 6, 5446–5452, 2013. Doi:10.1021/nn401429w.

44. Jiang Pu, Yohei Yomogida, Keng Ku Liu, Lain Jong Li, Yoshihiro Iwasa, and Taishi Takenobu, "Highly flexible MoS_2 thin-film transistors with ion gel dielectrics," *Nano Lett.*, 12, 4013–4017, 2012. Doi:10.1021/nl301335q.

45. Ji Feng, Xiaofeng Qian, Cheng-Wei Huang, and Ju Li, "Strain-engineered artificial atom as a broad-spectrum solar energy funnel," *Nat Photonics*, 6, 865–871, 2012. Doi:10.1038/nphoton.2012.285.

46. Jaesung Lee, Zenghui Wang, K. He, J. Shan, and P. Feng, "High frequency MoS_2 nanomechanical resonators," *ACS Nano*, 7, 6086–6091, 2013. Doi:10.1021/nn4018872.

47. Junjie Qi, Yann-Wen Lan, Adam Z. Stieg, Jyun-Hong Chen, Yuan-Liang Zhong, Lain-Jong Li, Chii-Dong Chen, Yue Zhang, and Kang L. Wang, "Piezoelectric effect in chemical vapour deposition-grown atomic-monolayer triangular molybdenum disulfide piezotronics," *Nat. Commun.*, 6, 7430, 2015. Doi:10.1038/ncomms8430.

48. Zheng Dongqi, Ziming Zhao, Rui Huang, Jiaheng Nie, Lijie Li, and Yan Zhang, "High-performance piezo-phototronic solar cell based on two-dimensional materials," *Nano Energy*, 32, 448–453, 2017. Doi:10.1016/j.nanoen.2017.01.005.

49. Y. Fu, Y. Wang, H. Zhao, Z. Zhang, B. An, C. Bai, Z. Ren, J. Wu, Y. Li, W. Liu, and P. Li, "Synthesis of ternary ZnO/ZnS/MoS_2 piezoelectric nanoarrays for enhanced photocatalytic performance by conversion of dual heterojunctions," *Appl. Surf. Sci.*, 556, 149695, 2021.

50. S. A. Han, T. H. Kim, S. K. Kim, K. H. Lee, H. J. Park, J. H. Lee, and S. W. Kim, "Point-defect-passivated MoS$_2$ nanosheet-based high performance piezoelectric nanogenerator," *Adv. Mater.*, 30, 21, 1800342, 2018.

51. W. Wu, L. Wang, Y. Li, F. Zhang, L. Lin, S. Niu, D. Chenet, X. Zhang, Y. Hao, T. F. Heinz, and J. Hone, "Piezoelectricity of single-atomic-layer MoS 2 for energy conversion and piezotronics," *Nature*, 514, 7523, 470–474, 2014.

52. Li Song, Lijie Ci, Hao Lu, Pavel B. Sorokin, Chuanhong Jin, Jie Ni, Alexander G. Kvashnin, Dmitry G. Kvashnin, Jun Lou, Boris I. Yakobson, and Pulickel M. Ajayan, "Large scale growth and characterization of atomic hexagonal boron nitride layers," *Nano Lett.*, 10, 8, 3209–3215, 2010. Doi:10.1021/nl1022139.

53. Chun Li, Yoshio Bando, Chunyi Zhi, Yang Huang, and Dmitri Golberg, "Thickness-dependent bending modulus of hexagonal boron nitride nanosheets," *Nanotechnology*, 20, 385707, 2009. Doi:10.1088/0957-4484/20/38/385707.

54. Yin Liu, Yuncan Pan, Deqiang Yin, Shufeng Song, Liyang Lin, Mingxia Zhang, Xueli Qi, and Jianyao Yao, "Mechanical properties and thickness-determined fracture mode of hexagonal boron nitride nanosheets under nanoindentation simulations," *Comput. Mater. Sci.*, 186, 110047, 2021. Doi:10.1016/j.commatsci.2020.110047.

55. Kun Chen, Xi Wan, and J. B. Xu, "Epitaxial stitching and stacking growth of atomically thin transition-metal dichalcogenides (TMDCs) heterojunctions," *Adv. Funct. Mater.*, 27, 19, 1603884, 2017. Doi:10.1002/adfm.201603884.

56. Michael S. Bresnehan, Matthew J. Hollander, Maxwell Wetherington, and Michael LaBella, "Integration of hexagonal boron nitride with quasi-freestanding epitaxial graphene: toward wafer-scale, high-performance devices," *ACS Nano*, 6, 6, 5234–5241, 2012. Doi:10.1021/nn300996t.

57. Cheng Ding, Yuehua Dai, Fei Yang, and Xiaoyan Chu, "A molecular dynamics study of the mechanical properties of the graphene/hexagonal boron nitride planar heterojunction for RRAM," *Mater. Today Commun.*, 26, 101653, 2021. Doi:10.1016/j.mtcomm.2020.101653.

58. Gwan-Hyoung Lee, Young-Jun Yu, Xu Cui, Nicholas Petrone, Chul-Ho Lee, Min Sup Choi, Dae-Yeong Lee, Changgu Lee, Won Jong Yoo, Kenji Watanabe, Takashi Taniguchi, Colin Nuckolls, Philip Kim, and James Hone, "Flexible and transparent MoS$_2$ field-effect transistors on hexagonal boron nitride-graphene heterostructures," *ACS Nano*, 7, 9, 7931–7936, 2013. Doi:10.1021/nn402954e.

59. Kiho Cho, Jiong Yang, and Yuerui Lu, "Phosphorene: an emerging 2D material," *J. Mater. Res.*, 32, 15, 2839–2487, 2017. Doi:10.1557/jmr.2017.71.

60. Zenghui Wang, Hao Jia, Xuqian Zheng, Rui Yang, Zefang Wang, G. J. Ye, X. H. Chen, Jie Shan, and Philip X.-L. Feng, "Black phosphorus nanoelectromechanical resonators vibrating at very high frequencies," *Nanoscale*, 7, 3, 877–884, 2015. Doi:10.1039/c4nr04829f.

61. Jin-Wu Jiang and Harold S. Park, "Negative poisson's ratio in single-layer black phosphorus," *Nat. Commun.*, 5, 4727, 2014. Doi:10.1038/ncomms5727.

62. Jin-Wu Jiang and Harold S. Park, "Mechanical properties of single-layer black phosphorus," *J. Phys. D-Appl. Phys.*, 47, 2014. Doi:10.1088/0022-3727/47/38/385304.

63. Damien Hanlon, Claudia Backes, Evie Doherty, Clotilde S. Cucinotta, Nina C. Berner, Conor Boland, Kangho Lee, Andrew Harvey, Peter Lynch, Zahra Gholamvand, Saifeng Zhang, Kangpeng Wang, Glenn Moynihan, Anuj Pokle, Quentin M. Ramasse, Niall McEvoy, Werner J. Blau, Jun Wang, Gonzalo Abellan, Frank Hauke, Andreas Hirsch, Stefano Sanvito, David D. O'Regan, Georg S. Duesberg, Valeria Nicolosi, and Jonathan N. Coleman, "Liquid exfoliation of solvent-stabilized few-layer black phosphorus for applications beyond electronics," *Nat. Commun.*, 6, 8563, 2015. Doi:10.1038/ncomms9563.

64. A. Morita, "Semiconducting black phosphorus," *Appl. Phys. A*, 39, 4, 227–242, 1986. Doi:10.1007/BF00617267.

65. Zhen-Dong Sha, Qing-Xiang Pei, Zhiwei Ding, Jin-Wu Jiang, and Yong-Wei Zhang, "Mechanical properties and fracture behavior of single-layer phosphorene at finite temperatures," *J. Phys. D: Appl. Phys.*, 48, 39, 395303, 2015. Doi:10.1088/0022-3727/48/39/395303.

66. Zhaoyao Yang, Junhua Zhao, and Ning Wei, "Temperature-dependent mechanical properties of monolayer black phosphorus by molecular dynamics simulations," *Appl. Phys. Lett.*, 107, 2, 023107, 2015. Doi:10.1063/1.4926929.

67. Yuanyue Liu, Fangbo Xu, Ziang Zhang, Evgeni S. Penev, and Boris I. Yakobson, "Two-dimensional mono-elemental semiconductor with electronically inactive defects: the case of phosphorus," *Nano Lett.*, 14, 12, 6782–6786, 2014. Doi:10.1021/nl5021393.

68. Yongqing Cai, Qingqing Ke, Gang Zhang, Boris I. Yakobson, and Yong-Wei Zhang, "Highly itinerant atomic vacancies in phosphorene," *J. Am. Chem. Soc.*, 138, 32, 10199–10206, 2016. Doi:10.1021/jacs.6b04926.

69. Ning Liu, Jiawang Hong, Ramana Pidaparti, and Xianqiao Wang, "Fracture patterns and the energy release rate of phosphorene," *Nanoscale*, 8, 10, 5728–5736, 2016. Doi:10.1039/C5NR08682E.

70. Zenghui Wang, Hao Jia, Xuqian Zheng, Rui Yang, Zefang Wang, G. J. Ye, X. H. Chen, Jie Shan, and Philip X.-L. Feng, "Black phosphorus nanoelectromechanical resonators vibrating at very high frequencies," *Nanoscale*, 7, 3, 877–884, 2015. Doi:10.1039/C4NR04829F.

71. Michael Naguib, Murat Kurtoglu, Volker Presser, Jun Lu, Junjie Niu, Min Heon, Lars Hultman, Yury Gogotsi, and Michel W. Barsoum, "Two-dimensional nanocrystals produced by exfoliation of Ti₃AlC₂," *Adv. Mater.*, 23, 37, 4248–4253, 2011. Doi:10.1002/adma.201102306.

72. Michael Naguib, Olha Mashtalir, Joshua Carle, Volker Presser, Jun Lu, Lars Hultman, Yury Gogotsi, and Michel W. Barsoum, "Two-dimensional transition metal carbides," *ACS Nano*, 6, 2, 1322–1331, 2012. Doi:10.1021/nn204153h.

73. Michael Naguib, Vadym N. Mochalin, Michel W. Barsoum, and Yury Gogotsi, "25th anniversary article: MXenes: a new family of two-dimensional materials," *Adv. Mater.*, 26, 7, 992–1005, 2014. Doi:10.1002/adma.201304138.

74. Michael Naguib, Jeremy Come, Boris Dyatkin, Volker Presser, Pierre-Louis Taberna, P. Simon, Michel Barsoum, and Yury Gogotsi, "MXene: a promising transition metal carbide anode for lithium-ion batteries," *Electrochem. Commun.*, 16, 61–64, 2012. Doi:10.1016/j.elecom.2012.01.002.

75. Mohammad Khazaei, Masao Arai, Taizo Sasaki, Mehdi Estiliand, and Yoshio Sakka, "Trends in electronic structures and structural properties of MAX phases: a first-principles study on M_2AlC (M = Sc, Ti, Cr, Zr, Nb, Mo, Hf, or Ta), M_2AlN, and hypothetical M_2AlB phases," *J. Phys.: Condens. Matter*, 26, 50, 505503 (1–12), 2014. Doi:10.1088/0953-8984/26/50/505503.

76. Wei Jin, Shiyun Wu, and Zhiguo Wang, "Structural, electronic and mechanical properties of two-dimensional janus transition metal carbides and nitrides," *Phys. E: Low-Dimens. Syst. Nanostructures*, 103, 307–313, 2018. Doi:10.1016/j.physe.2018.06.024.

77. Muhammad Kashif Aslam and Maowen Xu, "A mini-review: MXene composites for sodium/potassium-ion batteries," *Nanoscale*, 12, 15993–16007, 2020, Doi:10.1039/D0NR04111D.

78. Mohammad Khazaei, Ahmad Ranjbar, Masao Arai, Taizo Sasaki, and Seiji Yunoki, "Electronic properties and applications of MXenes: a theoretical review," *J. Mater. Chem. C*, 5, 10, 2488–2503, 2017. Doi:10.1039/C7TC00140A.

79. Deniz Çakır, F. M. Peeters, and Cem Sevik, "Mechanical and thermal properties of h-MX2 (M = Cr, Mo, W; X = O, S, Se, Te) monolayers: a comparative study," *Appl. Phys. Lett.*, 104, 20, 203110, 2014. Doi:10.1063/1.4879543.

80. Karel-Alexander N. Duerloo, Mitchell T. Ong, and Evan J. Reed, "Intrinsic piezoelectricity in two-dimensional materials," *J. Phys. Chem. Lett.*, 3, 2871–2876, 2012. Doi:10.1021/jz3012436.

81. Farzaneh Memarian, A. Fereidoon, and M. Darvish Ganji, "Graphene young's modulus: molecular mechanics and DFT treatments," *Superlattices Microstruct.*, 85, 348–356, 2015. Doi:10.1016/j.spmi.2015.06.001.

82. Soo Min Kim, Allen Hsu, Min Ho Park, Sang Hoon Chae, Seok Joon Yun, Joo Song Lee, Dae-Hyun Cho, Wenjing Fang, Changgu Lee, Toma´s Palacios, Mildred Dresselhaus, Ki Kang Kim, Young Hee Lee, and Jing Kong, "Synthesis of large-area multilayer hexagonal boron nitride for high material performance," *Nat. Commun.*, 6, 8662, 2015. Doi:10.1038/ncomms9662.

83. Andres Castellanos-Gomez, Menno Poot, Gary A. Steele, Herre S. J. van der Zant, Nicolás Agraït, and Gabino Rubio-Bollinger, "Elastic properties of freely suspended MoS_2 nanosheets," *Adv. Mater.*, 24, 772–775, 2012. Doi:10.1002/adma.201103965.

84. Alexey Lipatov, Haidong Lu, Mohamed Alhabeb, Babak Anasori, Alexei Gruverman, Yury Gogotsi, and Alexander Sinitskii, "Elastic properties of 2D $Ti_3C_2T_x$ MXene monolayers and bilayers," *Sci. Adv.*, 4, 6, eaat0491, 2018. Doi:10.1126/sciadv.aat0491.

85. Apratim Khandelwal, Karthick Mani, Manohar Harsha Karigerasi, and Indranil Lahiri, "Phosphorene—the two-dimensional black phosphorous: properties, synthesis and applications," *Mater. Sci. Eng.: B*, 221, 17–34, 2017. Doi:10.1016/j.mseb.2017.03.011.

86. Ming Xin, Jiean Li, Zhong Ma, Lijia Pan, and Yi Shi, "MXenes and their applications in wearable sensors," *Front. Chem.*, 8, 297, 2020. Doi:10.3389/fchem.2020.00297.

87. Michael Naguib, Vadym N. Mochalin, Michel W. Barsoum, and Yury Gogotsi, "Two-dimensional materials: 25th anniversary article: MXenes: a new family of two-dimensional materials," *Adv. Mater.*, 26, 7, 992–1005, 2014. Doi:10.1002/adma.201470041.

88. Zhongheng Fu, Ning Wang, Dominik Legut, Chen Si, Qianfan Zhang, Shiyu Du, Timothy C. Germann, Joseph S. Francisco, and Ruifeng Zhang, "Rational design of flexible two-dimensional MXenes with multiple functionalities," *Chem. Rev.*, 119, 23, 11980–12031, 2019. Doi:10.1021/acs.chemrev.9b00348.

89. Vadym N. Borysiuk, Vadym N. Mochalin, and Yury Gogotsi, "Bending rigidity of two-dimensional titanium carbide (MXene) nanoribbons: a molecular dynamics study," *Comput. Mater. Sci.*, 143, 418–424, 2018. Doi:10.1016/j.commatsci.2017.11.028.

90. Shaohong Luo, Shashikant Patole, Shoaib Anwer, Baosong Li, Thomas Delclos, Oleksiy Gogotsi, Veronika Zahorodna, Vitalii Balitskyi, and Kin Liao, "Tensile behaviors of $Ti_3C_2T_x$ (MXene) films," *Nanotechnology*, 31, 39, 395704, 2020. Doi:10.1088/1361-6528/ab94dd.

91. Babak Anasori, Maria Lukatskaya, and Yury Gogotsi, "2D metal carbides and nitrides (MXenes) for energy storage," *Nat. Rev. Mater.*, 2, 2, 1–17, 2017. Doi:10.1038/natrevmats.2016.98.

92. Masashi Okubo, Akira Sugahara, Satoshi Kajiyama, and Atsuo Yamada, "MXene as a charge storage host," *Acc. Chem. Res.*, 51, 3, 591–599, 2018. Doi:10.1021/acs.accounts.7b00481.

93. Jinxiu Zhao, Lei Zhang, Xiao-Ying Xie, Xianghong Li, Yongjun Ma, Qian Liu, Wei-Hai Fang, Xifeng Shi, Ganglong Cui, and Xuping Sun, "$Ti_3C_2T_x$ (T = F, OH) MXene nanosheets: conductive 2D catalysts for ambient electrohydrogenation of N_2 to NH_3," *J. Mater. Chem. A*, 6, 47, 24031–24035, 2018. Doi:10.1039/C8TA09840A.

94. Chuanfang (John) Zhang and Valeria Nicolosi, "Graphene and MXene-based transparent conductive electrodes and supercapacitors," *Energy Storage Mater.*, 16, 102–125, 2019. Doi:10.1016/j.ensm.2018.05.003.

95. Yi-Zhou Zhang, Jehad K. El-Demellawi, Qiu Jiang, Gang Ge, Hanfeng Liang, Kanghyuck Lee, Xiaochen Dong, and Husam N. Alshareef, "MXene hydrogels: fundamentals and applications," *Chem. Soc. Rev.*, 49, 7229–7251, 2020. Doi:10.1039/d0cs00022a.

96. Zheng Ling, Chang E. Ren, Meng-Qiang Zhao, Jian Yang, James M. Giammarco, Jieshan Qiu, Michel W. Barsoum, and Yury Gogotsi, "Flexible and conductive MXene films and nanocomposites with high capacitance," *Proc. Natl. Acad. Sci.*, 111, 47, 16676–16681, 2014. Doi:10.1073/pnas.1414215111.

97. Hyosung An, Touseef Habib, Smit Shah, Huili Gao, Miladin Radovic, Micah J. Green, and Jodie L. Lutkenhaus, "Surface-agnostic highly stretchable and bendable conductive MXene multilayers," *Sci. Adv.*, 4, 3, eaaq0118, 2018. Doi:10.1126/sciadv.aaq0118.

98. Seymur Cahangirov, Mehmet Topsakal, Aktürk Ethem, and Hasan Sahin, "Two- and one-dimensional honeycomb structures of silicon and germanium," *Phys. Rev. Lett.*, 102, 23, 236804, 2009. Doi:10.1103/PhysRevLett.102.236804.

99. Gexin Liu, X. L. Lei, Musheng Wu, and Bo Xu, "Is silicene stable in O2?—First-principles study of O2 dissociation and O2-dissociation—induced oxygen atoms adsorption on free-standing silicene," *EPL (Europhysics Letters)*, 106, 4, 47001, 2014. Doi:10.1209/0295-5075/106/47001.

100. Gexin Liu, X. L. Lei, Musheng Wu, Bo Xu, and Chuying Ouyang, "Comparison of the stability of free-standing silicene and hydrogenated silicene in oxygen: a first principles investigation," *J. Phys. Condens. Matter*, 26, 35, 355007, 2014. Doi:10.1088/0953-8984/26/35/355007.

101. Hasan Sahin, S. Cahangirov, Mehmet Topsakal, E. Bekaroglu, Aktürk Ethem, Tuğrul Senger, and Salim Ciraci, "Monolayer honeycomb structures of group-IV elements and III-V binary compounds: first-principles calculations," *Phys. Rev. B*, 80, 15, 155453, 2009. Doi:10.1103/PhysRevB.80.155453.

102. Andres Castellanos-Gomez, Menno Poot, Albert Amor-Amorós, Gary A. Steele, Herre S.J. van der Zant, Nicolás Agraït, and Gabino Rubio-Bollinger, "Mechanical properties of freely suspended atomically thin dielectric layers of mica," *Nano Res.*, 5, 8, 550–557, 2012. Doi:10.1007/s12274-012-0240-3.

103. Pascal Pochet, Brian C. McGuigan, Johann Coraux, and Harley T. Johnson, "Toward Moiré engineering in 2D materials via dislocation theory," *Appl. Mater. Today*, 9, 240–250, 2017. Doi:10.1016/j.apmt.2017.07.007.

8 2D Materials for Energy Harvesting

Astakala Anil Kumar, Sharmila Kumari Arodhiya, Shashank Priya, Ashok Kumar, and Shyam Sundar Pattnaik

CONTENTS

8.1 INTRODUCTION

The global energy demand is increasing progressively due to the rapid industrialization and population growth. Besides, the energy-harvesting devices operating on the fossil fuels/ nonrenewable energy sources cause adverse effects to the environment. Consequently, the demand for the clean energy technologies operating on the renewable energies, like solar, wind, and thermal, are gaining importance. Though the existing energy harvesting systems have been efficiently employed for energy harvesting, they are not yet competitive or on par with fossil fuels mainly due to the high production costs [1, 2]. It is an important task to develop inexpensive energy harvesting devices possessing improved efficiency to cater the societal energy needs. The layered two-dimensional materials such as graphene and other transition metal chalcogenides possess non -centrosymmetric crystal structures. They have extensively been used in the fabrication of electromechanical systems and energy harvesting devices. The perovskite solar cells centered on mixed metal halide perovskite materials attracted attention due to their bandgap tunability, high absorption coefficient

(~10^5 cm^{-1}), and large diffusion length [3]. The 2D materials such as graphene, transition metal dichalcogenides, nitrides, MXenes, and Xenes revealed application in the fabrication of various energy harvesting devices such as thermoelectric, perovskite solar cells, and piezoelectric systems. Since the first report published on the usage of organo-inorganic perovskites in 2009, nowadays the perovskite solar cells reached a stable power conversion efficiency of about 24% [4].

In the recent past, there are more reports on the combination of the emerging two-dimensional materials like graphene, transition metal dichalcogenides, and other emerging materials with the hybrid perovskites [5, 6]. The two-dimensional materials possess outstanding physical, chemical, and electronic properties that enhance the overall performance of the perovskite cell. The 2D materials are comprised of covalently bonded thin sheets, which can be molded into detached monolayers of desired thickness [7]. Among 2D materials, graphene is the earliest studied 2D material. It consists of carbon located in the symmetrical hexagons with sp^2 hybridization. It has two-dimensional honeycomb structure and available physically in the form of flakes of few layers of carbon atoms [8, 9]. The two-dimensional graphene that possesses extraordinary properties—such as high electrical conductivity, electron mobility, bandgap tunability, stability, high thermal stability, high mechanical strength—makes it a viable candidate for the energy-harvesting applications [10–12]. The various applications of the 2D materials are shown in Figure 8.1.

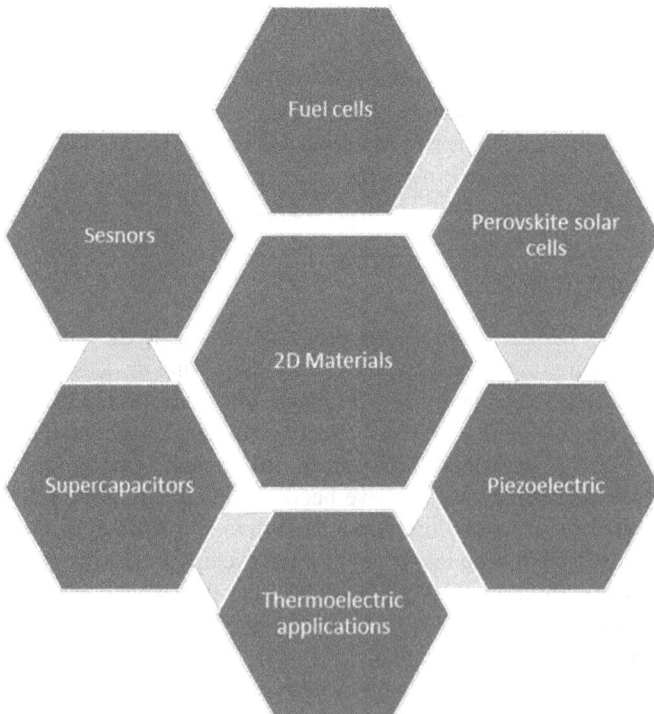

FIGURE 8.1 Various applications of 2D materials.

Recently, significant research is being done on the development of various materials for the energy-harvesting applications. Various 2D materials in general attracted a huge consideration in the fabrication process due to their unique electronic, mechanical, and optical properties [13]. The transition metal dichalcogenides are the most researched 2D materials after graphene in the energy-harvesting applications [14]. The transition metal dichalcogenides have general formula of MX_2, where M is a transitional metal ion, such as Mo, W, Ti, etc., and X is a chalcogenide atom, such as sulfur (S), selenium (Se), or tellurium (Te). The transition metal chalcogenides possess layered structure and are flexible as compared to graphene/ graphene-based derivatives [15, 16]. The transition metal chalcogenides such as MoS_2, $MoSe_2$, WS_2, WSe_2 have been investigated extensively for the energy-harvesting applications [17–19]. The bandgap tunability, high surface area, flexibility are the attributes of transition metal chalcogenides, which makes these materials interesting for the carrier transport applications in energy-harvesting applications [20]. In the current chapter, the recent advances in the 2D layered materials for their application in the energy-harvesting devices such as perovskite solar cells, thermoelectric, and piezoelectric systems will be discussed.

8.2 EMERGING APPLICATIONS OF TWO-DIMENSIONAL (2D) MATERIALS

8.2.1 PEROVSKITE SOLAR CELLS (PSCS)

Perovskite solar cells (PSCs) are the third-generation solar cells. These are a viable alternative to the existing silicon solar cells. Unlike silicon solar cells, the PSCs have different materials for different functionalities, such as absorption, separation, and collection of charge carriers. The organo-inorganic halide perovskite such as $MAPbX_3$, $FAPbX_3$ (where Methylammonium, MA^+, formamidinium, FA^+, and X = Cl, Br and I) act as a light absorber where the separation of electron and holes will initiate. The replacement of Pb with materials such as Sn, Ge, or other divalent materials is gaining importance due to toxicity of lead [21, 22]. The perovskite structured materials such as $MAPbI_3$ possess unique properties, such as lower binding energy, higher absorption coefficient of the order 10^4 cm^{-1}, and higher value of dielectric constant [23]. Once the perovskite solar cell is illuminated, the perovskite layer deposited on the surface of the electron transport layer will absorb the sunlight and generate the electron hole pairs [24, 25]. The carrier diffusion length of 100 nm makes the $MAPbI_3$ as the most researched material that has shown the higher power conversion efficiencies [26]. The schematic representation of the different layers in perovskite solar cell is represented in the Figure 8.2

Typically, the perovskite solar cell consists of the mesoporous TiO_2 layer as an electron transport layer, which is deposited on the FTO (fluorine doped tin oxide) substrate, Spiro- OMeTAD acts as an hole transport layer, and $MAPbI_3$ functions as a light absorber. The electron hole pairs are separated, and the electrons are collected at the electron transport layer, and the holes are collected at the metal contacts (usually Au or Ag) deposited on the hole transport layer [27–29]. The potential drop across the different layers affects the performance of perovskite solar cell. Incorporation

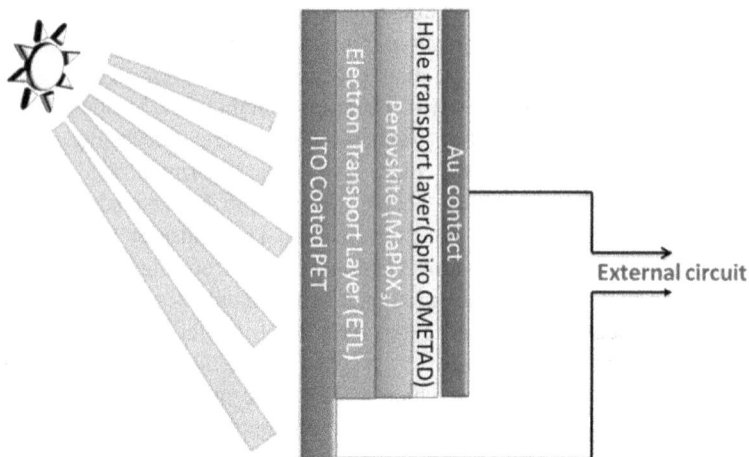

FIGURE 8.2   Schematic diagram of perovskite solar cell.

of the interface/buffer layers in between the electron transport layers of perovskite solar cell results in the enhancement of device performance. These interface/buffer layers tune the properties of the device without affecting the material choice [30–32]. The various methodologies, such as self-assembling layers, inorganic materials, polymers, and the deposition of thin films using several nanomaterials, are used to engineer the performance of PSC [33, 34]. According to the recent scientific reports, the usage of two-dimensional materials such as graphene derivatives and transitional metal dichalcogenides as the buffer layer in the PSC has gained prominence. The unique physical and chemical properties along with the optoelectronic tunability made these two-dimensional materials unique as compared to the other materials. The synthesis/deposition of these materials is facile and easily processable. The solar cell parameters such as incident photon to current efficiency (IPCE) and open circuit voltage (V_{oc}) increase considerably with the incorporation of buffer/interfacial layers in between the adjacent layers of perovskite solar cells.

8.2.1.1   Graphene-Related Materials

Among two-dimensional materials, graphene is the novel two-dimensional carbon derivative. It possesses good optical transparency and higher electron mobility, which makes it of enormous interest among scientific community [35–36]. In the recent past, owing to the interesting physical and chemical properties, graphene and its derivatives have been extensively utilized in the perovskite solar cells. The graphene can be utilized in the device fabrication in different forms, such as zero, one, two, and three dimensions as graphene quantum dots, graphene nanobelt, graphene sheet, and grapheme, respectively, as per requirements in the perovskite solar cells. In addition, graphene can also be fitted/ decorated with the other inorganic materials to enhance the optoelectronic properties [37]. In literature, it was reported that the inducement of graphene in the PSC as a buffer layer in between the adjacent layers of perovskite solar cell results in the enhancement of the performance of the perovskite solar cells [38].

Mariani et al. [39] reported the fabrication of PSC utilizing the carbon-based counter electrodes and studied the variation of parameters with the addition of graphene between the interface of HTM and electron transport layer. The perovskite solar cell was fabricated with the TiO_2 as electron transport layer, the $MAPbI_3$ as light absorber, and Spiro-OMeTAD as hole transport layer. The counter electrode was made with the conducting carbon paste unlike the metal contact used in the traditional perovskite solar cells. The highest power conversion efficiency (PCE) of 15.8% was achieved for the small area of 0.09 cm^2, and PCE of 14.1% is achieved for the large area of 1 cm^2. Further, the PCE of 13.9% was obtained for the device manufactured with the area of 4 cm^2. The PCE obtained is the highest reported with the p-i-n configuration without using the conventional metal surfaces as counter electrode. Mahmoudi et al. [40] utilized the *p*-type inorganic hole transport material (NiO) in contrast to the conventionally used Spiro-OMeTAD. The usage of inorganic materials in PSC results in the enhancement of stability as the inorganic compounds are more stable than the organic materials. Further, the perovskite solar cell was fabricated using the $MAPbI_{3-x}Cl_x$/NiO-graphene as photo absorber, hole transport material (HTM), and the NiO as the buffer layer in between the inorganic and organic layer. The device with the NiO interface depicts the highest power conversion efficiency of 20.8%. Interestingly, in contrast to the conventional PSC with organic functional layers, the device with inorganic hole transport that is kept in the ambience for a duration of 310 days shows the decrement in PCE of about 3% (conventional PSC 2%). Chen et al. [41] reported the PSC with the α-Fe_2O_3 as electron transport layer. Although α-Fe_2O_3 possesses better stability as compared to the conventionally used TiO_2, its poor interaction with the perovskite results in the lower efficiency of the device fabricated with α-Fe_2O_3 as electron transport layer. Further, with the inducement of N, S co-doped graphene quantum dots based buffer layer in between the electron transport layer (ETL) and perovskite layer results in the enhancement of power conversion efficiency from 14 to 19.2%, which is the highest PCE reported with α-Fe_2O_3. Besides, the interfacial engineering adopted with the incorporation of buffer layer in between the adjacent layers of the perovskite solar cell results in the enhancement of stability and durability of device as compared to the conventional device fabricated with TiO_2 as ETL. The mechanism/working of perovskite solar cell is shown in Figure 8.3.

Ahmed et al. [42] fabricated PSC with ZnO as electron transport layer. The ZnO is a viable alternative to the conventionally used TiO_2 as it has better electrical and optical conductivity. The presence of oxygen traps on the surface of the ZnO results in the decomposition of $MAPbI_3$ to MAI and PbI_2. The decomposition of the perovskite is minimized by using the graphene quantum dots buffer layer between ETL and perovskite layer. The inclusion of the buffer layer enhances the electron conductivity between the adjacent layers, which results in the enhancement of the power conversion efficiency of the device. The device fabricated with ZnO graphene quantum dots based ETL shows an enhanced power PCE of 17%. Chandrasekhar et al. [43] fabricated the PSC with nitrogen doped graphene and ZnO nanocomposite-based electron transport layer. The perovskite solar cell was fabricated on the FTO substrate

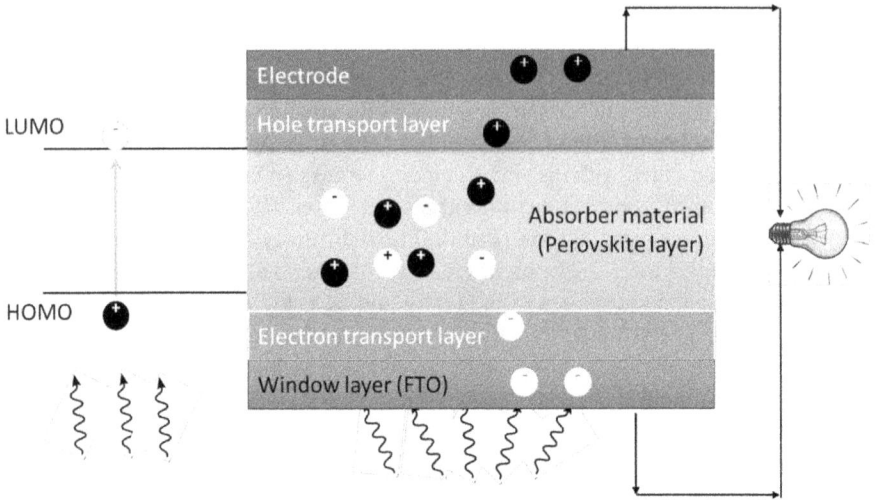

FIGURE 8.3   Working of a perovskite solar cell.

with nitrogen doped graphene and ZnO nanocomposite as electron transport layer, MAPbI$_3$ as light absorber, Spiro-MEOTAD as HTM, and Ag contact on the rear end acts as counter electrode. Several PSCs were fabricated with varying the concentration of nitrogen doped graphene in the nanocomposite mixture. The perovskite solar cell fabricated with the optimized concentration of 0.8 wt% of nitrogen doped graphene showed the highest power conversion efficiency of 16.82%. Guo et al. [44] synthesized Ni doped graphene nanocomposites using novel liquid phase impregnation method. The novel HTM-free perovskite solar cell was fabricated using the synthesized Ni doped graphene nanocomposites as electron transport layer, followed by the deposition of MAPbI$_3$ layer that is used as an optical window. The carbon electrode coated on the top of the light absorber acts as a counter electrode. The optimized perovskite solar cell with the concentration of Ni at 4×10^{-4} mol/g yields the highest power conversion efficiency of about 12.39%. The device may be further optimized to engineer the performance of the HTM-free perovskite solar cell.

8.2.1.2 Dichalcogenides

In perovskite solar cells, the ETL plays a crucial role in determining the device performance as it acts as a bridge for the photogenerated electrons. In conventional perovskite solar cells, the mesoporous TiO$_2$ was utilized as ETL and was known for better power conversion efficiencies. The presence of cation/oxygen vacancies on the surface of the metal oxide based ETL led the researchers to invade the other materials for the enhancement of device performance. The transition metal dichalcogenides will serve as an alternative for the traditionally used oxide semiconductor to conquer the role of an ETL in the perovskite solar cell. The transition metal halides possess the general formula MX$_2$, where the cation M is a transition metal ion, and X belongs to the chalcogenide family. The layered transition metal dichalcogenides are the viable candidates to graphene as they have similar properties, such as flexibility, optical transparency, and the dimensions are also as comparable to graphene. The absence of the surface lattice defect and atomistic thickness makes the transition metal dichalcogenides a viable candidate as an ETL in perovskite solar cells. Among the transition metal dichalcogenides, the MoS_2, WS_2, $MoSe_2$, and WSe_2 are the materials explored for the usage of carrier transport in the perovskite solar cell as these possess high electron mobility, layered structures, optical transparency, tunability of bandgap with the incorporation of external dopants in the lattice, and involve the facile synthesis process. Particularly, the MoS_2 has become more prominent in perovskite solar cells as it can have dual usage in the perovskite solar cell as both ETL and HTL due to its ambipolar properties. Malek et al. [45] successfully synthesized MOS_2 nanosheets by ultrasonic spray pyrolysis method. The perovskite solar cell was fabricated with the synthesized MoS_2 nanosheets. The optical window layer was composed of the triple cations namely Cs, MA, and FA $Cs_{0.05}$ [$MA_{0.13}$ $FA_{0.87}$]$_{0.95}Pb(I_{0.87}Br_{0.13})_3$ with Spiro-OMeTAD as hole transport layer and Au contacts on the rear end of the device as the counter electrode. The electron transport layer was fabricated using different substrate temperatures ranging from 180 to 250°C. The device fabricated with the sample sintered at 220°C provided the highest power conversion efficiency of 3.36%. The performance of the device can be engineered by modifying the physical, chemical, and morphological properties of the electron transport layer. Mahmood et al. [46] synthesized the MoS_2 nanosheets by facile and cost-effective hydrothermal method. The electron transport layer was fabricated by electro spraying the synthesized MoS_2 nanosheets on to the surface of FTO substrates, followed by the perovskite $MAPbI_3$ as light absorber, Spiro-OMeTAD as hole transport layer, and Au contact on the rear side of perovskite solar cells as the counter electrode. The highest power conversion efficiency of 16.17% was obtained with the electro spraying of MoS_2 on the FTO substrate, which is higher as compared to the spin coated electron transport layer.

The fabrication perovskite solar cell using the molybdenum selenide as hole transport layer is reported by Chen et al. [47]. The inverted planar structured perovskite solar cell is fabricated with the deposition of $MoSe_2$ on the FTO substrate as hole transport layer, followed by the deposition of perovskite structured $MAPbI_3$ as light absorber, fullerene (C60) as electron transport layer. The buffer layer of bathocuproine was introduced in between the electron transport layer and the counter electrode (Au) on the other end of the perovskite solar cell. On illumination with the 1.5 AM the optimized efficiency of 8.23% was obtained for the various domain sizes of $MAPbI_3$ in the range of 23–27 nm. Similarly, Chang et al. [48] reported the similar perovskite structure with $MoSe_2$ as hole transport layer, $MAPbI_3$ as light absorber, polymethyl methacrylate (PMMA) as electron transport layer, and Al doped ZnO as transparent conductive oxide, and the C60 as the buffer layer between the electron transport layer and the Al doped ZnO. The counter electrode was developed on the surface of Al doped ZnO with silver nanostructures. The unique modified PMMA ETM films have been prepared and characterized at various thicknesses of about 25 nm, where the highest power conversion efficiency of 16.9% is reported, which is the highest reported efficiency with the configuration and p-type $MoSe_2$ as hole transport layer. The device may be engineered further to enhance the performance of the fabricated perovskite solar cells.

8.2.2 THERMOELECTRIC MATERIALS

The thermoelectric generators convert heat energy directly into electrical energy without the emission of harmful gases and with noiseless operation because of the absence of any moving part. The thermoelectric materials are widely used in power generator, cooling devices, and sensors [49–52]. Thermoelectric generator can be the best solution for the utilization of waste heat generated from industries, vehicles, and also from human bodies. But low conversion efficiency (peak efficiency of 9.6% and system efficiency of 7.4%) is the main disadvantage of thermoelectric generator [53]. In the past years, it was very challenging to decouple the thermoelectric parameters to enhance the efficiency. The nano-structuring technique decouples the thermal conductivity and electrical conductivity due to quantum confinement. The working of the typical thermoelectric device is shown in the Figure 8.4

8.2.2.1 Oxide Thermoelectrics

A thermoelectric material for large-scale application should not only possess high figure of merit (ZT) but also it should have high abundance and easy availability. Therefore, the use of oxides as thermoelectric material can be promising.

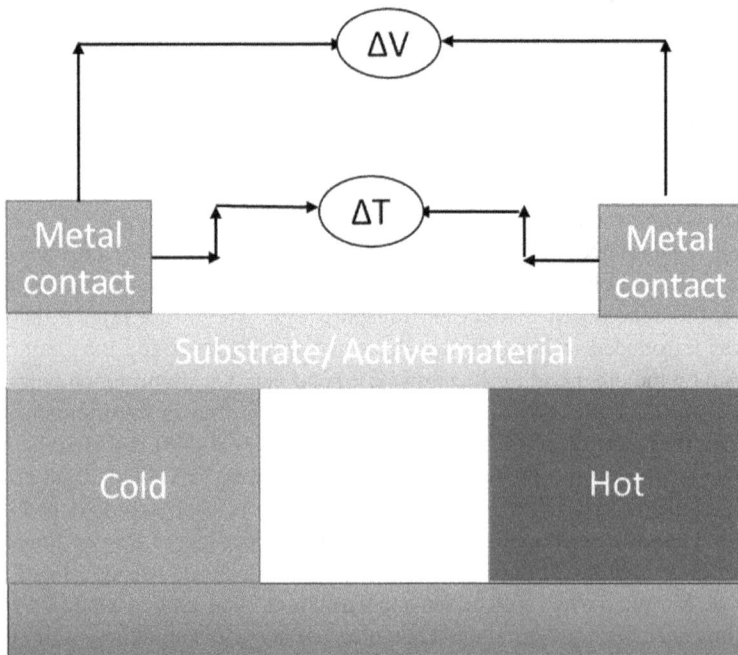

FIGURE 8.4 Working of a thermoelectric device.

Furthermore, oxides possess more chemical and thermal stability in comparison to other thermoelectric materials [54–56].

$NaCo_2O_4$ is quite prominent as a high temperature thermoelectric material. Several papers have reported the thermoelectric performance of $NaCo_2O_4$. Terasaki et al. [57] reported high thermoelectric power in single crystal of $NaCo_2O_4$. Fuzita et al. [58] reported ZT greater than one at 800 K and polycrystalline of $NaCo_2O_4$ showed ZT~0.8 at 1,000 K [59]. Flexible $Na_{1.4}Co_2O_4$ thin film prepared by self-flux method and painted on a printing paper showed a power factor of 159–223 $\mu Wm^{-1} K^{-2}$ in the temperature range of 303–522 K, which is comparable to conductive polymers [60]. $Ca_2Co_2O_5$ single crystalline whisker showed the figure of merit in the range of 1.2–2.7 at ≥873 K [61]. M. Ardyanian et al. prepared ZnO-SnO binary thin film on glass substrate by using spray pyrolysis method and reported a maximum power factor of 4.06×10^{-5} $Wm^{-1} K^{-2}$ for the molar ratio [Sn] / [Zn] = 10% at 490 K [62]. The $Zn_{0.98}Al_{0.02}O$ was reported as a promising thermoelectric material because of its high electrical conductivity, Seebeck coefficient, and melting point. It showed a power factor of 1.5×10^{-3} Wm^{-2} at 975 K [63–65]. Besides, the energy filtering process increases the carrier mobility. A major increment in electrical conductivity of single-layered ZnO:Mg and bilayer ZnO:Mg/ZnO was reported due energy filtering. It led to a much-improved Seebeck coefficient and hence power factor in comparison to ZnO [66]. Further, the ABO_3 type perovskites are the efficient thermoelectric oxide materials. Their thermoelectric performance can further be improved by two-dimensional electron gas techniques. A superlattice composed of undoped $SrTiO_3$ and Nb doped $SrTiO_3$ on $LaAlO_3$ single crystal substrate was fabricated by pulse laser deposition method, which attained a ZT of ~2.4, considering the thermal conductivity similar to those of bulk single crystal [67–69]. Table 8.1 shows the thermoelectric properties of CaCaO based oxides.

TABLE 8.1
Thermoelectric Properties of Ca-Co-O Ceramics

Composition and structure	ZT	T (K)	S(µV/K)	σ (S/cm)	κ (W/mK)	Ref.
$Ca_9Co_{12}O_{28}$	0.047	700	118	84	1.73	[70]
$Ca_2Co_2O_5$ single crystalline whisker	1.2–2.7	≥873	200		≈1	[61]
$(Ca_2CoO_3)_{0.7}CoO_2$ single crystal	0.87	973	240	430	3	[71]
Bi- and Na- substituted $Ca_3Co_4O_9$	0.32	1000	~200	~130	1.7	[72]
Textured $Ca_2Co_2O_5$		973	178	~100		[73]
La-doped textured $Ca_3Co_4O_9$	0.26	975	150	213	2.06	[74]
$(Ca_{2.8}Ag_{0.05}Lu_{0.15})\ Co_4O_9$ with nano-inclusion	0.6	1100	230	140	1.4	[75]
$[(Ca, Yb)_2CoO_3]_{0.62}[CoO_2]$	0.45	1000	240	45	0.6	[76]
$Ca_3Co_4O_{9+}\delta$	0.37	800	~165		1.25	[77]
Ga-doped $Ca_3Co_4O_{9+}\delta$	0.25	1000	170	125	1.5	[78]
La, Fe-doped $Ca_3Co_4O_9$	0.32	1000	225	80	1.3	[79]
Ba-doped $[Ca_2CoO_3]_{0.62}[CoO_2]$	0.33	1000	205	120	1.65	[80]

8.2.2.2 Two-Dimensional Chalcogenides

In the past, because of its reasonable price and performance, the only thermoelectric material used for commercial purpose was Bi_2Te_3 with ZT 1.35 and 0.9 for n- and p-type, respectively [81–82]. Despite the low conductivity of n-type Bi_2Te_3, it has been widely used in thermoelectric refrigerator in the temperature range of 180–450. For higher temperature power generation applications, PbTe and SiGe (ZT ~1) have been used [83]. The high temperature thermoelectric materials are widely used in space craft power production. Chalcogenides used as thermoelectric materials are divided into three categories depending on their working temperature range. For a temperature less than 150°C Bi_2Te_3, for 150–500°C TAGS [$(AgSbTe_2)1-x(GeTe)x$], and above 500°C SiGe based materials are used [84, 85]. Figure 8.4 shows the basic construction of a single couple thermoelectric module consisting p- and n-type legs. Lagraneur et al. [86] fabricated a thermoelectric generator with three phases involving different thermoelectric material for different temperature range. For less than 250°C, n and p-type Bi_2Te_3, for 250–500°C TAGS and PbTe, for 500–700°C skutterudite p-type $CeFe_3RuSb$ and n-type $CoSb_3$. These materials were chosen because of their highest figure of merit in the corresponding temperature range. A flat thermocouple solution was used along with these segmented thermoelectric materials between the heat source and sink to increase the possibility of achieving the thickness and coefficient of thermal expansion for various regions. Wilbrecht et al. [87] fabricated a two-phase TE generator using Bi_2Te_3 for low temperature and Mg_2SiSn/MnSi for moderate temperature. It achieved 44.2% higher efficiency but 2.5 kW less electrical power than single phase generator (3.2 kW).

The transition metal chalcogenides (TMDs), like MX_2 and CdI_2 (M = Mo, W and X = S, Se, Te), are the layered materials, and each layer is held by van der Waals interaction. These exhibit high Seebeck coefficient and low thermal conductivity. CdI_2 type 2D TMDs such as TiX_2, ZrX_2, HfX_2 possess lower thermal conductivity in comparison to MX_2 type TMDs [88]. The thermal conductivity for $ZrSe_2$ is 1.2 $Wm^{-1} K^{-1}$, and for $HfSe_2$ it's 1.8 $Wm^{-1} K^{-1}$, which are less than Mo based TMDs and comparable to bulk PbTe and Bi_2Te_3 [89–93]. The indirect band structure in multilayer converts into direct bandgap in monolayer hence leads to variation in electronic properties. Bilayer 2D MoS_2 exhibits a maximum power factor of 8.5 $mWm^{-1} K^{-2}$ at room temperature. The improved transport properties TMDs are because of their valley degeneracies and large effective mass, which is further modified by quantum confinement in 2D material [88, 94–96].

8.2.2.3 Polymer Layered Thermoelectric

Conductive polymers, because of their low thermal conductivity, higher availability, and easy synthesis process, are used to fabricate the thermoelectric devices for room temperature applications. Though comparative low electrical conductivity and Seebeck coefficient limit their use in thermoelectric applications [82, 97], these are highly flexible to be used in wearable thermoelectric devices (WTD). By using WTD, the heat released by human body can be directly converted into electricity and be used for small applications, such as wrist watch, mobile charging, etc. Furthermore, the thermoelectric properties of conductive polymers can be improved by doping, de-doping [98], solvent treatment [99–102], sulfuric acid, and crystallization [103]. Power factor of (polypyrrole) PPy/grapheme/polyaniline (PANI) was reported 52.5 μW/mK^2, which is much higher than that of pristine PPy and PANI [104]. Poly(3,4-ethylenedioxythiophene) (PEDOT)consisting (poly-styrenesulfonate) (PSS) have been studied intensively due to its low-cost, high transparency, water-dispensability, and excellent process ability [105]. The electrical conductivity can further be improved by post-treatment of PEDOT: PSS thin film with the co-solvent of water and an organic solvent such as ethanol, isopropyl alcohol, acetonitrile, acetone, and tetra-hydro furan. The increment in electrical conductivity depends upon ratio of water to organic solvent, temperature, and dielectric constant of the solvent [106]. The thin film of ethylene glycol mixed PEDOT: PSS has a ZT of 0.28 and dimethyl sulfoxide (DMSO) mixed PEDOT: PSS of 0.42 at room temperature [107]. Furthermore, Huang et al. reported the power factor of PEDOT: PSS thin film annealed at 220°C, which was 162.5 times greater than without annealed films [108]. Chalcogens that show the highest figure of merit can be mixed with PEDOT: PSS to produce highly efficient flexible thermoelectric material. Park et al. [109] fabricated a flexible thermoelectric device based on Ag$_2$Se nano-wire/PEDOT: PSS nano-composite film. For synthesizing this film, first the Se nano-wires coated with PEDOT: PSS were mixed with Ag precursor to prepare PEDOT: PSS coated Ag$_2$Se nanowires, and in a separate beaker, PEDOT:PSS was mixed in dimethyl sulfoxide (DMSO). The Ag$_2$Se nanowires coated with PEDOT: PSS were dispersed in the solution of PEDOT: PSS and DMSO. This final solution was than deposited on the glass substrate via drop cast method. The film consisting of 50 wt% nanowires showed the highest power factor of 327.15 μW/mK^2 at 300 K. The thermoelectric properties were decreased only by a factor of ~5.9% after bending of 1,000 times. The device fabricated using this film generated output voltage of 7.6 mV for the temperature difference of 20 K.

8.2.3 OTHER APPLICATIONS

8.2.3.1 Dye-Sensitized Solar Cells

Dye-sensitized solar cells (DSSC) are the third-generation solar cells that are extensively investigated due to their low operational cost and facile fabrication methodologies. Similar to the perovskite solar cell, the DSSC consists of the electron transport layer, light absorber, hole transport layer and the counter electrode to harvest the solar energy [110–112]. The conventional DSSC consists of TiO$_2$, N719 dye, iodide/triiodide electrolyte and platinum as ETL, light absorber, HTM, and counter electrode, respectively [113]. Various components of the DSSC are shown in the Figure 8.5.

The platinum that has been conventionally used as counter electrode material is to be replaced as the availability of platinum is scarce, which, in turn, increases the fabrication cost of the device. The two-dimensional materials such as graphene was effectively used as the counter electrode material in DSSC, which showed the power conversion efficiency close to 13% [114]. The success of graphene as the counter electrode material in DSSC implied the researchers to focus on the other two-dimensional materials, such as MoS$_2$, WS$_2$, MoSe$_2$, WSe$_2$ [115–117]. The power conversion

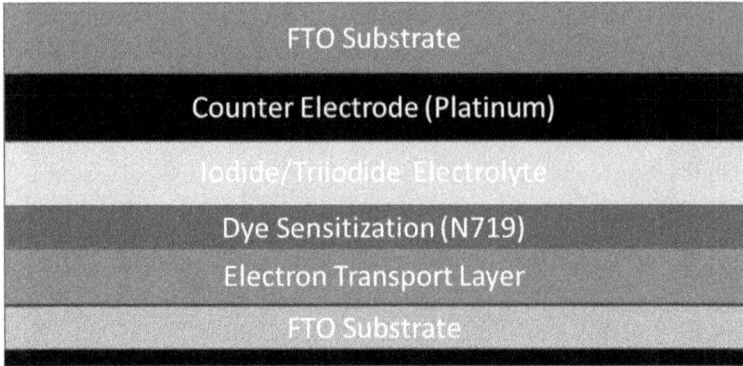

FIGURE 8.5 Schematic diagram of various components of DSSC.

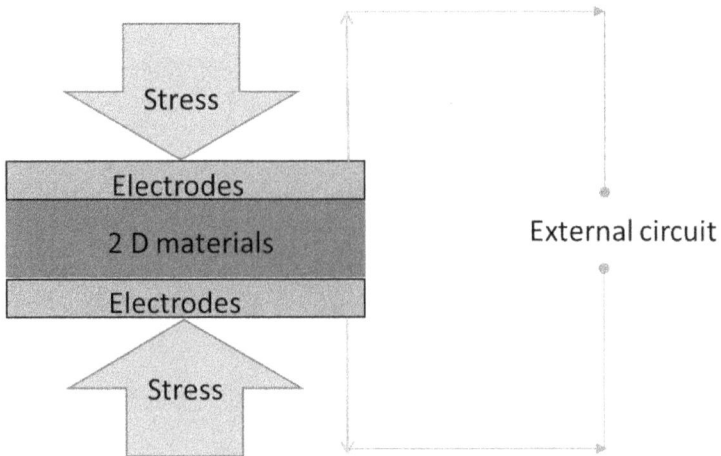

FIGURE 8.6 Working of a piezoelectric device.

efficiencies and the stability of the DSSC reported with the two-dimensional material are better as compared to conventionally used platinum counter electrodes. Currently, researchers are focused on the long-term stability of transition metal chalcogenides based DSSC in different environmental conditions.

8.2.3.2 Piezoelectric Applications

Piezoelectricity refers to the development of electrical charge on application of applied stress. In the absence of external strain, the charge distribution is symmetric, and the net electric dipole moment is zero. With the application of external stress, the charges are displaced, and the charge distribution is no longer symmetric. It gives rise to net polarization [118]. The schematic representation of the typical piezoelectric device is shown in the Figure 8.6

The two-dimensional materials have been of greater interest among the scientific community as these are known to have central symmetry and are predicted to have intrinsic piezoelectricity. The two-dimensional materials such as graphene/other transition metal dichalcogenides were extensively investigated due to their unique properties, such as intrinsic piezoelectricity, larger surface area, and facile synthesis process [119]. Compared to conventionally used ferroelectric materials, the two-dimensional layered materials have the advantages such as bandgap, tunability, and flexibility [120].

8.2.4 EMERGING 2D LAYERED MATERIALS FOR ENERGY HARVESTING

Two-dimensional layered materials like graphene and graphene-based materials, transition metal dichalcogenides, transition metal carbide, hexagonal boron nitride, etc. are widely used in energy harvesting applications specifically in piezoelectric and triboelectric devices due to their unique properties [121, 122]. Graphene has single-layer honeycomb lattice structure and zero bandgap (semiconductor) showing extraordinary chemical and physical properties. Hexa-boron nitride (h-BN) also has a structure like graphene but shows different properties (insulator). The 2D layered materials cover almost all types of electrical properties from insulator to superconductor. Graphene possesses a high value of electron mobility as well as thermal conductivity [123, 124]. Graphene is 100 times more conducting than Cu and has 100 times more electron mobility in comparison to single crystalline Si [122].

Graphite was used as an anode in lithium-ion batteries because of its large abundance, low-cost, and good theoretical capacity of 372 mAh/g. Han et al. found that graphene has much larger theoretical capacity (740mAh/g) than that of graphite [126]. TMDs also have larger theoretical capacity than that of graphite (670 mAh/g for MoS_2) [127]. Hence several research papers have been reported on better performance of lithium ion and sodium ion batteries using graphene and TMDs as an anode [64]. Besides, graphene is used for H_2storage and TMDs for generation of H_2. Few layers of graphene can store ~ 5wt% of H_2. Zhou et al. reported ~ 16 wt% storage capacity of graphene covered with Li ions [128].

Moreover, 2D layered materials have been used as supercapacitors and nanogenerators. Huang et al. [129] fabricated a supercapacitor based on TMDs with specific capacitance of 576 F/g at 5 mV/s. Goe et al. [130] fabricated a supercapacitor based on h-BN with a specific capacitance of 140 F/g at 2 A/g. A specific capacitance of 310 F/g at 2mV/s was reported for graphene-MnO_2 composite [131].

Metal organic frameworks (MOFs) are another option for energy storage and conversion such as photocatalytic hydrogen evolution, fuel cells, batteries, and supercapacitor due to their excellent gas storage capacity [132]. Further, due to nanotechnology and advanced characterization methods, new approaches are possible to rationally design and, hence, modify the material efficiency [133–135]. MOFs are porous materials and can be used direct as well as a substrate to integrate functional metals, metal oxides, semiconductor, and complexes to generate nanostructure [136, 137]. In 2003, the discovery of MOF-5 with hydrogen storage capacity has attracted the attention of researchers as hydrogen and methane adsorbent [138, 139]. The hydrogen storage of MOF-210 with highest surface area with total H_2gravimetric uptake is of 17.6 wt% at 80 bar and 77 K [140]. The absorption capacity of MOFs depends upon the volume of the pores. When benzene ring of NOTT-101 is replaced with pyrimidine ring of UTSA-76, an unexpected enhancement was observed in volumetric uptake [257 cm^3 (STP) cm^{-3} (298 K and 65 bar)] and with working capacity of 197 cm^3 (SPT) cm^{-3} [141]. MOF derivatives have been used as a photocatalyst in water splitting and carbon dioxide reduction. The products generated in CO_2 reduction, like CO, HCOOH, MeOH, and CH_4 are used as energy carriers for industrial applications [142, 143]. Fuel cells are very popular for energy storage and conversion technology as they convert fuel (e.g., hydrogen, natural gas, and methanol) into electricity. The MOFs, MOF composite and derivatives are used in fuel cells as an electrolyte and electrode catalyst. MOFs also used in oxygen reduction reaction, supercapacitor, LI, $Li-O_2$, Li-S batteries, electrolysis of water, etc. [132].

Boron nitride (BN) is used as fillers in pyroelectric devices due to low-cost and non-toxicity to improve the thermal transfer. Q. Wang et al. reported 65.6% increment in harvested power of BN filed PMN-PMS-PZT in comparison to unfiled PMN-PMS-PZT [84]. Because of its wide bandgap (6.2 eV), high chemical and thermal stability h-BN layers are used in solar cells as interfacial layers [144].

8.3   CONCLUSIONS AND FUTURE PROSPECTUS

The research on the development of various 2D materials for the energy harvesting device applications is on the high appreciation in the recent years due to their unique layered structure and optoelectronic properties. The various 2D materials such as graphene, graphene-related materials, transition metal dichalcogenides, MXenes, Xenes were synthesized and used as an active material in the various energy harvesting devices such as perovskite solar cells, thermoelectric, and piezoelectric device applications. In continuation, most of the 2D materials were not investigated due to the difficulty in the material synthesis. The various parameters for the synthesis methods for the 2D materials required to be optimized. It was found that the thermal conductivity of the 2D materials possess inherent thermal conductivity, which is a limitation to the performance of the thermoelectric device. The research needs to be focused on the structural modification and to tune the thermal conductivity as the thin films of various 2D materials have huge application in the various mechanical and optoelectronic systems. The deposition of uniform/layered thin films with 2D material is still challenging. The synthesis of 2D materials with better yield using the facile synthesis routes is the current active challenge among the scientific community. The several key phenomenon changes in the 2D materials due to the application of the strain, temperature gradient, and the photoinduced changes are to be investigated. Besides, the influence of the external parameters on the band structure and the bandgap tunability need to be investigated.

REFERENCES

1. Chawla, Rashmi, Poonam Singhal and Amit Kumar Garg. Photovoltaic review of all generations: Environmental impact and its market potential. *Trans Electr Electron Mater*, 1–21 (2020).

2. Mariotti, Nicole, Matteo Bonomo, Lucia Fagiolari, Nadia Barbero, Claudio Gerbaldi, Federico Bella and Claudia Barolo. Recent advances in eco-friendly and cost-effective materials towards sustainable dye-sensitized solar cells. *Green Chem*, 22(21), 7168–7218 (2020).

3. Burra, G.K., Ghosh, D.S. and Tiwari, S. X-ray and raman study of CH3NH3PbI3 perovskite nanocrystals. *J Mater NanoSci*, 8(1), 16–22 (2021).

4. Nair, S., Patel, S.B. and Gohel, J.V. Recent trends in efficiency-stability improvement in perovskite solar cells. *Mater Today Energy*, 17, 100449 (2020).

5. Khan, K., Tareen, A.K., Aslam, M., Wang, R., Zhang, Y., Mahmood, A., Ouyang, Z., Zhang, H. and Guo, Z. Recent developments in emerging two-dimensional materials and their applications. *J Mater Chem C*, 8(2), 387–440 (2020).

6. Ricciardulli, A.G., Yang, S., Smet, J.H. and Saliba, M. Emerging perovskite monolayers. *Nat Mater*, 1–12 (2021).

7. Bhunia, S., Deo, K.A. and Gaharwar, A.K. 2D covalent organic frameworks for biomedical applications. *Adv Funct Mater*, 30(27), 2002046 (2020).

8. Tuček, J., Błoński, P., Ugolotti, J., Swain, A.K., Enoki, T. and Zbořil, R. Emerging chemical strategies for imprinting magnetism in graphene and related 2D materials for spintronic and biomedical applications. *Chem Soc Rev*, 47(11), 3899–3990 (2018).

9. Kang, J., Huang, S., Jiang, K., Lu, C., Chen, Z., Zhu, J., Yang, C., Ciesielski, A., Qiu, F. and Zhuang, X. 2D Porous Polymers with sp2-carbon connections and sole sp2-carbon skeletons. *Adv Funct Mater*, 30(27), 2000857 (2020).

10. Upadhyay, S.N., Satrughna, J.A.K. and Pakhira, S. Recent advancements of two-dimensional transition metal dichalcogenides and their applications in electrocatalysis and energy storage. *Emergent Mater*, 1–20 (2021).

11. Kumar, P., Boukherroub, R. and Shankar, K. Sunlight-driven water-splitting using two-dimensional carbon based semiconductors. *J Mater Chem A*, 6(27), 12876–12931 (2018).

12. Wines, D., Ersan, F. and Ataca, C. Engineering the electronic, thermoelectric, and excitonic properties of two-dimensional group-III nitrides through alloying for optoelectronic devices (B1—x Al x N, Al1—x Ga x N, and Ga1—x In x N). *ACS Appl Mater Interfaces*, 12(41), 46416–46428 (2020).

13. Macha, M., Marion, S., Nandigana, V.V. and Radenovic, A. 2D materials as an emerging platform for nanopore-based power generation. *Nat Rev Mater*, 4(9), 588–605 (2019).

14. Zhang, H., Cheng, H.M. and Ye, P. 2D nanomaterials: Beyond graphene and transition metal dichalcogenides. *Chem Soc Rev*, 47(16), 6009–6012 (2018).

15. Ponomarenko, V.P., Popov, V.S., Popov, S.V. and Chepurnov, E.L. Photo-and nanoelectronics based on two-dimensional materials. Part I. Two-dimensional materials: Properties and synthesis. *J Commun Technol Electron*, 65(9), 1062–1104 (2020).

16. Chen, E., Xu, W., Chen, J. and Warner, J.H. 2D layered noble metal dichalcogenides (Pt, Pd, Se, S) for electronics and energy applications. *Mater Today Adv*, 7, 100076 (2020).

17. Hemanth, N.R., Kim, T., Kim, B., Jadhav, A.H., Lee, K. and Chaudhari, N.K. Transition metal dichalcogenide-decorated MXenes: Promising hybrid electrodes for energy storage and conversion applications. *Mater Chem Front*, 5(8), 3298–3321 (2021).

18. Samadi, M., Sarikhani, N., Zirak, M., Zhang, H., Zhang, H.L. and Moshfegh, A.Z. Group 6 transition metal dichalcogenide nanomaterials: Synthesis, applications and future perspectives. *Nanoscale Horiz*, 3(2), 90–204 (2018).

19. James Singh, K., Ahmed, T., Gautam, P., Sadhu, A.S., Lien, D.H., Chen, S.C., Chueh, Y.L. and Kuo, H.C. Recent advances in two-dimensional quantum dots and their applications. *Nanomaterials*, 11(6), 1549 (2021).

20. Chen, W., Hou, X., Shi, X. and Pan, H.,. Two-dimensional Janus transition metal oxides and chalcogenides: Multifunctional properties for photocatalysts, electronics, and energy conversion. *ACS Appl Mater Interfaces*, 10(41), 35289–35295 (2018).

21. Fu, Y., Zhu, H., Chen, J., Hautzinger, M.P., Zhu, X.Y. and Jin, S.,. Metal halide perovskite nanostructures for optoelectronic applications and the study of physical properties. *Nat Rev Mater*, 4(3), 169–188 (2019).

22. Zhou, C., Lin, H., Lee, S., Chaaban, M. and Ma, B.,. Organic-inorganic metal halide hybrids beyond perovskites. *Mater Res Lett*, 6(10), 552–569 (2018).

23. Chen, Y., Lao, Z., Sun, B., Feng, X., Redfern, S.A., Liu, H., Lv, J., Wang, H. and Chen, Z.,. Identifying the ground-state NP sheet through a global structure search in two-dimensional space and its promising high-efficiency photovoltaic properties. *ACS Mater Lett*, 1(3), 375–382 (2019).

24. Kumar, M., Raj, A., Kumar, A. and Anshul, A.,. An optimized lead-free formamidinium Sn-based perovskite solar cell design for high power conversion efficiency by SCAPS simulation. *Opt Mater*, 108, 110213 (2020).

25. Tang, K.W., Li, S., Weeden, S., Song, Z., McClintock, L., Xiao, R., Senger, R.T. and Yu, D.,. Transport modeling of locally photogenerated excitons in halide perovskites. *J Physic Chem Lett*, 12(16), 3951–3959 (2021).

26. Xia, C.Q., Peng, J., Poncé, S., Patel, J.B., Wright, A.D., Crothers, T.W., Uller Rothmann, M., Borchert, J., Milot, R.L., Kraus, H. and Lin, Q.,. Limits to electrical mobility in lead-halide perovskite semiconductors. *J Phys Chem Lett*, 12(14), 3607–3617 (2021).

27. Shao, S. and Loi, M.A.,. The role of the interfaces in perovskite solar cells. *Adv Mater Interfaces*, 7(1), 1901469 (2020).

28. Roghabadi, F.A., Ahmadi, N., Ahmadi, V., Di Carlo, A., Aghmiuni, K.O., Tehrani, A.S., Ghoreishi, F.S., Payandeh, M. and Fumani, N.M.R.,. Bulk heterojunction polymer solar cell and perovskite solar cell: Concepts, materials, current status, and opto-electronic properties. *Solar Energy*, 173, 407–424 (2018).

29. Ansari, M.I.H., Qurashi, A. and Nazeeruddin, M.K.,. Frontiers, opportunities, and challenges in perovskite solar cells: A critical review. *J Photochem Photobiol C: Photochem Rev*, 35, 1–24 (2018).

30. Dagar, J., Castro-Hermosa, S., Lucarelli, G., Cacialli, F. and Brown, T.M.,. Highly efficient perovskite solar cells for light harvesting under indoor illumination via solution processed SnO2/MgO composite electron transport layers. *Nano Energy*, 49, 290–299 (2018).

31. Bang, S.M., Shin, S.S., Jeon, N.J., Kim, Y.Y., Kim, G., Yang, T.Y. and Seo, J.,. Defect-tolerant sodium-based dopant in charge transport layers for highly efficient and stable perovskite solar cells. *ACS Energy Letters*, 5(4), 1198–1205 (2020).

32. Yokoyama, T., Nishitani, Y., Miyamoto, Y., Kusumoto, S., Uchida, R., Matsui, T., Kawano, K., Sekiguchi, T. and Kaneko, Y. Improving the open-circuit voltage of Sn-based perovskite solar cells by band alignment at the electron transport layer/perovskite layer interface. *ACS Appl Mater Interfaces*, 12(24), 27131–27139 (2020).

33. Maio, A., Pibiri, I., Morreale, M., Mantia, F.P.L. and Scaffaro, R. An overview of functionalized graphene nanomaterials for advanced applications. *Nanomater*, 11(7), 1717 (2021).

34. Shao, S. and Loi, M.A.,. The role of the interfaces in perovskite solar cells. *Adv Mater Interfaces*, 7(1), 1901469 (2020).

35. Bolotsky, A., Butler, D., Dong, C., Gerace, K., Glavin, N.R., Muratore, C., Robinson, J.A. and Ebrahimi, A. Two-dimensional materials in biosensing and healthcare: From in vitro diagnostics to optogenetics and beyond. *ACS Nano*, 13(9), 9781–9810 (2019).

36. Kumar, K.S., Choudhary, N., Jung, Y. and Thomas, J.,. Recent advances in two-dimensional nanomaterials for supercapacitor electrode applications. *ACS Energy Lett*, 3(2), 482–495 (2018).

37. Zhang, Z., Lin, P., Liao, Q., Kang, Z., Si, H. and Zhang, Y.,. Graphene-Based Mixed-Dimensional van der Waals Heterostructures for Advanced Optoelectronics. *Adv Mater*, 31(37), 1806411 (2019).

38. Saranin, D., Pescetelli, S., Pazniak, A., Rossi, D., Liedl, A., Yakusheva, A., Luchnikov, L., Podgorny, D., Gostischev, P., Didenko, S. and Tameev, A. Transition metal carbides (MXenes) for efficient NiO-based inverted perovskite solar cells. *Nano Energy*, 82, 105771 (2021).

39. Mariani, P., Najafi, L., Bianca, G., Zappia, M.I., Gabatel, L., Agresti, A., Pescetelli, S., Di Carlo, A., Bellani, S. and Bonaccorso, F. Low-temperature graphene-based paste for large-area carbon perovskite solar cells. *ACS Appl Mater Interfaces*, 13, 22368–22380 (2021).

40. Mahmoudi, T., Wang, Y. and Hahn, Y.B.,. Highly stable perovskite solar cells based on perovskite/NiO-graphene composites and NiO interface with 25.9 mA/cm2 photocurrent density and 20.8% efficiency. *Nano Energy*, 79, 105452 (2021).

41. Chen, H., Luo, Q., Liu, T., Tai, M., Lin, J., Murugadoss, V., Lin, H., Wang, J., Guo, Z. and Wang, N.,. Boosting multiple interfaces by co-doped graphene quantum dots for high efficiency and durability perovskite solar cells. *ACS Appl Mater Interfaces*, 12(12), 13941–13949 (2020).

42. Ahmed, D.S., Mohammed, M.K. and Majeed, S.M.,. Green synthesis of eco-friendly graphene quantum dots for highly efficient perovskite solar cells. *ACS Appl Ener Mater*, 3(11), 10863–10871 (2020).

43. Chandrasekhar, P.S., Dubey, A. and Qiao, Q.,. High efficiency perovskite solar cells using nitrogen-doped graphene/ZnO nanorod composite as an electron transport layer. *Solar Energy*, 197, 78–83 (2020).

44. Guo, M., Wei, C., Liu, C., Zhang, K., Su, H., Xie, K., Zhai, P., Zhang, J. and Liu, L.,. Composite electrode based on single-atom Ni doped graphene for planar carbon-based perovskite solar cells. *Mater Design*, 209, 109972 (2021).

45. Abd Malek, N.A., Alias, N., Saad, S.K.M., Abdullah, N.A., Zhang, X., Li, X., Shi, Z., Rosli, M.M., Abd Aziz, T.H.T., Umar, A.A. and Zhan, Y. Ultra-thin MoS$_2$ nanosheet for electron transport layer of perovskite solar cells. *Opt Mater*, 104, 109933 (2020).

46. Mahmood, K., Khalid, A., Ahmad, S.W., Qutab, H.G., Hameed, M. and Sharif, R., 2020. Electrospray deposited MoS$_2$ nanosheets as an electron transporting material for high efficiency and stable perovskite solar cells. *Solar Energy*, 203, 32–36.

47. Chen, L.C., Tseng, Z.L., Chen, C.C., Chang, S.H. and Ho, C.H.,. Fabrication and characteristics of CH3NH3PbI3 perovskite solar cells with molybdenum-selenide hole-transport layer. *Appl Phy Express*, 9(12), 122301 (2016).

48. Chang, L.B., Tseng, C.C., Lee, J.H., Wu, G.M., Jeng, M.J., Feng, W.S., Chen, D.W., Chen, L.C., Lee, K.L., Popko, E. and Jacak, L.,. Preparation and characterization of MoSe2/CH3NH3PbI3/PMMA perovskite solar cells using polyethylene glycol solution. *Vacuum*, 178, 109441 (2020).

49. Snyder, G.J. and Toberer, E.S. Complex thermoelectric materials, *Nat Mater*, 7(2), 105–114 (2008).

50. Rull-Bravo, M., Moure, A., Fernandez, J.F. and Martin-Gonzalez, M. Skutterudites as thermoelectric materials: Revisited. *RSC Adv*, 5(52), 41653–41667 (2015).

51. Sundarraj, P., Maity, D., Roy, S.S. and Taylor, R.A. Recent advances in thermoelectric materials and solar thermoelectric generators- a critical review. *RSC Adv*, 4(87), 46860–46874 (2014).

52. Tan, G., Zhao L. -D. and Kanatzidis, M. G. Rationally designing high- performance bulk thermoelectric materials. *Chem Rev*, 116,19, 12123–12149 (2016).

53. Kraemer, D., Jie, Q., McEnaney, K., Cao F. Liu, W. and Weinstein, L. Concentration solar thermoelectric generator with peak efficiency of 7.4%. *Nat Energy* (2016).

54. Vineis, C. J., Shakouri, A., Marjumdar A. and Kanatzidis, M. G. Nanostructured thermoelectric: Big efficiency gains from small features. *Adv Mater*, 22, 3970 (2010).

55. Kanatzidis, M. G. Nanostructured thermoelectric: The new paradigm? *Chem Mater*, 22, 648 (2010).

56. Minnich, A. J., Dresselhaus, MS., Ren, Z. F. and Chen, G. Bulk nanostructured thermoelectric materials: Current research and future prospects. *Energy Environ Sci*, 2, 466 (2009).

57. Terasaki, I., Sasago, Y. and Uchinokura, K. Large thermoelectric power in $NaCo_2O_4$ single crystals. *Phys Rev* B56, R12685–12687 (1997).

58. Fujita, K. Mochida, T. and Nakamura, K. High- temperature thermoelectric properties of $NaCo_2O_4$, single crystals. *Jpn J Appl Phys Pt, 1*, 40, 4644–4647 (2001).

59. Ohtaki, M. Nojiri, Y. and Maeda, E. *Improved thermoelectric performance of sintered $NaCo_2O_4$ with enhanced 2-dimensional microstructure*. Proc. 19th International Conference on Thermoelectrics (ICT2000), Wales, 190–195 (August 2000).

60. Tian, Z., Wang, J., Yaer, X., Jun Kang, H., Wang, X. H., Liu, H. M., Yang, D. Z. and Wang, T.M. Pencil painting like preparation for flexible thermoelectric material of high performance p-type $Na_{1.4}Co_2O_4$ and novel n-type $Na_xCo_2O_4$. *Journal of Materiomics*, 7, 1153–1160 (2021). https://doi.org/10.1016/j.jmat.2021.01.006

61. Funahashi, R., Matsubara, I., Ikuta, H., Takeuchi, T., Mizutani, U. and Sodeoka, S. An oxide single crystal with high thermoelectric performance in air. *Jpn J Appl Phys Pt, 2*, 39, L1127 (2000).

62. Ardyanian, M., Moeini, M. and AzimiJuybari, H. Thermoelectric and photoconductivity properties of zinc oxide–tin oxide binary systems prepared by spray pyrolysis. Thin solid film, 552, 39–45 (2014).

63. Yabuta, H., Kaji, N., Hayashi, R., Kumomi, H., Nomura, K., Kamiya, T. and Hiraono, M. Hideo hosono. *Appl Phys Lett*, 97, 072111 (2010).

64. Ohtaki, M., Tsubota, T., Eguchi, T., Arai, H. High-temperature thermoelectric properties of (Zn1– x Al x) O. *J Appl Phys*, 79, 1816 (1996).

65. Chena, Q., Maa, H., Lia, X., Wanga Y., Liua B., Wanga C., Jia G., Wanga J., Changa L., Zhangb Y., Jiaa, X., Synergistic Optimization of the Thermoelectric Properties of Zn0.98Al0.02O using High-Pressure and High-Temperature treatment, *J. Alloys Compd*, 156124 (2020)

66. Pham, A. T. T., Vo, P. T. N., Ta, H. K. T., Lai, H. T., Tran, V. C., Doan, T. L. H., Duong, A. T., Lee, C. T., Nair, P. K., Zulueta, Y. A., Phan, T. B. and Luu, S. D. N. Improved power factor achieved by energy filtering in ZnO:Mg/ZnO hetero-structures, *Thin Solid Films*, 721, 1385537 (2021).

67. Ohta, H., Kim, S. W., Mune, Y., Mizoguchi, T., Nomura, K., Ohta, S., Nomura, T., Yakanishi, Ikuhara, Y., Hirano, M., Hosono, H. and Koumoto, K. Giant thermoelectric Seebeck coefficient of a two dimensional electron gas in $SrTiO_3$. *Nat Mater*, 6, 129 (2007).

68. Mune, Y., Ohta, H., Koumoto, K., Mizoguchi, T. and Ikuhara, Y. Enhanced Seebeck coefficient of quantum-confined electron in $SrTiO_3$/ $SrTi_{0.8}Nb_{0.2}O_3$ superlattice. *Appl Phys Lett*, 91, 192105 (2007).

69. Lee, K. H., Mune, Y., Ohta, H. and Koumoto, K. Thermal stability of giant thermoelectric Seebeck coefficient for $SrTiO_3$/ $SrTi_{0.8}Nb_{0.2}O_3$ superlattice at 900K. *Appl Phys Express*, 1, 015007 (2008).

70. Li, S. Funahashi, R., Matsubara, I., Ueno, K. and Yamada, H. High temperature thermoelectric properties of oxide $Ca_9Co_{12}O_{28}$. *J Mater Chem*, 9, 1659–1660 (1999).

71. Shikanao, M. and Funahashi, R. Electrical and thermal properties of single-crystalline $(Ca_2CoO_3)_{0.7}$ CoO_2 with a $Ca_3Co_4O_9$ structure. *Appl Phys Lett*, 82, 1851–1853 (2003).

72. Xu, G., Funahashi, R., Shikano, M., Matsubara, I. and Zhou, Y. Thermoelectric properties of the Bi-and Na-substituted $Ca_3Co_4O_9$ system. *Appl Phys Lett*, 80, 3760–3762 (2002).

73. Lan, J., Lin, Y. H., Li, G. J., Xu, S., Liu, Y., Wen Nan, Ce. and Zhao, S. J. High-temperature electrical transport behaviors of the layered $Ca_2C_2O_5$-based ceramics. *Appl Phys Lett*, 192104(1–3), (2010).

74. Lin, Y. H., Lan, J., Shen, Z., Liu, Y., Nan, C. W. and Li, J. F. High-temperature electrical transport behaviors in textured $Ca_3Co_4O_9$-based polycrystalline ceramics. *Appl Phys Lett*, 94, 072107 (2009).

75. Nong, N. V., Pryds, N., Linderoth, S. and Ohtaki, M. Thermoelectric Performance of p-Type Layered Oxide $Ca_3Co_4O_9$ delta Through Heavy Doping and Metallic Nanoinclusions. *Adv Mater*, 23, 2484–2490 (2011).

76. Wang, Y., Sui, Y., Li, F. and Liu, X. Thermoelectrics in misfit-layered oxides [(Ca, Ln)$_2CoO_3]_{0.62}$ [CoO_2]: From bulk to nano. *Nano Energy*, 1(3), 456–465 (2012).

77. Wu, N., Holgate, T. C., Nong, N. V., Pryds, N. and Linderoth, S. Effects of synthesis and spark plasma sintering conditions on the thermoelectric properties of $Ca_3Co_4O_{9+\delta}$. *J Electronic Materials*, 42(7), 2134–2142 (2013).

78. Tian, R., Donelson, R., Ling, C. D., Blanchard, P. E. R., Zhang, T., Chu, D., Tan, T. T. and Li, S. Ga substitution and oxygen diffusion kinetics in $Ca_3Co_4O_{9+\delta-}$ based thermoelectric oxides. *J Phys Chem C*, 117(26), 13382–13387 (2013).

79. Butt, S., Liu, Y. C., Lan, J. L., Shehzad, K., Zhan, B., Lin, Y. and Nan, C. W. High-temperature thermoelectric properties of La and Fe co-doped Ca–Co–O misfit-layered cobaltites consolidated by spark plasma sintering. *Journal of Alloys and Compounds*, 588, 277–283 (2014).

80. Butt, S., Ren, Y., Farooq, M. U., Zhan, B., Sagar, R. U. R., Lin, Y. and Nan, C. W. Enhanced thermoelectric performance of heavy-metals (M: Ba, Pb) doped misfit-layered ceramics: $(Ca_{2-}xMxCoO_3)_{0.62}$ (CoO_2). *Energy Conversion and Management*, 83, 35–41 (2014).

81. Lan, Y., Minnich, A. J., Chen, G. and Ren, Z. Enhancement of thermoelectric figure-of-merit by a bulk nanostructuring approach. *Adv Funct Mater*, 20, 357–376 (2010).

82. Jouhara, H., Zabnienska-Gora, A., Khordehgah, N., Doraghi, Q., Ahmad, L., Norman, L., Axcell, B., Wrobel, L. and Dai, S. Thermoelectric generator (TEG) technologies and applications. *Inernational Journal of Thermofluids*, 9, 100063 (2021).

83. Terry Hendricks, Choate, *Engineering scoping study of thermoelectric generator system for industrial waste heat recovery*. U.S. Department of Energy Office of Scientific and Technical Information, 2006.

84. Romanjek, K., Vesin, S., Aixala, L., Baffie, T., Bernard-Granger, G. and Dufourcq, J. High performance silcon germanium-based thermoelectric modules for gas exhaust energy scavenging. *J Electron Mater*, 44, 2192–2202 (2015).

85. Zoui, M. A., Bentouba, S., Stocholm, J.G. and Bourouis, M. A Review on thermoelectric generator: Progress and application. *Energies*, 13, 3606 (2020).

86. LaGraneur, J., Crane, D., Hung, S., Mazar, B., and Eder, A. *Automotive waste heat conversion to electric power using skutterudite*. TAGS, PbTe and BiTe, 25th International Conference on Thermoelectrics, 343 (2006).

87. Wilbrecht, S. and Beitelschmidt, M. The potential of a cascaded TEG system for waste heat usage in railway vehicles. *J Electron Mater*, 47, 3358–3369 (2018).

88. Chen, Z.-G., Han, G., Yang, L., Cheng, L., L. and Zou, L. Nanostructured thermoelectric materials: Current research and future challenge. *Prog Nat Sci*, 22(6), 535–549 (2012).

89. Huang, H. and Fan, X. Research advances of typical two dimensional layered thermo-electric materials. *Res Appl Mater Sci*, 2, 1–11 (2020).

90. Kumar, S. and Schwingenschlogl, U. Thermoelectric response of bulk and monolayer MoSe2 and WSe2. *Chem Mater*, 27, 1278–1284 (2015).

91. Jin, Z., Liao, Q., Fang, H., Liu, Z., Liu, W., Ding, Z., Luo, T. and Yang, N. A revisit to high thermoelectric performance of single-layer MoS 2. *Sci Rep*, 5, 18342 (2015).

92. Sahoo, S., Gaur, A. P., Ahmadi, M., Guinel, M. J.-F. and Katiyar, R. S. Temperature-dependent Raman studies and thermal conductivity of few-layer MoS_2. *J Phys Chem C*, 117, 9042–9047 (2013).

93. Hellman, O. and Broido, D. A. Phonon thermal transport in Bi2Te3 from first principles. *Phys Rev B*, 90, 134309 (2014).

94. Qiu, B., Bao, H., Ruan, X., Zhang, G. and Wu, Y. In molecular dynamics simulations of lattice thermal conductivity and spectral phonon mean free path of PbTe: Bulk and nanostructures, Comput. *Mater Sci*, 278–285 (2012).

95. Hippalgaonkar, K., Wang, Y., Ye, Y., Qiu, D. Y., Zhu, H., Wang, Y., Moore, J., Louie, S. G. and Zhang, X. High thermoelectric power factor in two-dimensional crystals of MoS_2. *Phys Rev B*, 95, 115407 (2017).

96. Wickramaratne, D., Zahid, F. and Lake, R. K. Electronic and thermoelectric properties of few-layer transition metal dichalcogenides. *J Chem Phys* 140, 124710 (2014).

97. Huang, W., Luo, X., Gan, C. K., Quek, S. Y. and Liang, G. Theoretical study of thermoelectric properties of few-layer MoS_2 and WSe2. *Phys Chem Chem Phys*, 16, 10866–10874 (2014).

98. Park, D., Kim, M. and Kim, J. Conductive PEDOT: PSS-based organic/inorganic flexible thermoelectric films and power generators. *Polymers*, 13, 210 (2021).

99. X. Li, C. Liu, W. Zhou et al. Roles of polyethylenimineethoxylated in efficiently tuning the thermoelectric performance of poly(3,4-ethylenedioxythiophene)-rich nanocrystal films, *ACS Appl Mater Interfaces*, 11(8), 8138–8147 (2019).

100. Zhang, S., Fan, Z., Wang, X. and Zhang, Z. Enhancement of the thermoelectric properties of PEDOT: PSS via one-step treatment with cosolvents or their solutions of organic salts. *J Mater Chem A*, 6(16), 7080–7087 (2018).

101. Lee, S. H., Park, H., Kim, S., Son, W., Cheong, I. W. and Kim, J. H. Transparent and flexible organic semiconductor nanofilms with enhanced thermoelectric efficiency. *J Mater Chem A*, 2(20), 7288–7294 (2014).

102. Zhao J. and Tan, D. A strategy to improve the thermoelectric performance of conducting polymer nanostructures. *J Mater Chem C*, 5(1), 47–53 (2017).

103. Xie, J., Zhao, C., Lin, Z. and Gu, P. Nanostructured conjugated polymers for energy-related applications beyond solar cells. *Chem Asian J*, 11(10), 1489–1511 (2016).

104. Kim, N., Kee, S., Lee S. H. et al. Highly conductive PEDOT: PSS nanofibrils induced by solution-processed crystallization. *Adv Mater*, 26(14), 2268–2272 (2014).

105. Wang, Y., Yang, J., Wang, L., Du, K., Yin, Q. and Yin, Q. Polypyrrole/graphene/polyaniline ternary nanocomposite with high thermoelectric power factor. *ACS Appl Mater. Interfaces*, 9, 20124–20131 (2017).

106. Fei-Peng Du, Nan-Nan Cao, Yun-Fei Zhang, Ping Fu, Yan-Guang Wu, Zhi-Dong Lin, Run Shi, Abbas Amini and Chun Cheng. PEDOT: PSS/graphene quantum dots films with enhanced thermoelectric properties via strong interfacial interaction and phase separation, *Sci Rep*, 8(1–12), 6441 (2018).

107. Yijie Xia and Jianyong Ouyang. PEDOT: PSS films with significantly enhanced conductivities induced by preferential solvation with cosolvents and their application in polymer photovoltaic cells. *J Mater Chem*, 21, 4927–4936 (2011).

108. Kim, G-H., Shao, L., Zhang, K. and Pipe, K. P. Engineered doping of organic semiconductors for enhanced thermoelectric efficiency. *Nat Mater*, 12, 729–723 (2013).

109. Huang, X., Deng, L., Liu, F., Zhang Q. and Chen, G. Effect of crystalline microstructure evolution on thermoelectric performance of PEDOT: PSS films. *Energy Mater Adv*, 2021, 10 (2021).

110. Aftabuzzaman, M., Lu, C. and Kim, H.K. Recent progress on nanostructured carbon-based counter/back electrodes for high-performance dye-sensitized and perovskite solar cells. *Nanoscale*, *12*(34), 17590–17648 (2020).

111. Joseph, I., Louis, H., Unimuke, T.O., Etim, I.S., Orosun, M.M. and Odey, J. An overview of the operational principles, light harvesting and trapping technologies, and recent advances of the dye sensitized solar cells. *Appl Solar Energy*, 56(5), 334–363 (2020).

112. Bera, S., Sengupta, D., Roy, S. and Mukherjee, K. Research on dye sensitized solar cell: A review highlighting the progress in India. *J Phys Energy* 3, 032013 (2021).

113. Ling, C.K., Aung, M.M., Abdullah, L.C., Ngee, L.H. and Uyama, H. A short review of iodide salt usage and properties in dye sensitized solar cell application: Single vs binary salt system. *Solar Energy*, 206, 1033–1038 (2020).

114. Muchuweni, E., Martincigh, B.S. and Nyamori, V.O. Recent advances in graphene-based materials for dye-sensitized solar cell fabrication. *RSC Advances*, 10(72), 44453–44469 (2020).

115. Tapa, A.R., Xiang, W. and Zhao, X., Metal chalcogenides (M x E y; E= S, Se, and Te) as counter electrodes for dye—Sensitized solar cells: An overview and guidelines. *Adv Energy Sustainab Res*, 2100056.

116. Dey, S. Atomically controlled two-dimensional heterostructures: Synthesis, characterization and applications. In *Functional Properties of Advanced Engineering Materials and Biomolecules* (pp. 201–235). Springer, Cham, 2021.

117. Fatima, N., Tahir, M.B., Noor, A., Sagir, M., Tahir, M.S., Alrobei, H., Fatima, U., Shahzad, K., Ali, A.M. and Muhammad, S. Influence of van der waals heterostructures of 2D materials on catalytic performance of ZnO and its applications in energy: A review. *I J Hydr Energy* 46, 25413–25423 (2021).

118. Goel, S. and Kumar, B. A review on piezo-/ferro-electric properties of morphologically diverse ZnO nanostructures. *J Alloys Compd*, 816, 152491 (2020).

119. Liang, Q., Zhang, Q., Zhao, X., Liu, M. and Wee, A.T. Defect engineering of two-dimensional transition-metal dichalcogenides: Applications, challenges, and opportunities. *ACS Nano*, 15(2), 2165–2181 (2021).

120. Miao, J. and Wang, C., Avalanche photodetectors based on two-dimensional layered materials. *Nano Res*, 14(6), 1878–1888 (2021).

121. Park, D., Kim M. and Kim, J. Conductive PEDOT: PSS-based organic/inorganic flexible thermoelectric films and power generators. *Polymers*, 13,210 (2021).

122. Yang, P-K. and Lee, C.-P. 2D-layered nanomaterials for energy harvesting and sensing applications, monolayers of emerging nanomaterials. *Intech Open*, 1–14 (2019). http://dx.doi.org/10.5772/intechopen.85791

123. Han, S.A., Sohn, A. and Kim, S-W. Recent advanced in energy harvesting and storage applications with two-dimensional layered materials. *FlatChem*, 6, 37–47 (2017).

124. Bolotin, K. I., Sikes, K. J., Jiang, Z., Klima, M., Fudenberg, G., Hone, J., Kim P. and Stormer, H. L. Ultrahigh electron mobility in suspended graphene. *Solid State Commun*, 146, 351–355 (2008).

125. Balandin, A. A., Ghosh, S., Bao, W., Calizo, I., Teweldebrhan, D., Miao, F. and Lau, C. N. Superior thermal conductivity of single-layer graphene. *Nano Lett*, 8, 902–907 (2008).

126. Han, S., Wu, D. Q., Li, S., Zhang, F. and Feng. X. L. Graphene: a two-dimensional platform for lithium storage. *Small*, 9 (2013) 1173–1187.

127. Ding, S., Zhang, D., Chen, J. S. and Lou, X. W. Facile synthesis of hierarchical MoS_2 microspheres composed of few-layered nanosheets and their lithium storage properties. *Nanoscale*, 4 (2012) 95–98.

128. Zhou, W., Zhou, J., Shen, J., Ouyang, C. and Shi. S. First-principles study of high-capacity hydrogen storage on graphene with Li atoms. *J Phys Chem Solids*, 73 (2012) 245–251.

129. Huang, K.-J., Zhang, J.-Z., Shi, G.-W. and Liu, Y.-M. Hydrothermal synthesis of molybdenum disulfide nanosheets as supercapacitors electrode material. *ElectrochimicaActa*, 132 (2014) 397–403.

130. Gao, T., Gong, L., Wang, Z., Yang, Z., Pan, W., He, L., Zhang, J., Ou, E., Xiong, Y. and Xu, W. Boron nitride/reduced graphene oxide nanocomposites as supercapacitors electrodes. *Mater Lett*, 159 (2015) 54–57.

131. Yana, J., Fana, Z., Weia, T., Qianb, W., Zhanga, M. and Weib, F. Fast and reversible surface redox reaction of graphene–MnO$_2$ composites as supercapacitor. *Carbon*, 48 (2010) 3825–3833.

132. Wang, H., Zhu, Q.-L., Zou, R. and Xu, Q. Metal-organic frameworks for energy applications. *Chem*, 2, 50–80 (2017).

133. Zhang, Z., Chen, Y., Xu, X., Zhang, J., Xiang, G., He, G. and Wang, X. Well defined metal-organic framework hollow nanocages. *Angew Chem Int Ed Engl*, 53, 429–433 (2014).

134. Lu, G., Li, S., Guo, Z., Farha, O.K., Hauser, B.G., Qi, X., Wang, Y., Wang, X., Han, S., Liu, X. et al. Imparting functionality to a metal-organic framework material by controlled nanoparticle encapsulation. *Nat Chem*, 4, 310–316 (2012).

135. Zhu, Q.-L. and Xu, Q. Metal-organic framework composites. *Chem Soc Rev*, 43, 5468–5512 (2014).

136. Sun, J.-K. and Xu, Q. Functional materials derived from open framework templates/precursors: Synthesis and applications. *Energy Environ Sci*, 7, 2071–2100 (2014).

137. Rosi, N.L., Eckert, J., Eddaoudi, M., Vodak, D.T., Kim, J., O'Keeffe, M. and Yaghi, O.M. Hydrogen storage in microporous metal-organic frameworks. *Science*, 300, 1127–1129 (2003).

138. Schoedel, A., Ji, Z. and Yaghi, O.M. The role of metal-organic frameworks in a carbon-neutral energy cycle. *Nat Energy*, 1, 16034–16046 (2016).

139. Furukawa, H., Ko, N., Go, Y.B., N. Aratani, S.B. Choi, E. Choi, A.O. Yazaydin, R.Q. Snurr, M. O'Keeffe, J. Kim and O.M. Yaghi. Ultrahigh porosity in metal-organic frameworks. *Science*, 329, 424–428 (2010).

140. Li, B., Wen, H.-M., Wang, H., Wu, H., Tyagi, M., Yildirim, T., Zhou, W. and Chen, B. A porous metal-organic framework with dynamic pyrimidine groups exhibiting record high methane storage working capacity. *J Am Chem Soc*, 136, 6207–6210 (2014).

141. Dhakshinamoorthy, A., Asiri, A.M. and Garc, H. Metal-organic framework (MOF) compounds: Photocatalysts for redox reactions and solar fuel production. *Angew Chem Int Ed Engl*, 55, 5414–5445 (2016).

142. Kornienko, N., Zhao, Y., Kley, C.S., Zhu, C., Kim, D., Lin, S., Chang, C.J., Yaghi, O.M. and Yang, P. Metal-organic frameworks for electrocatalytic reduction of carbon dioxide. *J Am Chem Soc*, 137, 14129–14135 (2015).

143. Wanga, Q., Bowenc, C. R., Lewisc, R., Chend, J., Leia, W., Zhange, H., Lia, M-Y. and Jianga, S. Hexagonal boron nitride nanosheets doped pyroelectric ceramic composite for high-performance thermal energy harvesting, *Nano Energy*, 60, 144–152, 2019.

144. Miyata, N., Moriki, K., Mishima, O., Fujisawa, M. and Hattori, T. Optical constants of cubic boron nitride. *Phys Rev B*, 40 12028–12029 (1989).

9 2D Nanomaterials for Energy Applications

Sayli Pradhan and Neetu Jha

CONTENTS

9.1 INTRODUCTION

9.1.1 GRAPHENE

Graphene is a one-atom-thick two-dimensional layered material with sp^2- bonded carbon. It has become one of the most important carbon materials in the multiple areas like materials science and condensed matter physics.[1,2] Motivated by its high theoretical specific surface area ($2,620$ m^2 g^{-1}), high mechanical flexibility, ultrathinness, good electrical conductivity, and high theoretical capacitance (e.g., 550 F g^{-1}), much attentions have been paid to the application of graphene for energy storage devices,[3] in particular, high-energy batteries and high-power supercapacitors (SCs).[4–9] At present, various methods have been developed to effectively synthesize graphene via top-down or bottom-up strategies. In general, top-down strategies are mainly based on the exfoliation

DOI: 10.1201/9781003247890-9

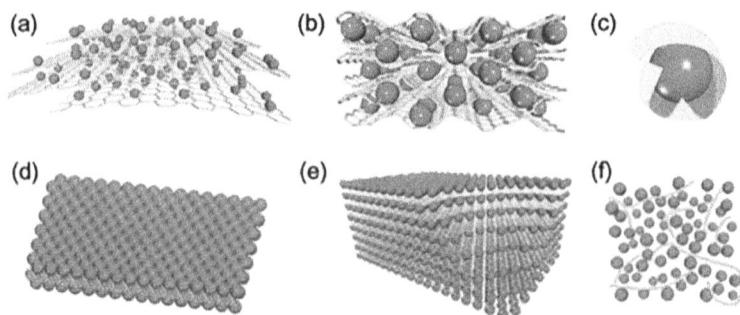

FIGURE 9.1   (Color online) Schematic of structural models of graphene/metal oxide hybrids. (a) Anchored model: oxide NPs anchored on graphene. (b) Wrapped model: oxide wrapped by graphene. (c) Encapsulated model: oxide encapsulated by graphene. (d) Sandwich-like model: an oxide/graphene/oxide sandwich-like structure. (e) Layered model: an alternating layered structure of oxide and graphene. (f) Mixed model: graphene and oxide mechanically mixed.

Source: Reprinted with permission, copyright 2012, Elsevier[5]

of graphite,[10] such as micromechanical cleavage, oxidation-exfoliation-reduction, intercalation exfoliation, and solid exfoliation (e.g., ball milling[11]). And bottom-up strategies include organic synthesis from small structurally-defined molecules (e.g., polyphenylene[12]) and chemical vapor deposition (CVD).[13] Notably, the graphene products vary distinctly from different synthesis methods in terms of number of layers, sizes, functional groups, wrinkles, defects (in-plane holes or heteroatom doping), and shapes (e.g., graphene nanoribbons), which could be remarkably reflected by huge differences in apparent properties (e.g., SSA, apparent density) and device performances. For instance, mechanically exfoliated graphene sheets (GSs) possess perfect crystallinity without abundant defects. It exhibits low sheet resistance (R_s) of around 400 Ωsq^{-1} at room temperature,[14] much higher than that of reduced graphene oxide (rGO) sheets by wet chemical routes, in which many defects and oxygen containing functional groups are left in GS, resulting in low conductivities mainly ranging from 298 to 0.045 S m^{-1}.[15] Furthermore, the layers and defects of GSs could be precisely prepared by CVD method via adjusting growth parameter (e.g., temperature, time, catalyst); however, in the case of oxidation-exfoliation-reduction, the control of layers and defects seems to be impossible. Therefore, by using appropriately physical or chemical method, one can create desirable defects to adjust the band structure of graphene-based materials (e.g., porous graphene and doped graphene sheets) and functionalize graphene with organic/inorganic matters for performance enhancement.[16]

9.1.2   Hexagonal Boron Nitride

h-BN is a semiconducting material from group III-V compound consisting of layers of sp^2- hybridized alternating boron and nitrogen atoms in a honeycomb lattice, with the layers coupled by van der Waals forces. Isostructural to graphene, h-BN nanosheets (h-BNNS) are also known as "white graphene." Its layered structure, along with high

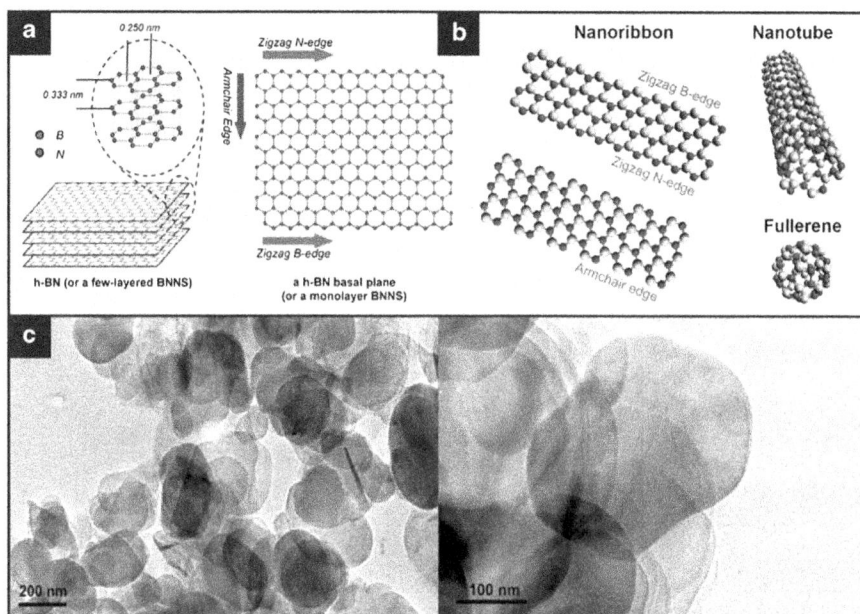

FIGURE 9.2   An overview of hexagonal boron nitride structures.

Source: Figures reproduced with permission[26]

chemical and thermal stability, allows the applications as lubricants and protective coatings. Pristine h-BN has a wide bandgap energy of 5.9 eV[17], enabling it to emit deep ultraviolet light at around 200 nm.[18] Unlike graphene, which is gapless and has rich electronic properties, the wide bandgap renders h-BN an insulating property and consequently hinders its applications in the fields of energy conversion and storage. There have been a lot of research activities focusing on the modification of h-BN properties so that it can be used in various applications. Heteroatom doping and functional group grafting have been proved effective. Depending on the incorporated dopants and functional groups (H, F, OH, CH_3, CHO, CN, NH_2, etc.), the bandgap of h-BN can be theoretically varied from 0.3 eV to 3.1 eV.[19] Experimentally doping h-BN with carbon can reduce its bandgap from 5.9 eV to 2.6 eV and therefore make it a promising nonmetal photocatalyst.[20] Carbon and oxygen co-doping can modify h-BN's optical, electronic, and surface properties, granting it interesting fluorescent, electrochemical energy storage and gas sorption capabilities.[21–23] Edge-hydroxylation, an effective functionalization method, which can be realized by the reaction between commercial h-BN and hot steam, gives the material enhanced catalytic oxidative dehydrogenation of alkane.[24,25]

9.1.3 Transition Metal Dichalcogenides (TMDCs)

TMD monolayers are thin 2D semiconductors in the form of MX_2, where M is a transition metal from group 4–10 (e.g., Mo or W) and X is a chalcogen (e.g., S, Se, or Te).[27]

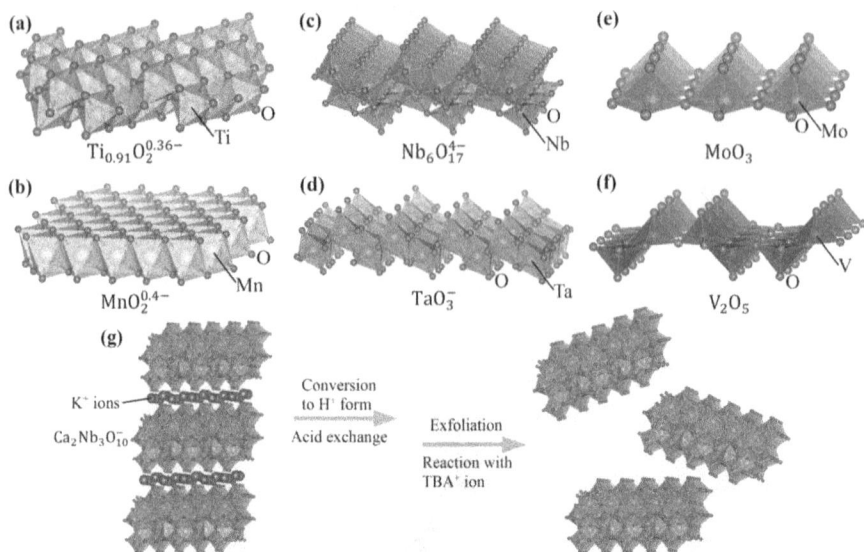

FIGURE 9.3 Representative structures of selected oxide nanosheets whose bulk counterparts are naturally layered. (a) $Ti_{0.91}O_2^{0.36-}$, (b) $MnO_2^{0.4-}$, (c) $Nb_6^{1}O_{17}^{4-}$, (d) TaO_3^-, (e) MoO_3, (f) V_2O_5. (g) Schematic illustration of the exfoliation of layered $KCa_2Nb_3O_{10}$ into $Ca_2Nb_3O_{10}^-$ nanosheets using the cation exchange-assisted liquid exfoliation method.

Source: Reproduced with permission, copyright 2014, Springer Nature[31]

One transition metal atomic layer is sandwiched between two layers of chalcogen atoms. Covalent bonds exist in individual layers and van der Waals forces dominate between layers. TMD monolayers can provide large exposed surface and abundant surface sites for electrostatic adsorption or surface redox reactions. Besides, some TMDs have multiple phases with diverse electrical properties, ranging from insulating, semiconducting, to metallic.[28] These unique properties have attracted interest over their use in capacitive energy storage applications. Among various TMD materials, MoS_2 is currently the most popular electrode materials for SCs due to its high specific capacitance (~1000 F g^{-1}).[29,30]

9.2.4 PHOSPHORENE

Phosphorene is a monolayer or several-layered black phosphorus (BP), similar to graphene, which can be synthesized from bulk BP using many methods.[32–34] BP is the most stable allotropy of phosphorus in ambient conditions. Under certain conditions, it can be mutually transformed from white phosphorus or red phosphorus. Bridgman accidentally obtained BP during an investigation of the effect of high pressure on the melting point of ordinary white phosphorus.[35] The crystal structure of BP in normal conditions consists of puckered layers parallel to the ac plane.[36,37] Some of the great advantages of BP are the direct bandgap tunability from 0.3 eV (in bulk) to 1.5 eV

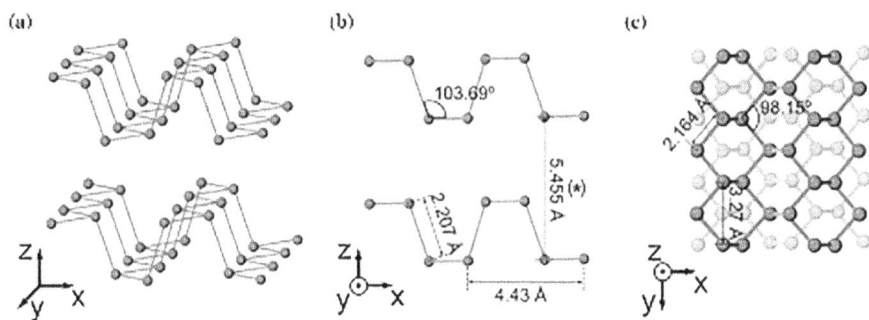

FIGURE 9.4 Black phosphorus structure. (a) 3D representation. (b) Lateral view. (c) Top view.

Source: Reproduced from with permission from 2D materials[32]

(in monolayer),[38–42] enhanced charge carrier mobility, and anisotropic in-plane properties. These properties make BP a special candidate for research associated with nanophotonic devices, incorporating 2D layered materials and moderate bandgap-based nano-electronic applications. Such a band structure fills a much needed gap for field-effect transistor (FET) applications. BP and phosphorene were reported to have several more advantageous properties than other 2D materials in application as a negative material for batteries. In particular, black phosphorus has a folded structure, and each P atom reacts with three Li or Na atoms to form Li_3P and Na_3P, giving it a very high theoretical specific capacity of 2596 mAh g^{-1},[43] which is far more than the graphite negative electrode (372 mAh g^{-1}) with its excellent energy storage properties. These outstanding performances have attracted many people to focus their attention on BP as a negative material for energy storage.[44]

9.2.5 MXENES

A large group of early transition metal carbides, nitride, or carbonitride (MXenes) have been identified as a new class of two-dimensional (2D) materials.[45] The general formula of MXene is $M_{n+1}X_nT_x$, where M stands for an early transition metal (Ti, Mo, Cr, Nb, V, Sc, Zr, Hf, or Ta), X represents carbon or nitrogen, n is usually an integral number between 1 and 3, and T_x is the surface termination, such as hydroxyl, oxygen, or fluorine. These surface terminations render hydrophilicity to MXenes and also have a significant influence on their Fermi level density of the states, thereby electronic properties.[46,47] MXenes are synthesized by selectively removing the "A" layer from their parent MAX phases, and "ene" was added at the last to show its 2D nature that is similar to graphene. Since its discovery in 2011,[48] MXenes have been particularly attractive as electrode materials for energy storage applications because of their unique structures, including (1) the inner conductive transition metal carbide layer that enables efficient electron transportation and (2) a transitional metal oxide-like surface that acts as active sites for fast redox reactions. Although there are more than 20 different MXenes that have been synthesized,[49–52] most research to date has

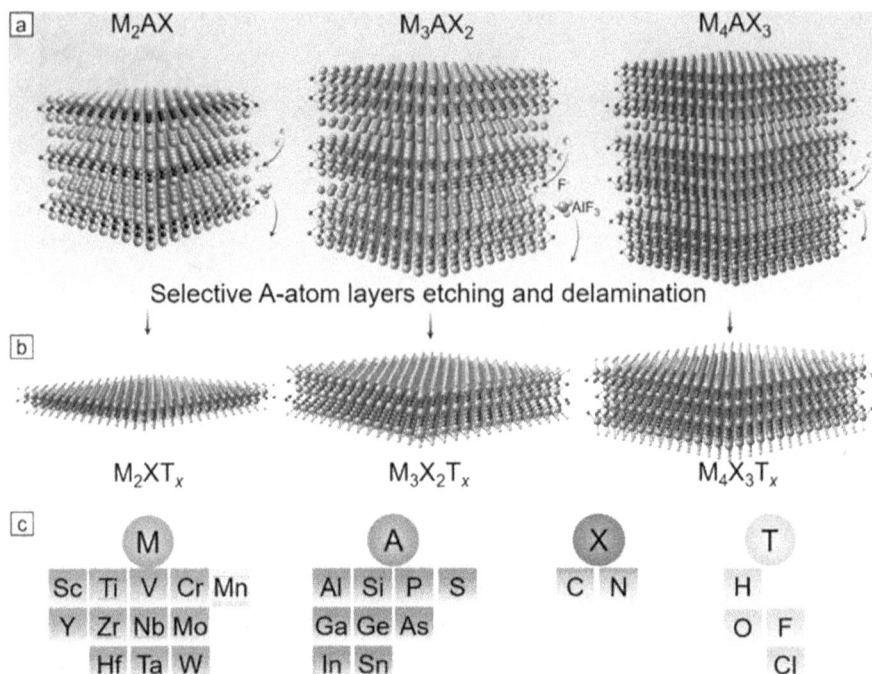

FIGURE 9.5 MXenes precursors and their synthesis. (a) Three types of mono-M MAX phases M_2AX, M_3AX_2, M_4AX_3 and the selective etching process of the A group layers (red atoms). (b) MXenes are formed after selective etching and formation of surface terminations (yellow atoms) labeled as T. (c) Possible elements for M, A, X, and T in MAX and MXene phases.

been focused on titanium carbide (Ti_3C_2) for MSC application due to its ultrahigh conductivity (2.4×10^4 S cm^{-1})[53] and volumetric capacitance (1500 F/cm^3) along with ultrahigh rate capability (10 V/s) in acidic media.[54] Besides their high volumetric capacitance, MXenes are particularly favorable for MSCs because of their 2D nature, which enhances the mechanical stability and shortens the ion diffusion paths between the positive and negative electrodes.[55]

9.2 PROPERTIES OF 2D NANOMATERIALS COMPARED TO BULK

Removal of van der Waals interactions—A layered bulk material consists of many covalently bonded planes held together by weak van der Waals interactions. When a force is applied to a material, these van der Waals forces can be easily overcome and the material breaks, making it seem weak. Conversely, the covalent bonds that hold the atoms together in the layers are actually very strong. A monolayer will only have covalent bonds. By removing the "weak links" from the material, it appears to become much stronger. For example, graphene has a tensile strength 1,000 times greater than graphite, and while a graphite pencil can be easily broken, graphene is over 100 times stronger than steel.

An increase in the ratio of surface-area-to-volume—The surface-area-to-volume ratio of a material defines how much of it is exposed to its environment. This is important for chemical reactions—the more reactant that is in contact with the material, the faster the reaction can occur, so 2D materials tend to be more reactive than their bulk counterparts. It also makes 2D materials more sensitive to their surroundings, an effect that is exploited for sensors based on 2D materials.

Confinement of electrons in a plane—The electronic and optical properties of a material depend upon its electronic band structure. This describes how electrons move through the material and is a result of the periodicity of its crystal structure. When a material goes from bulk to 2D, the periodicity is removed in the direction perpendicular to the plane, which can greatly change the band structure. The modified band structures are responsible for the extremely high conductivity of graphene and the fluorescence of monolayer MoS_2.

Another effect of dimensional confinement is reduced dielectric screening between electrons and holes in semiconductors. When there is less material to screen the electric field, there will be an increase in Coulomb interaction and more strongly bound excitons, making them more stable than excitons found in bulk materials. If the excitons are confined in a plane that is thinner than their Bohr radius (as is the case for many 2D semiconductors), quantum confinement will result in an increase in their energy compared to bulk excitons, changing the wavelength of light they absorb and emit.

Their energy can be tuned somewhat by changing the number of layers in the 2D material (i.e., a bilayer structure will absorb/emit lower energy light than a monolayer). However, this can also affect the band structure, resulting in changes to other properties as well (for example, bilayer MoS_2 becomes non-emissive compared to a monolayer due to changes in electronic band structure).

9.3 ENERGY STORAGE APPLICATIONS OF 2D NANOMATERIALS

9.3.1 Li Ion Batteries

From introduction of rechargeable Li-battery concept (1970s), LIBs received huge attention as a significant energy storage device in the last few decades. LIBs with their rechargeable nature is a significant energy storage devices and possess almost zero environmental impact with long lifetime and elevated energy density than other conventional rechargeable batteries. However, the utmost attainable energy density of LIBs is 250Wh kg^{-1}, limiting their extensive utilization in recently promising uses where high-energy battery systems are needed (battery-powered electric vehicles (EVs)).[57] LIBs have been an important part in transportable electronic devices because of its elevated Li-storage ability, extraordinary cyclability, and elevated reliability. Therefore, LIBs, using chemistry of $LiCoO_2$ graphite with their cathode anode, having an estimated simulated energy density of ~380Wh kg^{-1}, and experimentally calculated in between 120 and 150Wh kg^{-1}, are the most important nominee for hybrid and all electrical automobile for convenience use. These obtained rates are better than other batteries, for instance, Pb-acid (~30Wh kg^{-1}) and Ni-hydride (45–68 Wh kg^{-1}).[58] The LIB energy density

activity rely on the chemical and physical nature of cathode and anode (e.g., conventional LIBs use graphite as anode). The low specific capacitance (C_{sp}) (i.e., total Ah obtainable, when a battery is discharged at a distinct discharge current, per unit weight) of graphite (372 mAh g^{-1})[59] formulated it's significance to discover alternate anodes.[60] Moreover, graphite Possess low Csp (372 mAh g^{-1}) and poor reaction rate while used in heavy-duty vehicles. On the other hand, with rising power and energy requirements for state-of-the-art electronic devices, accessible LIBs are not sufficient because of their low energy density, sluggish charging rate, short life cycle, and safety scrutinization. Till now, large research work has been done to discover new ideal anode materials to substitute graphite. So many kinds of storage are introduced, including significant storage for instance hydroelectric power and condensed air, plus fly wheels and electrochemical energy storage (LIBs, redox-FBs, and SCs).[61]

9.3.2 Zinc Based Batteries

9.3.2(a) Zinc Ion Batteries

The zinc ion battery (ZIB) is considered as one of the most important alternative battery chemistries to date. However, one of the challenges in ZIB development is the limitation of materials that can exhibit satisfactory Zn^{2+} storage. Transition metal dichalcogenides (TMDs) are widely investigated in energy-related applications due to their distinct physical and chemical properties. In particular, the wide interlayer spacings for these TMDs are particularly attractive as viable Zn^{2+} storage sites.[62] Using MoS_2 as a model system demonstrated experimentally and theoretically that even hosts with an originally poor Zn^{2+} diffusivity can allow fast Zn^{2+} diffusion. Through simple interlayer spacing and hydrophilicity engineering, the Zn^{2+} diffusivity is boosted by three orders of magnitude, effectively enabling the otherwise barely active MoS_2 to achieve a high capacity of 232 mAh g^{-1}, which is ten times that of its pristine form.[63]

The two-dimensional structural features of covalent organic frameworks (COFs) promotes the electrochemical storage of cations like H^+, Li^+, and Na^+ through both faradaic and non-faradaic processes. However, the electrochemical storage of cations like Zn^{2+} ion is still unexplored although it bears a promising divalent charge. Herein, for the first time, we have utilized hydroquinone linked β-ketoenamine COF acting as a Zn^{2+} anchor in an aqueous rechargeable zinc ion battery. The charge-storage mechanism comprises of an efficient reversible interlayer interaction of Zn^{2+} ions with the functional moieties in the adjacent layers of COF (−182.0 kcal mol^{-1}). Notably, due to the well-defined nanopores and structural organization, a constructed full cell displays a discharge capacity as high as 276 mA h g^{-1} at a current rate of 125 mA g^{-1}.[64] A zincophile interphase based on 3D-printed g-C_3N_4 modulating interface to concurrently achieve homogeneous zinc nucleation and a dendrite-free growth. The Zn/C_3N_4 anode affords lower the energy barrier and more homogeneously charged distribution to facilitate highly reversible Zn plating/stripping. The symmetric Zn/C_3N_4 cell presents appreciably low voltage hysteresis and superior cycling stability compared to the bare Zn. Furthermore, the Zn/C_3N_4/AC supercapacitor and Zn/C_3N_4/MnO_2 battery show long cycle stability.[65]

The traditional method to fabricate a MXene-based energy storage device starts from etching MAX phase particles with dangerous acid/alkali etchants to MXenes, followed by device assembly. This is a multistep protocol and is not environmentally friendly. Herein, an all-in-one protocol is proposed to integrate synthesis and battery fabrication of MXene. By choosing a special F-rich electrolyte, MAX V_2AlC is directly exfoliated inside a battery, and the obtained V_2CT_X MXene is in situ used to achieve an excellent battery performance. This is a one-step process with all reactions inside the cell, avoiding any contamination to external environments. Through the lifetime, the device experiences three stages of exfoliation, electrode oxidation, and redox of V_2O_5. While the electrode is changing, the device can always be used as a battery, and the performance is continuously enhanced. The resulting aqueous zinc ion battery achieves outstanding cycling stability (4,000 cycles) and rate performance (97.5 mAh g^{-1} at 64 A g^{-1}), distinct from all reported aqueous MXene-based counterparts with pseudocapacitive properties, and outperforming most vanadium-based zinc ion batteries with high capacity. This work sheds light on the green synthesis of MXenes, provides an all-in-one protocol for MXene devices, and extends MXenes' application in the aqueous energy storage field.[66]

9.3.2(b) Zinc Air Batteries

Metal-air battery, represented by zinc air battery (ZAB), is recognized as a promising secondary battery technology with great energy density. Since the catalyst in the air-electrode must simultaneously catalyze oxygen evolution reaction (OER) and oxygen reduction reaction (ORR), mass transfer structure of the air-electrode catalyst would significantly affect the ZAB performance. A facile selectively etching strategy is therefore proposed to in situ adjusting the mass transfer structure of the CoFe OER/ORR bifunctional catalyst encapsulated in graphene cage. This graphene-encapsulated CoFe catalyst with optimized mass transfer channel exhibits superior OER and ORR activity to IrO_2 and Pt/C, respectively. The corresponding ZAB performance is therefore tripled compared with that equipped with Pt/C-IrO_2 catalysts in terms of maximum power density (206 mW/cm²). Together with the satisfied stability, this facilely synthesized CoFe catalyst shows promising application potential in metal-air batteries.

Electronically coupled metallic hybrids of NiFe layered double hydroxide nanosheet/Ti_3C_2 MXene quantum dots deposited on a nitrogen-doped graphene surface (LDH/MQD/NG) for high-performance flexible zinc air batteries (ZABs). As verified from the Mott-Schottky and Nyquist plots, as well as spectroscopic, electrochemical, and computational analyses, the electronic and chemical coupling of LDH/MQD/NG modulates the local electronic and surface structure of the active LDH to provide metallic conductivity and abundant active sites, leading to significantly improved bifunctional activity and electrocatalytic kinetics. The rechargeable ZABs with LDH/MQD/NG hybrids are superior to the previous LDH-based ZABs, demonstrating a high-power density (113.8 mW cm⁻²) and excellent cycle stability (150 h at 5 mA cm⁻²). Moreover, the corresponding quasi solid-state ZABs are completely flexible and practical, affording a high power density of 57.6 mW cm⁻² even in the bent state, and in real-life operation of tandem cells for powering various electronic devices.[67]

The construction of an efficient oxygen reduction reaction and oxygen evolution reaction (ORR/OER) bifunctional electrocatalyst is of great significance but still remains a giant challenge for high-performance metal-air batteries. In this study, uniform FeS/Fe_3C nanoparticles embedded in a porous N,S-dual doped carbon honeycomb-like composite (*abbr.* $FeS/Fe_3C@NS-C-900$) have been conveniently fabricated by pyrolysis of a single-crystal Fe-MOF, which has a low potential gap ΔE of ca. 0.72 V, a competitive power density of 90.9 mW/cm^2, a specific capacity as high as 750 mAh/g, and excellent cycling stabilities over 865 h (1,730 cycles) at 2 mA/cm^2 when applied as a cathode material for rechargeable zinc air batteries. In addition, the two series-linked zinc air batteries successfully powered a 2.4 V LED light as a real power source. The efficient ORR/OER bifunctional electrocatalytic activity and long-term durability of the obtained composite might be attributed to the characteristic honeycomb-like porous structure with sufficient accessible active sites, the synergistic effect of FeS and Fe_3C, and the N,S co-doped porous carbon, which provides a promising application potential for portable electronic zinc air battery related devices.[68]

MnO_2 nanorods on hexagonal boron nitride (h-BN) and their composite with high surface area carbon by a chemical method. The optimized nanocomposite catalyst (MnBN/C-75) exhibits a substantial higher onset potential (E_{onset} = 0.9 V vs. RHE) and limiting kinetic current density (J_L = 5.6 mA cm^{-2}) during the oxygen reduction reaction (ORR) compared to other reported h-BN-based metal-supported or metal-free electrocatalysts. Moreover, this catalyst shows a ~4-electron transfer pathway with a low peroxide (HO_2^-) intermediate yield during electroreduction of oxygen, indicating a single step, first-order kinetics as a commercial Pt/C catalyst. Besides, the mass activity of 222 mA mg^{-1} calculated at 0.6 V for the MnBN/C-75 catalyst is ~21 times higher than that of MnBN (10.4 mA mg^{-1}) and slightly lower than Pt/C (239 mA mg^{-1} at 0.9 V). Importantly, the MnBN/C-75 nanocomposite reveals a smaller deviation in half-wave potential ($\Delta E_{1/2}$ = 18 mV) compared to the Pt/C catalyst ($\Delta E_{1/2}$ = 50 mV) even after 5k potential cycling under similar conditions. The relatively lower ionic diffusion and charge transfer resistance at the electrode/electrolyte interface by the MnBN/C-75 electrode support to our claim regarding a higher electrocatalytic activity. Thus, the presence of Mn^{3+} ions in the form of MnOOH (during composite formation) along with both h-BN support and KB carbon at the electrode surface contributes immensely in boosting the electrocatalytic activity. Thus, it could be a promising electrocatalyst if employed in the cathode compartment of low-temperature fuel cells to lead faster ORR kinetics.[69]

9.3.3 SODIUM ION BATTERIES

It is a rechargeable battery analogous to Li ion battery but using sodium instead of Li.[70] Usually, insertion of Na^+ or K^+ ions in extended Ti_3C_2 layers are studied, which increases the difference between vertical planes.[71,72] Several studies are focused on 2D nanostructure anode materials for application in SIBs. Depending on the distinct Na storage method, four different types of nanomaterials are presented, including carbonaceous materials, sodium alloys/compounds, phosphides/oxides/sulfides/selenides, and titanium-based materials. Based on the theoretical work for interlayer expansion without and with functionalized group like O and F decorated- Ti_3C_2 MXenes with 413.0, 367.7 mAh g^{-1} and 151.2 mAh g^{-1} Na storage capacity, respectively,[73] with approximately small volume changes (−0.5% to +1.6%) in sodiation and desodiation. On the other hand, the experimental results based on theoretical estimation show that the Na half cells have small reversible capacity for MXene materials, and Ti_3C_2 delivered a 100 mAh g^{-1} Csp @ $i = 20$ mA g^{-1} over 100 cycles.[74] Afterward, MoS_2-inserted $Ti_3C_2T_x$ layers were introduced, with enhanced 250.9 mAh g^{-1} high C_{sp} of over 100 cycles, and rate performance with 162.7 mAh g^{-1} capacitance at 1 Ag^{-1}.[75] In an extra half Na-ion cell,[76] free standing Ti_3C_2 MXene/CNTs porous films with volumetric capacity of 345 mAh cm^{-3} @ 100 mA g^{-1} over 500 cycles (175 mAh g^{-1} charging capacitance @ 20 mA g^{-1} after 100 cycles) are achieved. Bak et al.[77] demonstrated 2D V_2CT_x MXene as anode material, with 78 mAh g^{-1} reproducible capacity up to 100 cycles at 20 mA g^{-1} a current density. The 3D MXenes, Ti_3CT_x, V_2CT_x, Mo_2CT_x films also demonstrate improved catalytic efficiency with reversible capacitance of 295, 310, 290 mAh g^{-1} at 2.5°C (1C = 200 mA g^{-1}) over 1,000 cycles, enhanced compared to multilayer MXenes or their hybrids.[78] Similarly, K^+ intercalated Ti_3C_2 MXene nanoribbons also show outstanding Na/K storage, e.g., reversible capacities of 168, 136 mAh g^{-1} @ 20 mA g^{-1} and 84 mAh g^{-1}, 78 mA hg^{-1} @ 200 mA g^{-1} in SIBs and PIBs, correspondingly.[79] Moreover, Sb_2O_3/ MXene ($Ti_3C_2T_x$) hybrid materials provide 295 mAh g^{-1} rate performance at 2 mA g^{-1} and 472 mAh g^{-1} increased cycling performance subsequent to 100 cycles @ 100 mA g^{-1}.[80] Presently, the most significant issue is slow progress of research for electrode materials, attuned separator, and electrolyte with outstanding Na-ion diffusion, etc. For full battery system, some productive and informative efforts are performed to ensure the possibility of its realistic application. Since the huge interlayer TMDC spacing, i.e., $MoSe_2$ (as compared to MoS_2), it is suitable for accommodating cations like Na.[81] Wang anode and cathode material Ti_3C_2/CNTs and $Na_{0.44}$ MnO_2, respectively, powering for ~25 min a 2.5 V LED 0.041 mW h^{-1} electrical energy.[76] The first charging/discharging capacitance in full cell was 270 and 286 mAh cm^{-3} of $Ti_3C_2T_x$/CNT-SA electrode volume, correspondingly, and preserves volumetric 242 mAh cm^3 discharge capacitance more than 60 cycles @ $i = 50$ mA g^{-1}, with elevated columbic efficiency (99%). In one more complete SIB, 50 mAh g^{-1} capacitance with highest cell voltage (3.5 V) is also achieved via hard carbon as negative electrode.[82] Briefly, poor anode rate capability with low capacitance is still an obstacle for SIBs. Further simulation evaluation and experimental development are needed to build up more technological data, to more regularly deal with these situations in the near future. Since the huge interlayer TMDC spacing, i.e., $MoSe_2$ (as compared to MoS_2), it is suitable for accommodating cations like Na.[81] Wang et al. formed 2D $MoSe_2$ as large as 200 nm for SIBs,[83] with solid state type diffusion process; the electrochemical performance was simply showing that Na-ions can reversibly diffuse all over the nanoparticles. They also theoretically stimulated the activation barriers for the surface and interlayer reactions to be 0.34 and 1.31 eV, correspondingly, and also reported that the battery performance is because of the high SSA of the $MoSe_2$ sample (62.3 m^2 g^{-1}). Zhang et al. decorated 2D $MoSe_2$ on MWCNTs.[84] The presence of MWCNTs increased the specific capacity up to 20% and enhanced the cyclability for longtime performance. Shi et al. customized the $MoSe_2$ composition via doping S synthesizing $Mo(Se_{0.85}S_{0.15})_2$, but the battery performance was analogous to MoSe$_2$.[85] Kang et al. formed yolk-shell-shaped MoSe$_2$ microspheres. In the vertically formed MoSe$_2$ nanosheets, the counter-ions could directly diffuse along the 2D structure of MoSe$_2$. Therefore, the battery performance was outstanding in contrast with other reported SIBs and LIBs, showing familiar flat plateau during both charging and discharging. The similar group also reported fullerene-like MoSe$_2$, presenting an outstanding structural stability in the SIBs' performance.[86] As the Na incorporation resultant change in structure to form metallic Mo and Na$_2$Se (similar case for LIBs), carbon nanotubes built the fullerene-like cage via keeping the electroactive materials intact. So the electrode kept its capacitance till 50 cycles.[87] Yajun Zhao et al., for the first time, studied CoFe LDH with nitrates in its interlayer as an intercalation-type anode material for SIBs. As compared to well-established conversion reaction for MHOs in LIBs, an extraordinary intercalation/de-intercalation mechanism for Na$^+$ storage was exposed. The CoFe-NO$_3$-LDH anode has first 498 mAh g^{-1} discharge capacity and 209 mAh g^{-1} reversible capacity after 200 cycles at 1 A g^{-1}. The excellent Na$^+$ storage efficiency is credited to the pillared interlayer nitrates expanded in between the CoFe-LDH interlayer distance, helping the intercalation/de-intercalation of Na$^+$-ions. Moreover, the LDH material topochemical transformation property assured high structural stability even with the change in valence of Co$_{2+}$/Co$_{2-x}$ and Fe$_{2+}$/Fe$_{3+}$ in their host layers. Additionally, the CoFe-NO$_3$-LDH interlayer space obviously shows that the LDH is a good Na$^+$-ion conductor.[88] Bare oxide anodes usually suffer from bad rate capability and poor cyclability due to their very large changes in volume in the Na insertion-extraction process. To overcome such issues, one method is to use 2D nanosheets because it can give short paths for ion transport, e.g., ultrathin NiO nanosheets (thickness about 4–5 nm) synthesized by facile solvothermal method followed by annealing in air. For Na storage, the NiO nanosheets deliver a high reversible specific capacity of 299 mA h g^{-1} at a current density of 1 A g^{-1}, and the capacity still remains as high as 266 mA h g^{-1} after the 100th cycle. Recently, Yu et al. studied a two-step method for 2D TMO nanosheet synthesis with variable pore sizes via GOs as a sacrificial template.[89,90] This approach has been demonstrated for the synthesis of various 2D TMO nanosheets, counting simple oxides for instance Co$_3$O$_4$, Fe$_2$O$_3$, Mn$_2$O$_3$, and mixed oxides, for example, ZnMn$_2$O$_4$ (ZMO), ZnCo$_2$O$_4$ (ZCO), NiCo$_2$O$_4$ (NCO), and CoFe$_2$O$_4$ (CFO). Particularly, 2D TMO nanosheets show greatly enhanced rate capability and cycle stability for both Li- and Na-ion storage because of the increased SSA and interfaces, facile interfacial transport, and small diffusion paths.[91] One more plan is to combine the oxide with a 2D support, for example, graphene, carbon cloth, and stainless steel mesh that can improve the conductivity. A variety of oxides (e.g., Fe$_2$O$_3$, Fe$_3$O$_4$, Co$_3$O$_4$, NiO, CuO, MoO$_3$, Mn$_3$O$_4$, SnO$_2$, and SnO) have been studied as SIB anodes. Based on those studies, the Sn-based oxides, for example, SnO$_2$ and SnO, attracted particular awareness due to their high hypothetical reversible capacity, reasonable operating voltage, excellent performance, and low-cost. For instance, Wang et al. studied SnO$_2$ nanosize particles (5 nm) anchor on 2D-rGO and the synthesized electrode with capacitance of 302 mAh g^{-1} over 100 cycles at 160 mA g^{-1}[92]. In recent times, Pei et al. showed that 3D SnO$_2$ @G (SnO$_2$@3DG) has improved Na storage than its 2D equivalent (SnO$_2$@2DG), which is due to its high

SSA and 3D porous structural design.[93] The SnO_2@3DG based anode illustrated a superior reversible capacitance of 432 mAh g^{-1} after 200 cycles at 100 mA g^{-1}. Moreover, the controlled SnO atomic layer nanosheets were synthesized on carbon cloth. The thinnest SnO nanosheet anodes (2–6 SnO monolayers) showed good performance. Particularly, an initial charge and discharge capacity of 848 and 1072 mAh g^{-1} was calculated, correspondingly, at 0.1 A g^{-1}. Additionally, inspiring reversible capacity of 665 mAh g^{-1} after 100 cycles at 0.1 A g^{-1} and 452 mAh g^{-1} after 1,000 cycles at 1.0 A g^{-1} was calculated, with outstanding rate performance.[91]

9.3.4 SUPERCAPACITORS

Supercapacitors are one of the most promising energy storage devices, which have been attracting a lot of interest because of their advantages, including high power density, high capacitive retention, fast charge/discharge rate, and long cyclic life. The main constituents of supercapacitors are electrode and electrolyte materials, current collectors, separators, and sealants, which play an important role in the electrochemical performance of the supercapacitor devices. Among all, the selection of electrode materials and their design play an essential role in improving the capacitance performance of supercapacitors, in which the electrodes should be able to deliver a high specific surface area, corrosion resistance, conductivity, as well as a good thermal and chemical stability. Furthermore, they should be low-cost and environmentally friendly. Also, their ability to transfer the faradic charge is essential to increase the total capacitance value.[94] Another effective factor is the size of the pores, for instance, Y. Gogotsi et al.[95] have studied the influence of the pore size on the specific capacitance of carbide-derived carbon material, and they reported that increasing the pore size lesser than 2 nm can provide more capacitance and, consequently, higher energy density, whereas growing the pore size greater than 2 nm led to decreasing the specific capacitance. However, the smaller the pore, the greater the equivalent series resistance (ESR) and hence decreasing the power density. According to this, the selection of the electrode materials also depends upon their application. For instance, for the applications in which having more peak currents is essential, the electrode materials should have larger pores, while materials with smaller pores make the electrodes suitable for the purposes that require developed energy density.[94] Supercapacitors are categorized into the electrical double-layer capacitors (EDLCs) and the pseudocapacitors based on their energy storage mechanism. The charge storage mechanism in EDLCs is according to the electrosorption of ions and formation of an electrochemical double layer (EDL); however, the pseudocapacitors store the charges through the faradic reactions (redox reactions). The electrode materials for supercapacitors are classified based on three main categories, including carbon-based materials (e.g., carbon nanotubes (CNTs), activated carbon materials (ACs), and graphene), transition metal oxides (e.g., ruthenium oxide (RuO_2), manganese oxide (MnO_2), and nickel oxide (NiO)), and conducting polymers (e.g., polyaniline (PANI), polythiophene (PTh), and polypyrrole (PPy)). The charge storage mechanism for nanostructured carbon-based supercapacitor is based on the electrochemical double layer that formed between the electrode and the electrolyte interface, while conducting polymers and transition metal oxides are known as pseudocapacitive materials of their ability to provide the fast-reversible redox reactions.[96,97] Novel two-dimensional (2D) materials have shown many potentials to be used in different fields, such as environmental sensing,[98] water treatment,[99] and energy conversion/storage[100] because of their extraordinary optical, physical, chemical, and electronic properties. There are different 2D materials/2D based composites for electrodes in supercapacitors, such as MXenes, transition metal dichalcogenides (TMDs), and black phosphorus (BP).

MXenes

MXenes are a promising candidate in the field of energy storage, especially as electrodes for supercapacitors applications, because of their unique combination of hydrophilicity and metallic conductivity.[101] There are several examples of MXene, such as Ti_2CT_x, $Ti_3C_2T_x$, V_2CT_x, $Nb_4C_3T_x$, Nb_2CT_x, $Ta_4C_3T_x$, and Ti_2NT_x MXene, which can be used as electrode materials for supercapacitors due to their capability of storing the charges through pseudocapacitive mechanisms.[102,103] Qiu et al. have synthesized Ti_3C_2 MXene through different methods and reported a conductivity of 260 S cm^{-1}, which is two orders of magnitude greater than that of rGO with conductivity about 1 S cm^{-1}.[104] The electrical conductivity of Ti_3C_2 MXene films attained through the spin-cast method was reported up to 6500 S cm^{-1}, which is higher than most of the 2D materials.[100] The intercalation of multivalent, such as Mg^{2+}, Al^{3+}, or Zn^{2+}, and the large organic ions is one of the important parameters, whereas controlling the ion dynamics between MXene layers leads to achieving a high power in supercapacitors, which provides MXene-based supercapacitors with a low resistance to replace the electrolytic capacitors.[100] Zhang et al.[105] have prepared a conductive and highly deformable freestanding electrode by designing the composite of MXene and MnO_2 on carbon cloth. The resultant electrode showed an enhanced specific capacitance of about 511.2 F g^{-1} in comparison with MnO_2 nanorods/carbon cloth, which exhibited 302.3 F g^{-1}. Also, it was reported that the enhanced capacitance and high rate ability are because of the synergistic coupling of MnO_2 and MXene. Furthermore, the fabricated asymmetric flexible device made by MnO_2 nanorods, MXene, and carbon cloth has shown large energy and power density of 29.58 Wh kg^{-1} and 749.92 W kg^{-1}, respectively. Besides the advantages of having an ultrahigh conductivity and a hydrophilic surface, MXenes material also exhibited high mechanical and chemical stability.[101] The structural stability and conductivity of active electrode materials can be enhanced by means of efficient methods provided by interfacial engineering. For example,[106] Ti_3C_2 MXene was introduced to the flexible Ni_2Co-LDHs nanoarrays to enhance the mechanical stability of the structure. The battery-type supercapacitors fabricated by the resultant composite exhibited ultrahigh long-life stability as well as a good rate capability (126 mAh g^{-1} at 150 A g^{-1}), which was approximately 5.7 times higher than that of pure Ni_2Co- LDHs (22 mAh g^{-1}). The enhanced intrinsic performance is due to the excellent conductivity and the strong interfacial interactions of AL-Ti_3C_2 MXene composite (layered Ti_3C_2 MXene, which was obtained by etching the Al layers from the primary Ti_3AlC_2 MAX). Furthermore, the Ti_3C_2 MXene showed a high Young's modulus of 330 GPa that is greater than those of other prepared Nico-LDH nanosheets such as FeCo@NiCo-LDH reported by L. Gao.[107] The vanadium carbide (V_4C_3) MXenes were prepared through etching the intermediate aluminum layer from their precursors (vanadium aluminum carbide MAX phase) using HF. Etching the Al provides the path, which eases the mobility of the ions from the electrolyte into the electrode, thus increasing the electrochemically active sites. The resultant electrode showed a high specific capacitance (330 F g^{-1} at 5 mV s^{-1}) as well as high capacitive retention of 90% even after 3,000 cycles.[108] Furthermore, the pure $Ti_3C_2T_x$ and $Ti_3C_2T_x$/MWCNT composite films were tested along with rGO electrodes in asymmetric supercapacitors consisting of different electrolytes. The asymmetric supercapacitors with aqueous electrolyte provided the specific capacitances (48 F g^{-1} and 78 F cm^{-3}), which were lower than those values achieved by asymmetric cell based on $Ti_3C_2T_x$ electrodes. However, the reported potential window and, consequently, the energy density for the asymmetric device were higher than that of the symmetric cell. The gravimetric and volumetric cell capacitances of organic asymmetric supercapacitor reached to 30 F g^{-1} and 41 F cm^{-3}, respectively, which were the symmetric cells fabricated by $Ti_3C_2T_x$ and rGO. The potential window and, consequently, the energy density of the asymmetric supercapacitor with organic electrolytes delivered higher values in comparison with the asymmetric aqueous device.[109] Also, Rakhi et al.[110] have synthesized the nanocrystalline-MnO_2 directly on MXene nanosheets using a facile chemical method. The aqueous pseudocapacitor fabricated using MnO_2/Ti_{3-} C_2T_{x-}Ar composite demonstrated a higher specific capacitance (212 F g^{-1}) compared to pure $Ti_3C_2T_{x-}$Ar. Furthermore, the symmetric supercapacitors, fabricated using the resultant composite as electrode materials, showed higher capacity retention (approximately 88% after 10,000 cycles) compared to the symmetric capacitor containing the pure $-MnO_2$-based electrodes.

Transition Metal Dichalcogenides (TMDs)

Transition metal dichalcogenides (TMDs) are layered compounds with the crystal structure formulae of MX_2, in which M is a transition metal element, for instance, Ti, Mo, V, W, Re, Ta; and the term X can be any chalcogenide element, such as S, Se, Te. The TMD monolayers stack together by the van der Waals force, and each layer comprises three atomic layers, in which a transition metal layer is inserted between two chalcogen sheets.[111] The TMDs such as TiS_2, MoS_2, TaS_2, and $NbSe_2$ are the layered structure similar to the graphite and the first graphene descendants that are almost comparable to graphene in terms of their flexible, thin, and transparent structure.[112] However, in contrast with graphene, many 2D TMDs are semiconductor in nature, and they exhibit a high potential to be made into low power and ultrasmall transistors that are more efficient and successful than silicon-based transistors to deal with ever-shrinking devices.[113] TMDs have attracted a lot of interest as electrode materials due to their excellent physicochemical properties, such as large surface area, high catalytic activity, good semiconducting ability, low-cost, high chemical stability, high mechanical properties, their easy synthesis through various preparation methods,[114,115] such as mechanical exfoliation, direct sonication in solvents, chemical vapor deposition (CVD), etc. 2D TMDs are versatile materials applied in various application fields, including sensing photonics, electronics, and energy devices[114,115] such as sensors, optical displays, transistors, and supercapacitors due to their thin atomic profile, which leads to having ideal conditions to achieve higher electrostatic efficiency, a tunable electronic structure, mechanical strength, optical transparency, and sensor sensitivity.[116] For instance, MoS_2 nanosheets as direct band-gap materials are attractive semiconductors, in which two S atoms and one molybdenum (Mo) occupy two various sublattices of the hexagonal MoS_2 structure, and the Mo layer is inserted between two sulfur layers.[117] MoS_2 can be used in optoelectronic devices and flexible electronic as well as for composite films due to their superior elastic properties of single-layer and multilayer.[115] In the case of energy storage development, the 2D layered applications have been enormously increased due to their superior structure with interspace between layers and high surface area, which provides easier intercalation for ions and small molecules and leads to the reversible and quick redox reactions.[118] For example, the asymmetric supercapacitor was fabricated using AC as the negative electrode and Co@NiSe2 material with a nanowire structure as the positive electrode that was obtained through a simple hydrothermal method. The resultant electrode showed a high specific capacitance of 3167.6 Fg^{-1} at a current density of 1 A g^{-1}, as well as a high energy density of 50 Wh kg^{-1} at a power density of 779 W kg^{-1}. Also, 79.4% of initial capacitance remained after 4,000 cycles.[119] Besides all the advantages, the interactions between the 2D TMDs layers and their high surface energy lead to an increase in the restacking possibility of the layers, thus reducing the number of active sites.[120] One of the efficient solutions to address the mentioned disadvantages and increase the TMD activity is to synthesize the new functional hybrid materials, which consist of 2D TMDs and a highly conductive supporting material.[121,122] For instance, the hybridization of 2D TMDs with different carbon-based supporting materials to increase the active sites and ease the charge transferability has been reported in several numbers of research.[121,123,124] Over the last several years, the heterostructure hybrid of 2D TMDs with graphene sheets has offered significant potential for various applications due to their electronic and enhanced physicochemical properties. There are different approaches to synthesize the 2D TMDs/Gr hybrids with specific characteristics, such as high specific surface area, unique morphology, thickness, and electroactive site numbers, according to their application purposes, including (a) chemical vapor deposition synthesis,[125] (b) thermal and chemical reduction synthesis,[126] (c) electrochemical synthesis,[127] (d) microwave-assisted synthesis,[128] (e) hydrothermal and solvothermal synthesis,[129] etc. Furthermore, for the first time, the electrochemical supercapacitor performance of a VSe_2/RGO hybrid was reported, in which the hybrid supercapacitor was synthesized through the one-step hydrothermal method. The supercapacitor showed a specific capacitance of about 680 F g^{-1} at a normalized mass current of 1 A g^{-1}, as well as a high energy density of about 212 Wh kg^{-1} at a power density of about 3.3 kW kg^{-1}. Also, 81% of retention was reported almost even after 10,000 cycles.[130]

Black Phosphorus (BP)

Black phosphorus is a p-type semiconducting layered structure with narrow and direct bandgap.[131] BP has shown great potential owing to its high theoretical specific capacitance (2596 mAh g^{-1}), controllable bandgap (0.302.2 eV) by regulating the number of layers of the BP (phosphorene),[131] as well as the high carrier mobility as high as 1,000 cm^2/Vs[132] for building optoelectronic technologies.[131,133] Also, the high volumetric capacitance of 13.75F cm^{-3} has been obtained, exploiting liquid-exfoliated BP nanoflakes as electrode materials for flexible all-solid-state supercapacitor

(ASSP), which is higher than that of restacked graphene (1.0 F cm^{-3})[134]. Also, BP materials are a suitable candidate as electrode material for EDLCs. The typical electrochemical double-layer capacitance of BP material is due to the large space between BP layers, which leads to the rapid ion diffusion within the BP layers.[135,136] Hao et al.[137] have assembled a flexible asymmetric supercapacitor using the liquid-exfoliated BP nanoflakes that reached to a large charging-discharging rate of 10 V s^{-1} and provided a volumetric capacitance of 13.75 F cm^{-3}, E_d of 2.47 mWh cm^{-3}, P_d of 8.83 W cm^{-3}, and good cycling stability of over 30,000 cycles. Apart from the energy storage applications, the chemical activities and high surface area make them a suitable candidate to be used as absorbents, catalytic agent.[137] Yang et al.[136] have fabricated composite film using black phosphorus and CNT (ratio of 1:4) as electrodes for supercapacitor. The fabricated supercapacitor showed a volumetric specific capacitance of 41.1 F cm^{-3} at 0.005 V s^{-1}, a high energy density of 5.71 mWh cm^{-3}, and a high power density of 821.62 W cm^{-3} as well as superior mechanical flexibility. In addition, 91.5% of initial capacitance remained after 10,000 cycles.

9.3.5 HYBRID ENERGY STORAGE (HES) DEVICES

Perfect ESDs are those that have elevated energy and power. Although major progress has been made in raising energy amount of SCs, still their energy is low. The HES system intends to achieve similar efficiency to batteries regarding energy storage and in keeping the elevated power nature with good cycle life, like SCs. The HES devices can get through hybridization at material level (HDL/HML) or device, called internal and external hybrids, correspondingly. For device-based, EDLC and a battery might be included in tandem in modules and need power conversion methodology to separately handle both EES systems. Based on these necessities, HDL-based HES systems are multicomponent and usually experience fabrication difficulty, are not cost-efficient, and have high weight/volume. Recently, progress has been made for energy storage to achieve battery-level energy, SC-level power, and cyclability in a single device. The HML system becomes more advanced in improving the device performance. The HML system includes one battery-like charge accumulation and one capacitive (EDLC or PCs) charge storage procedure (Figure 9.6). For instance, in earlier reports, hybridization of an asymmetric SC based on activated carbon (SC electrode) with zinc (battery electrode), in a parallel combination. In this process, the activated carbon acts as a negative electrode of SCs, and Zn as negative electrode of secondary battery, and self-doped polyaniline acts as a collective positive electrode. As the working potential window of both material coincides, the HES devices can get high-energy nature like battery and the high-power ability like SCs. Hybrid ion capacitors in energy are intermediary in between batteries and SCs but, by theory, showing SC-like power ability and cyclability.[61]

9.3.6 FUEL CELLS

A number of 2D materials (e.g., graphene,[139,140] borophene,[141,142] and MXene[143,144]) have been recognized as materials with high surface area and great electric conductivity, which makes them suitable for their application in catalysis. MAB, a group

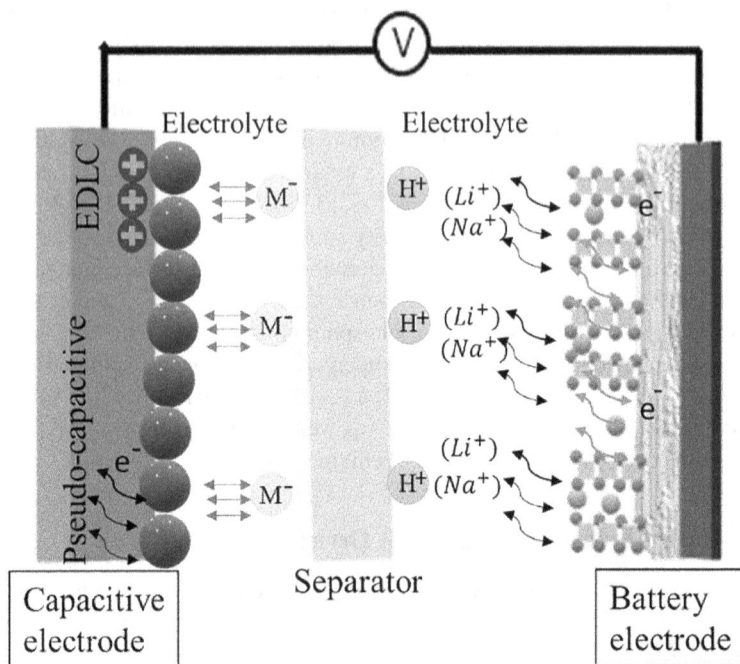

FIGURE 9.6 Illustration of HES systems with capacitive and battery-type electrode.

of 2D ceramic/metallic boride (instead of C or N in MXenes) materials with similar structure to MXenes, also considered as excellent candidates for electrocatalysis owing to the exceptionally high conductivity, rich interlayer porosity, high surface area, adjustable bandgap structure, etc. These materials were found to possess high heterogeneous electron transfer (HET) rates and promising electrocatalytic performances toward hydrogen evolution reaction (HER) and oxygen reduction reaction (ORR)[145] and other electrochemical energy storage systems.[146]

Two-dimensional (2D) materials such as graphene and hexagonal boron nitride (h-BN) can be used as robust and flexible encapsulation overlayers, which effectively protect metal cores but allow reactions to occur between inner cores and outer shells. Electrochemical (EC) tests combined with operando EC-Raman characterizations were performed to monitor the reaction process and its intermediates, which confirm that Pt-catalyzed electrocatalytic processes happen under few-layer h-BN covers. The confinement effect of the h-BN shells prevents Pt nanoparticles from aggregating and helps to alleviate the CO poisoning problem. Accordingly, embedding nanocatalysts within ultrathin 2D material shells can be regarded as an effective route to design high-performance electrocatalysts.[147]

The design of new cathode materials is vital to develop the high performance proton exchange membrane fuel cell (PEMFC). The adsorption properties of the species involved in the oxygen reduction reaction (ORR) and several ORR mechanisms on phosphorene were presented using the first-principle calculations. Results show that the H adsorption is more preferable than O_2, and the OOH spontaneously form from the adsorbed H and O_2. The HOOH from the hydrogenation of the OOH species could be further hydrogenated into OH and H_2O, and the OH would be finally hydrogenated into the (second) H_2O. The reaction barrier of rate-determining step is 1.02 eV. Moreover, the presented kinetic processes are confirmed by the thermodynamic simulation from Gibbs free energy calculations and molecular dynamics simulations. These results indicate that phosphorene as one of new 2D materials could be a promising metal-free cathode material for PEMFC.[148]

Development of bifunctional catalysts with low platinum (Pt) content for the ethanol oxidation reaction (EOR) and the oxygen reduction reaction (ORR) is highly desirable yet challenging. Herein, we present structural engineering of a series of two-dimensional/three-dimensional (2D/3D) hierarchical N-doped graphene-supported nanosized Pt_3Co alloys and Co clusters (PtCo@N-GNSs) via a hydrolysis-pyrolysis route. For the ORR, the optimal PtCo@N-GNS exhibits a high mass activity of 3.01 A mg_{Pt}^{-1}, which is comparable to the best Pt-based catalyst obtained through sophisticated synthesis. It also possesses excellent stability with minor decay after 50,000 cyclic voltammograms (CV) cycles in acidic medium. For the EOR, PtCo@N-GNS achieves the highest mass-specific and area-specific activities of 1.96 A mg_{Pt}^{-1} and 5.75 mA cm^{-2}, respectively, among all the reported EOR catalysts to date. The unique 2D/3D hierarchy, high Pt utilization, and valid encapsulation of nanosized Pt_3Co/ Co synergistically contribute to the robust ORR and EOR activities of the present PtCo@N-GNS. A direct ethanol fuel cell based on PtCo@N-GNS delivers a high open-circuit potential of 0.9 V, a stable power density of 10.5 mW cm^{-2}, and an excellent rate performance, implying the feasibility of the bifunctional PtCo@N-GNS.[149] A conventional carbon felt (CF) electrode was modified by $NiFe_2O_4$ ($NiFe_2O_4$@CF), MXene (MXene@CF), and $NiFe_2O_4$-MXene ($NiFe_2O_4$-MXene@CF) using facile dip-and-dry and hydrothermal methods. In these modified CF electrodes, the electrochemical performance considerably improved, while the highest power density (1,385 mW/m^2), which was 5.6, 2.8, and 1.4 times higher than those of CF, $NiFe_2O_4$@ CF, and MXene@CF anodes, respectively, was achieved using $NiFe_2O_4$-MXene@ CF. Furthermore, electrochemical impedance spectroscopy and cyclic voltammetry results confirmed the superior bio-electrochemical activity of a $NiFe_2O_4$-MXene@ CF anode in a MFC. The improved performance could be attributed to the low charge transfer resistance, high conductivity, and number of catalytically active sites of the $NiFe_2O_4$-MXene@CF anode. Microbial community analysis demonstrated the relative abundance of electroactive bacteria on a $NiFe_2O_4$-MXene@CF anodic biofilm rather than CF, MXene@CF, and $NiFe_2O_4$@CF anodes.[150]

MoS_2 is a promising H_2 evolution catalyst with excellent kinetics for driving the hydrogen evolution reaction (HER). The catalytic effects were suggested to stem from the sulfur edges of MoS_2 plates while the basal planes were catalytically inert.[151] Therefore the surface area and the crystallinity of the catalysts play an important role in the H_2 activation property. Promising results have been reported using MoS_2 nanoparticles with a high concentration of edges. As shown in Figure 9.4 (a), Dai's group has demonstrated a high HER efficiency using MoS_2-graphene oxide composites as the catalysts. MoS_2-graphene hybrid nanomaterials can be obtained by a one-step solvothermal reaction of $(NH_4)MoS_4$ and hydrazine in an N,N-dimethylformamide (DMF) solution of mildly oxidized graphene oxide at 200°C. The MoS$_2$ synthesized on graphene surface has abundant open edges, which facilitate the electrochemical reaction for H$_2$ generation.[127]

9.4 CONCLUSION

In this chapter, we try to explain the development of two-dimensional materials in terms of its morphology, surface area, and electrical conductivity, where we picked commonly employed materials like graphene, hexagonal boron nitride, transition metal chalcogenides, phosphorene, and MXene. We tried to explain their property dependence on the morphology and its importance in energy storage devices like Na/Li batteries, supercapacitors, and fuel cells. Along with this, we also focused on their synthesis mechanisms developed in literature and property dependence on these mechanisms.

REFERENCES

1. Geim, A. K. & Novoselov, K. S. The rise of graphene. *Nat Mater* (2007). doi:10.1038/nmat1849
2. Allen, M. J., Tung, V. C. & Kaner, R. B. Honeycomb carbon: A review of graphene. *Chem Rev* (2010). doi:10.1021/cr900070d
3. Yu, M., Li, R., Wu, M. & Shi, G. Graphene materials for lithium-sulfur batteries. *Energy Storage Mater* (2015). doi:10.1016/j.ensm.2015.08.004
4. Raccichini, R., Varzi, A., Passerini, S. & Scrosati, B. The role of graphene for electrochemical energy storage. *Nat Mater* (2015). doi:10.1038/nmat4170
5. Wu, Z. S. *et al.* Graphene/metal oxide composite electrode materials for energy storage. *Nano Energy* (2012). doi:10.1016/j.nanoen.2011.11.001
6. Sun, Y., Wu, Q. & Shi, G. Graphene based new energy materials. *Energy Environ Sci* (2011). doi:10.1039/c0ee00683a
7. Lv, W., Li, Z., Deng, Y., Yang, Q. H. & Kang, F. Graphene-based materials for electrochemical energy storage devices: Opportunities and challenges. *Energy Storage Mater* (2016). doi:10.1016/j.ensm.2015.10.002
8. Wu, H. *et al.* Graphene based architectures for electrochemical capacitors. *Energy Storage Mater* (2016). doi:10.1016/j.ensm.2016.05.003
9. Zheng, S. *et al.* Graphene-based materials for high-voltage and high-energy asymmetric supercapacitors. *Energy Storage Mater* (2017). doi:10.1016/j.ensm.2016.10.003
10. Ren, W. & Cheng, H. M. The global growth of graphene. *Nat Nanotechnol* (2014). doi:10.1038/nnano.2014.229
11. Jeon, I. Y. *et al.* Edge-carboxylated graphene nanosheets via ball milling. *Proc Natl Acad Sci U. S. A.* (2012). doi:10.1073/pnas.1116897109
12. Dössel, L., Gherghel, L., Feng, X. & Müllen, K. Graphene nanoribbons by chemists: Nanometer-sized, soluble, and defect-free. *Angew Chemie—Int Ed* (2011). doi:10.1002/anie.201006593
13. Chen, Z. *et al.* Three-dimensional flexible and conductive interconnected graphene networks grown by chemical vapor deposition. *Nat Mater* (2011). doi:10.1038/nmat3001
14. Wang, B. *et al.* Electric field effect in atomically thin carbon films. *Mater Today* (2010). doi:10.1063/1.2774096
15. Pei, S. & Cheng, H. M. The reduction of graphene oxide. *Carbon N. Y.* (2012). doi:10.1016/j.carbon.2011.11.010

16. Dong, Y., Wu, Z. S., Ren, W., Cheng, H. M. & Bao, X. Graphene: a promising 2D material for electrochemical energy storage. *Sci Bull* (2017). doi:10.1016/j.scib.2017.04.010

17. Watanabe, K., Taniguchi, T. & Kanda, H. Direct-bandgap properties and evidence for ultraviolet lasing of hexagonal boron nitride single crystal. *Nat Mater* (2004). doi:10.1038/nmat1134

18. Kubota, Y., Watanabe, K., Tsuda, O. & Taniguchi, T. Deep ultraviolet light-emitting hexagonal boron nitride synthesized at atmospheric pressure. *Science* (2007). doi:10.1126/science.1144216

19. Bhattacharya, A., Bhattacharya, S. & Das, G. P. Band gap engineering by functionalization of BN sheet. *Phys Rev B—Condens Matter Mater Phys* (2012). doi:10.1103/PhysRevB.85.035415

20. Huang, C. *et al.* Carbon-doped BN nanosheets for metal-free photoredox catalysis. *Nat Commun* (2015). doi:10.1038/ncomms8698

21. Liu, F. *et al.* Fluorescent carbon- and oxygen-doped hexagonal boron nitride powders as printing ink for anticounterfeit applications. *Adv Opt Mater* (2019). doi:10.1002/adom.201901380

22. Lei, W. *et al.* Large scale boron carbon nitride nanosheets with enhanced lithium storage capabilities. *Chem Commun* (2013). doi:10.1039/c2cc36998b

23. Weng, Q. *et al.* Preparation and hydrogen sorption performances of BCNO porous microbelts with ultra-narrow and tunable pore widths. *Chem—An Asian J* (2013). doi:10.1002/asia.201300940

24. Xiao, F. *et al.* Edge-hydroxylated boron nitride nanosheets as an effective additive to improve the thermal response of hydrogels. *Adv Mater* (2015). doi:10.1002/adma.201502803

25. Shi, L. *et al.* Edge-hydroxylated boron nitride for oxidative dehydrogenation of propane to propylene. *ChemCatChem* (2017). doi:10.1002/cctc.201700745

26. García-Miranda Ferrari, A., Rowley-Neale, S. J. & Banks, C. E. Recent advances in 2D hexagonal boron nitride (2D-hBN) applied as the basis of electrochemical sensing platforms. *Anal Bioanal Chem* (2021). doi:10.1007/s00216-020-03068-8

27. Chhowalla, M. *et al.* The chemistry of two-dimensional layered transition metal dichalcogenide nanosheets. *Nat Chem* (2013). doi:10.1038/nchem.1589

28. Novoselov, K. S., Mishchenko, A., Carvalho, A. & Castro Neto, A. H. 2D materials and van der Waals heterostructures. *Science* (2016). doi:10.1126/science.aac9439

29. Hu, X., Zhang, W., Liu, X., Mei, Y. & Huang, Y. Nanostructured Mo-based electrode materials for electrochemical energy storage. *Chem Soc Rev* (2015). doi:10.1039/c4cs00350k

30. Yang, W. *et al.* Carbon-MEMS-based alternating stacked MoS$_2$@rGO-CNT microsupercapacitor with high capacitance and energy density. *Small* (2017). doi:10.1002/smll.201700639

31. Osada, M. & Sasaki, T. Nanosheet architectonics: A hierarchically structured assembly for tailored fusion materials. *Polymer J* (2015). doi:10.1038/pj.2014.111

32. Andres, C.-G. *et al.* Isolation and characterization of few-layer black phosphorus. *2D Mater* (2014).

33. Samuel Reich, E. Phosphorene excites materials scientists. *Nature* (2014). doi:10.1038/506019a

34. Liu, H. *et al.* Phosphorene: An unexplored 2D semiconductor with a high hole mobility. *ACS Nano* (2014). doi:10.1021/nn501226z

35. Bridgman, P. W. Two new modifications of phosphorus. *J Am Chem Soc* (1914). doi:10.1021/ja02184a002

36. Morita, A. Semiconducting black phosphorus. *Appl Phys A Solids Surfaces* (1986). doi:10.1007/BF00617267

37. Cartz, L., Srinivasa, S. R., Riedner, R. J., Jorgensen, J. D. & Worlton, T. G. Effect of pressure on bonding in black phosphorus. *J Chem Phys* (1979). doi:10.1063/1.438523

38. Low, T. *et al.* Tunable optical properties of multilayer black phosphorus thin films. *Phys Rev B—Condens Matter Mater Phys* (2014). doi:10.1103/PhysRevB.90.075434

39. Tran, V., Soklaski, R., Liang, Y. & Yang, L. Layer-controlled band gap and anisotropic excitons in few-layer black phosphorus. *Phys Rev B—Condens Matter Mater Phys* (2014). doi:10.1103/PhysRevB.89.235319

40. Chen, Y. *et al.* Field-effect transistor biosensors with two-dimensional black phosphorus nanosheets. *Biosens Bioelectron* (2017). doi:10.1016/j.bios.2016.03.059

41. Li, D. *et al.* Black phosphorus polycarbonate polymer composite for pulsed fibre lasers. *Appl Mater Today* (2016). doi:10.1016/j.apmt.2016.05.001

42. Zhipei Sun. Optical modulators with two-dimensional layered materials. *Nature Photonics*, 10, 227 (2016). doi:10.1109/piers.2016.7735451

43. He, H., Wang, H., Tang, Y. & Liu, Y. Current studies of anode materials for sodium-ion battery. *Prog Chem* (2014). doi:10.7536/PC130919

44. Ren, X. *et al.* Properties, preparation and application of black phosphorus/phosphorene for energy storage: A review. *J Mater Sci* (2017). doi:10.1007/s10853-017-1194-3

45. Gogotsi, Y. & Anasori, B. The rise of MXenes. *ACS Nano* (2019). doi:10.1021/acsnano.9b06394

46. Hart, J. L. *et al.* Control of MXenes' electronic properties through termination and intercalation. *Nat Commun* (2019). doi:10.1038/s41467-018-08169-8

47. Zhang, C. (John) & Nicolosi, V. Graphene and MXene-based transparent conductive electrodes and supercapacitors. *Energy Storage Mater* (2019). doi:10.1016/j.ensm.2018.05.003

48. Naguib, M. *et al.* Two-dimensional nanocrystals: two-dimensional nanocrystals produced by exfoliation of Ti_3AlC_2 (Adv. Mater. 37/2011). *Adv Mater* (2011). doi:10.1002/adma.201190147

49. Kim, H., Wang, Z. & Alshareef, H. N. MXetronics: Electronic and photonic applications of MXenes. *Nano Energy* (2019). doi:10.1016/j.nanoen.2019.03.020

50. Zhong, Y. *et al.* Transition metal carbides and nitrides in energy storage and conversion. *Adv Sci* (2015). doi:10.1002/advs.201500286

51. Anasori, B., Lukatskaya, M. R. & Gogotsi, Y. 2D metal carbides and nitrides (MXenes) for energy storage. *Nat Rev Mater* (2017). doi:10.1038/natrevmats.2016.98

52. Naguib, M., Mochalin, V. N., Barsoum, M. W. & Gogotsi, Y. 25th anniversary article: MXenes: A new family of two-dimensional materials. *Adv Mater* (2014). doi:10.1002/adma.201304138

53. Ling, Z. *et al.* Flexible and conductive MXene films and nanocomposites with high capacitance. *Proc Natl Acad Sci U. S. A.* (2014). doi:10.1073/pnas.1414215111

54. Lukatskaya, M. R. *et al.* Ultra-high-rate pseudocapacitive energy storage in two-dimensional transition metal carbides. *Nat Energy* (2017). doi:10.1038/nenergy.2017.105

55. Jiang, Q. *et al.* Review of MXene electrochemical microsupercapacitors. *Energy Storage Mater* (2020). doi:10.1016/j.ensm.2020.01.018

56. Hong, W., Wyatt, B. C., Nemani, S. K. & Anasori, B. Double transition-metal MXenes: Atomistic design of two-dimensional carbides and nitrides. *MRS Bull* (2020). doi:10.1557/mrs.2020.251

57. Jung, K. N. *et al.* Rechargeable lithium-air batteries: A perspective on the development of oxygen electrodes. *J Mat Chem A* (2016). doi:10.1039/c6ta04510c

58. Divya, K. C. & Østergaard, J. Battery energy storage technology for power systems-An overview. *Electr Power Syst Res* (2009). doi:10.1016/j.epsr.2008.09.017

59. Bruce, P. G., Freunberger, S. a., Hardwick, L. J. & Tarascon, J.-M. Li-O_2 and Li-S batteries with high energy storage. *Nat Mater* (2011).

60. Ferrari, A. C. *et al.* Science and technology roadmap for graphene, related two-dimensional crystals, and hybrid systems. *Nanoscale* (2015). doi:10.1039/c4nr01600a

61. Khan, K. *et al.* Going green with batteries and supercapacitor: Two dimensional materials and their nanocomposites based energy storage applications. *Prog Solid State Ch* (2020). doi:10.1016/j.progsolidstchem.2019.100254

62. Lee, W. S. V., Xiong, T., Wang, X. & Xue, J. Unraveling MoS_2 and transition metal dichalcogenides as functional zinc-ion battery cathode: A perspective. *Small Methods* (2021). doi:10.1002/smtd.202000815

63. Liang, H. *et al.* Aqueous zinc-ion storage in MoS_2 by tuning the intercalation energy. *Nano Lett* **19**, 3199–3206 (2019).

64. Khayum, A. M. *et al.* Zinc ion interactions in a two-dimensional covalent organic framework based aqueous zinc ion battery. *Chem Sci* (2019). doi:10.1039/c9sc03052b

65. Liu, P. *et al.* Ultra-highly stable zinc metal anode via 3D-printed g-C3N4 modulating interface for long life energy storage systems. *Chem Eng J* (2021). doi:10.1016/j.cej.2020.126425

66. Li, X. *et al.* In situ electrochemical synthesis of MXenes without acid/alkali usage in/for an aqueous zinc ion battery. *Adv Energy Mater* (2020). doi:10.1002/aenm.202001791

67. Han, X. *et al.* Electronically coupled layered double hydroxide/ MXene quantum dot metallic hybrids for high-performance flexible zinc-air batteries. *InfoMat* (2021). doi:10.1002/inf2.12226

68. Li, Y. W. *et al.* Fe-MOF-derived efficient ORR/OER bifunctional electrocatalyst for rechargeable zinc-air batteries. *ACS Appl Mater Interfaces* (2020). doi:10.1021/acsami.0c11945

69. Patil, I. M., Swami, A., Chavan, R., Lokanathan, M. & Kakade, B. Hexagonal boron nitride-supported crystalline manganese oxide nanorods/carbon: A tunable nanocomposite catalyst for dioxygen electroreduction. *ACS Sustain Chem Eng* (2018). doi:10.1021/acssuschemeng.8b04241

70. Gao, Q. *et al.* Synergetic effects of K+ and Mg2+ ion intercalation on the electrochemical and actuation properties of the two-dimensional Ti3C2 MXene. *Faraday Discuss* (2017). doi:10.1039/c6fd00251j

71. Shi, C. *et al.* Structure of nanocrystalline Ti3 C2 MXene using atomic pair distribution function. *Phys Rev Lett* (2013). doi:10.1103/PhysRevLett.112.125501

72. Wu, S., Ge, R., Lu, M., Xu, R. & Zhang, Z. Graphene-based nano-materials for lithium-sulfur battery and sodium-ion battery. *Nano Energy* (2015). doi:10.1016/j.nanoen.2015.04.032

73. Yu, Y. X. Prediction of mobility, enhanced storage capacity, and volume change during sodiation on interlayer-expanded functionalized Ti3C2 MXene anode materials for sodium-ion batteries. *J Phys Chem C* (2016). doi:10.1021/acs.jpcc.5b10366

74. Kajiyama, S. *et al.* Sodium-ion intercalation mechanism in MXene nanosheets. *ACS Nano* (2016). doi:10.1021/acsnano.5b06958

75. Wu, Y. *et al.* MoS_2-nanosheet-decorated 2D titanium carbide (MXene) as high-performance anodes for sodium-ion batteries. *ChemElectroChem* (2017). doi:10.1002/celc.201700060

76. Xie, X. *et al.* Porous heterostructured MXene/carbon nanotube composite paper with high volumetric capacity for sodium-based energy storage devices. *Nano Energy* (2016). doi:10.1016/j.nanoen.2016.06.005

77. Bak, S. M. *et al.* Na-ion intercalation and charge storage mechanism in 2D vanadium carbide. *Adv Energy Mater* (2017). doi:10.1002/aenm.201700959

78. Zhao, M. Q. *et al.* Hollow MXene spheres and 3D macroporous MXene frameworks for Na-Ion storage. *Adv Mater* (2017). doi:10.1002/adma.201702410

79. Lian, P. *et al.* Alkalized Ti3C2 MXene nanoribbons with expanded interlayer spacing for high-capacity sodium and potassium ion batteries. *Nano Energy* (2017). doi:10.1016/j.nanoen.2017.08.002

80. Guo, X. *et al*. Sb2O3/MXene(Ti3C2Tx) hybrid anode materials with enhanced performance for sodium-ion batteries. *J Mater Chem A* (2017). doi:10.1039/c7ta02689g

81. Xie, D. *et al*. Facile fabrication of integrated three-dimensional C-MoSe2/reduced graphene oxide composite with enhanced performance for sodium storage. *Nano Res* (2016). doi:10.1007/s12274-016-1056-3

82. Dall'Agnese, Y., Taberna, P. L., Gogotsi, Y. & Simon, P. Two-dimensional vanadium carbide (MXene) as positive electrode for sodium-ion capacitors. *J Phys Chem Lett* (2015). doi:10.1021/acs.jpclett.5b00868

83. Jeon, C. W., Cheon, T., Kim, H., Kwon, M. S. & Kim, S. H. Controlled formation of MoSe2 by MoNx thin film as a diffusion barrier against Se during selenization annealing for CIGS solar cell. *J Alloys Compd* (2015). doi:10.1016/j.jallcom.2015.04.120

84. Zhang, Z., Yang, X., Fu, Y. & Du, K. Ultrathin molybdenum diselenide nanosheets anchored on multi-walled carbon nanotubes as anode composites for high performance sodium-ion batteries. *J Power Sources* (2015). doi:10.1016/j.jpowsour.2015.07.008

85. Shi, Z. T. *et al*. In situ carbon-doped Mo(Se0.85S0.15)2 hierarchical nanotubes as stable anodes for high-performance sodium-ion batteries. *Small* (2015). doi:10.1002/smll.201501360

86. Choi, S. H. & Kang, Y. C. Fullerene-like MoSe2 nanoparticles-embedded CNT balls with excellent structural stability for highly reversible sodium-ion storage. *Nanoscale* (2016). doi:10.1039/c5nr07733h

87. Eftekhari, A. Molybdenum diselenide (MoSe2) for energy storage, catalysis, and optoelectronics. *Appl Mater Today* (2017). doi:10.1016/j.apmt.2017.01.006

88. Zhao, Y. *et al*. Discovery of a new intercalation-type anode for high-performance sodium ion batteries. *J Mater Chem A* (2019). doi:10.1039/c9ta03753e

89. Peng, L. *et al*. Holey two-dimensional transition metal oxide nanosheets for efficient energy storage. *Nat Commun* (2017). doi:10.1038/ncomms15139

90. Chen, D. *et al*. Two-dimensional holey Co3O4 nanosheets for high-rate alkali-ion batteries: From rational synthesis to in situ probing. *Nano Lett* (2017). doi:10.1021/acs.nanolett.7b01485

91. Mao, J. *et al*. Two-dimensional nanostructures for sodium-ion battery anodes. *J Mater Chem A* (2018). doi:10.1039/c7ta10500b

92. Wang, Y. X. *et al*. Ultrafine SnO2 nanoparticle loading onto reduced graphene oxide as anodes for sodium-ion batteries with superior rate and cycling performances. *J Mater Chem A* (2014). doi:10.1039/c3ta13592f

93. Pei, L. *et al*. Ice-templated preparation and sodium storage of ultrasmall SnO2 nanoparticles embedded in three-dimensional graphene. *Nano Res* (2015). doi:10.1007/s12274-014-0609-6

94. Xie, L. *et al*. Hierarchical porous carbon microtubes derived from willow catkins for supercapacitor applications. *J Mater Chem A* (2016). doi:10.1039/c5ta09043a

95. Chmiola, J., Yushin, G., Dash, R. & Gogotsi, Y. Effect of pore size and surface area of carbide derived carbons on specific capacitance. *J Power Sources* (2006). doi:10.1016/j.jpowsour.2005.09.008

96. Li, J., Cheng, X., Shashurin, A. & Keidar, M. Review of electrochemical capacitors based on carbon nanotubes and graphene. *Graphene* (2012). doi:10.4236/graphene.2012.11001

97. Shi, F., Li, L., Wang, X. L., Gu, C. D. & Tu, J. P. Metal oxide/hydroxide-based materials for supercapacitors. *RSC Advances* (2014). doi:10.1039/c4ra06136e

98. Bo, Z. *et al*. Emerging energy and environmental applications of vertically-oriented graphenes. *Chem Soc Rev* (2015). doi:10.1039/c4cs00352g

99. Dervin, S., Dionysiou, D. D. & Pillai, S. C. 2D nanostructures for water purification: Graphene and beyond. *Nanoscale* (2016). doi:10.1039/c6nr04508a

100. Pomerantseva, E. & Gogotsi, Y. Two-dimensional heterostructures for energy storage. *Nat Energy* (2017). doi:10.1038/nenergy.2017.89

101. Ghidiu, M., Lukatskaya, M. R., Zhao, M. Q., Gogotsi, Y. & Barsoum, M. W. Conductive two-dimensional titanium carbide 'clay' with high volumetric capacitance. *Nature* (2015). doi:10.1038/nature13970

102. Djire, A. *et al.* Pseudocapacitive storage in nanolayered Ti2NTx MXene using Mg-Ion electrolyte. *ACS Appl Nano Mater* (2019). doi:10.1021/acsanm.9b00289

103. Anasori, B. *et al.* Two-dimensional, ordered, double transition metals carbides (MXenes). *ACS Nano* (2015). doi:10.1021/acsnano.5b03591

104. Wu, X., Wang, Z., Yu, M., Xiu, L. & Qiu, J. Stabilizing the MXenes by carbon nanoplating for developing hierarchical nanohybrids with efficient lithium storage and hydrogen evolution capability. *Adv Mater* (2017). doi:10.1002/adma.201607017

105. Zhou, H. *et al.* MnO2 nanorods/MXene/CC composite electrode for flexible supercapacitors with enhanced electrochemical performance. *J Alloys Compd* (2019). doi:10.1016/j.jallcom.2019.06.173

106. Lu, C. *et al.* Interface design based on Ti3C2 MXene atomic layers of advanced battery-type material for supercapacitors. *Energy Storage Mater* (2020). doi:10.1016/j.ensm.2019.11.021

107. Gao, L. *et al.* Mechanically stable ternary heterogeneous electrodes for energy storage and conversion. *Nanoscale* (2018). doi:10.1039/c7nr07789k

108. Syamsai, R. & Grace, A. N. Synthesis, properties and performance evaluation of vanadium carbide MXene as supercapacitor electrodes. *Ceram Int* (2020). doi:10.1016/j.ceramint.2019.10.283

109. Navarro-Suárez, A. M. *et al.* Development of asymmetric supercapacitors with titanium carbide-reduced graphene oxide couples as electrodes. *Electrochim Acta* (2018). doi:10.1016/j.electacta.2017.10.125

110. Rakhi, R. B., Ahmed, B., Anjum, D. & Alshareef, H. N. Direct chemical synthesis of MnO2 nanowhiskers on transition-metal carbide surfaces for supercapacitor applications. *ACS Appl Mater Interfaces* (2016). doi:10.1021/acsami.6b04481

111. Li, X. & Zhu, H. Two-dimensional MoS₂: Properties, preparation, and applications. *J Mater* (2015). doi:10.1016/j.jmat.2015.03.003

112. Zhang, H. Ultrathin two-dimensional nanomaterials. *ACS Nano* (2015). doi:10.1021/acsnano.5b05040

113. Das, S., Robinson, J. A., Dubey, M., Terrones, H. & Terrones, M. Beyond graphene: Progress in novel two-dimensional materials and van der waals solids. *Annu Rev Mater Res* (2015). doi:10.1146/annurev-matsci-070214–021034

114. Li, M. Y., Chen, C. H., Shi, Y. & Li, L. J. Heterostructures based on two-dimensional layered materials and their potential applications. *Mater Today* (2016). doi:10.1016/j.mattod.2015.11.003

115. Huang, X., Zeng, Z. & Zhang, H. Metal dichalcogenide nanosheets: Preparation, properties and applications. *Chem Soc Rev* (2013). doi:10.1039/c2cs35387c

116. Kang, J. *et al.* Graphene and beyond-graphene 2D crystals for next-generation green electronics. *Micro- and Nanotechnology Sensors, Systems, and Applications VI* (2014). doi:10.1117/12.2051198

117. Das, S., Kim, M., Lee, J. W. & Choi, W. Synthesis, properties, and applications of 2-D materials: A comprehensive review. *Crit Rev Solid State Mater Sci* (2014). doi:10.1080/10408436.2013.836075

118. Feng, J. *et al.* Metallic few-layered VS2 ultrathin nanosheets: High two-dimensional conductivity for in-plane supercapacitors. *J Am Chem Soc* (2011). doi:10.1021/ja207176c

119. Jiang, J. W. & Park, H. S. Mechanical properties of MoS$_2$/graphene heterostructures. *Appl Phys Lett* (2014). doi:10.1063/1.4891342

120. Muhammad Yousaf *et al.* Advancement in layered transition metal dichalcogenide composites for lithium and sodium ion batteries. *J Electr Eng* (2016). doi:10.17265/2328–2223/2016.02.003

121. Toth, P. S. *et al.* Asymmetric MoS$_2$/graphene/metal sandwiches: Preparation, characterization, and application. *Adv Mater* (2016). doi:10.1002/adma.201600484

122. Zhang, X. *et al.* MoS$_2$/Carbon nanotube core-Shell nanocomposites for enhanced non-linear optical performance. *Chem—A Eur J* (2017). doi:10.1002/chem.201604395

123. Liu, X. *et al.* Rotationally commensurate growth of MoS$_2$ on epitaxial graphene. *ACS Nano* (2016). doi:10.1021/acsnano.5b06398

124. Jiang, L. *et al.* Monolayer MoS$_2$-graphene hybrid aerogels with controllable porosity for lithium-ion batteries with high reversible capacity. *ACS Appl Mater Interfaces* (2016). doi:10.1021/acsami.5b10692

125. Li, X., Zhang, L., Zang, X., Li, X. & Zhu, H. Photo-promoted platinum nanoparticles decorated MoS$_2$@graphene woven fabric catalyst for efficient hydrogen generation. *ACS Appl Mater Interfaces* (2016). doi:10.1021/acsami.6b01903

126. Qin, W. *et al.* MoS$_2$-reduced graphene oxide composites via microwave assisted synthesis for sodium ion battery anode with improved capacity and cycling performance. *Electrochim Acta* (2015). doi:10.1016/j.electacta.2014.11.034

127. Li, Y. *et al.* MoS$_2$ nanoparticles grown on graphene: An advanced catalyst for the hydrogen evolution reaction. *J Am Chem Soc* (2011). doi:10.1021/ja201269b

128. MarriThese Authors Contributed Equally., S. R., Ratha, S., Rout, C. S. & Behera, J. N. 3D cuboidal vanadium diselenide embedded reduced graphene oxide hybrid structures with enhanced supercapacitor properties. *Chem Commun* (2017). doi:10.1039/c6cc08035a

129. Xia, X. H., Chao, D. L., Zhang, Y. Q., Shen, Z. X. & Fan, H. J. Three-dimensional graphene and their integrated electrodes. *Nano Today* (2014). doi:10.1016/j.nantod.2014.12.001

130. Zheng, Y. *et al.* Recent advances of two-dimensional transition metal nitrides for energy storage and conversion applications. *FlatChem* (2020). doi:10.1016/j.flatc.2019.100149

131. Wang, H. *et al.* Ultrathin black phosphorus nanosheets for efficient singlet oxygen generation. *J Am Chem Soc* (2015). doi:10.1021/jacs.5b06025

132. El-Kady, M. F. & Kaner, R. B. Scalable fabrication of high-power graphene micro-supercapacitors for flexible and on-chip energy storage. *Nat Commun* (2013). doi:10.1038/ncomms2446

133. Wu, S., Hui, K. S. & Hui, K. N. 2D Black Phosphorus: from Preparation to Applications for Electrochemical Energy Storage. *Adv Sci* (2018). doi:10.1002/advs.201700491

134. Yang, B. *et al.* Flexible black-phosphorus nanoflake/carbon nanotube composite paper for high-performance all-solid-state supercapacitors. *ACS Appl Mater Interfaces* (2017). doi:10.1021/acsami.7b13572

135. Hao, C. *et al.* Flexible all-solid-state supercapacitors based on liquid-exfoliated black-phosphorus nanoflakes. *Adv Mater* (2016). doi:10.1002/adma.201505730

136. Yasaei, P. *et al.* Stable and selective humidity sensing using stacked black phosphorus flakes. *ACS Nano* (2015). doi:10.1021/acsnano.5b03325

137. Giorgi, G., Fujisawa, J. I., Segawa, H. & Yamashita, K. Small photocarrier effective masses featuring ambipolar transport in methylammonium lead iodide perovskite: A density functional analysis. *J Phys Chem Lett* (2013). doi:10.1021/jz4023865

138. Noori, A., El-Kady, M. F., Rahmanifar, M. S., Kaner, R. B. & Mousavi, M. F. Towards establishing standard performance metrics for batteries, supercapacitors and beyond. *Chem Soc Rev* (2019). doi:10.1039/c8cs00581h

139. Sun, J. *et al.* Regulation of electronic structure of graphene nanoribbon by tuning long-range dopant-dopant coupling at distance of tens of nanometers. *J Phys Chem Lett* (2020). doi:10.1021/acs.jpclett.0c01839

140. Suter, J. L., Sinclair, R. C. & Coveney, P. V. Principles governing control of aggregation and dispersion of graphene and graphene oxide in polymer melts. *Adv Mater* (2020). doi:10.1002/adma.202003213

141. Mannix, A. J., Zhang, Z., Guisinger, N. P., Yakobson, B. I. & Hersam, M. C. Borophene as a prototype for synthetic 2D materials development. *Nat Nanotechnol* (2018). doi:10.1038/s41565-018-0157-4

142. Kiraly, B. *et al.* Borophene synthesis on Au(111). *ACS Nano* (2019). doi:10.1021/acsnano.8b09339

143. Tang, X., Guo, X., Wu, W. & Wang, G. 2D Metal carbides and nitrides (MXenes) as high-performance electrode materials for lithium-based batteries. *Adv Energy Mater* (2018). doi:10.1002/aenm.201801897

144. Xu, C. *et al.* MXene (Ti3C2Tx) and carbon nanotube hybrid-supported platinum catalysts for the high-performance oxygen reduction reaction in PEMFC. *ACS Appl Mater Interfaces* (2020). doi:10.1021/acsami.0c02446

145. Wang, J. *et al.* Discovery of hexagonal ternary phase Ti2InB2 and its evolution to layered boride TiB. *Nat Commun* (2019). doi:10.1038/s41467-019-10297-8

146. Rosli, N. F. *et al.* MAX and MAB Phases: Two-dimensional layered carbide and boride nanomaterials for electrochemical applications. *ACS Appl Nano Mater* (2019). doi:10.1021/acsanm.9b01526

147. Duan, Y. e. *et al.* Cobalt nickel boride nanocomposite as high-performance anode catalyst for direct borohydride fuel cell. *Int J Hydrogen Energy* (2021). doi:10.1016/j.ijhydene.2021.02.064

148. Lu, Z. *et al.* Phosphorene: A promising metal free cathode material for proton exchange membrane fuel cell. *Appl Surf Sci* (2019). doi:10.1016/j.apsusc.2019.02.013

149. Gao, J. *et al.* MOF-Derived 2D/3D hierarchical N-doped graphene as support for advanced Pt utilization in ethanol fuel cell. *ACS Appl Mater Interfaces* (2020). doi:10.1021/acsami.0c15493

150. Tahir, K. *et al.* Nickel ferrite/MXene-coated carbon felt anodes for enhanced microbial fuel cell performance. *Chemosphere* (2021). doi:10.1016/j.chemosphere.2020.128784

151. Kibsgaard, J., Chen, Z., Reinecke B. N. & Jaramillo, T. F. Engineering the surface structure of MoS$_2$ to preferentially expose active edge sites for electrocatalysis. *Nature Mater* (2012)

10 Role of 2D Materials in Environmental Monitoring

Renu Dhahiya, Moumita Saha, Ashok Kumar, Pankaj Sharma, Ram Sevak Singh, Varun Rai, and Kamalakanta Behera

CONTENTS

10.1 INTRODUCTION

The improvement in the living standards of living beings and the Industrial Revolution has led to environmental degradation to its peak. Today the most important concerns are to protect our environment for the betterment of human beings and other living creatures. However, due to several anthropogenic activities, the environment is loaded with various types of direct and indirect pollution-causing substances, such as heavy metals, toxic gases, volatile organic compounds, persistent organic

DOI: 10.1201/9781003247890-10

pollutants, and pesticides. These pollutants have a hazardous impact on the ecosystem and ultimately on human beings.[1,2] Heavy metal is the most released pollutant in water and soil, which chiefly includes copper, iron, mercury, arsenic, etc. Owing to their nonbiodegradability, heavy metal ions accumulate in the living system due to biomagnifications. Moreover, these ions are carcinogenic and cause several other diseases, such as damage to kidneys, lungs, liver, esophagus, stomach, prostate, and skin.[3] Another major source of environmental degradation is poisonous gases such as oxides of nitrogen (NO_x), methane gas (CH_4), hydrogen sulfide (H_2S), carbon monoxide, and carbon dioxides. These poisonous gases are released from various point and nonpoint sources, such as power plants, transport vehicles, factories, agricultural activities, forest fires, volcanic eruptions, and waste treatments. These gases cause serious ailments such as lung cancer, heart disease, and both acute and chronic respiratory diseases, such as asthma. The release of these gases also results in global warming and ozone degradation.[4] The next class of environmental pollutants includes pesticides. These are the chemicals that are meant to control the growth of weeds and undesirable plants that decline the crop yield. After the green revolution, the use of chemicals in agriculture has boomed to a new height. These chemicals include organophosphates, organochlorines, carbamates, and thiocarbamates. These chemical pesticides along with their benefits have various hazards also, which includes the degradation of aquatic plants and animals and several death-causing diseases in human beings. Antibiotics, which include tetracyclines, aminoglycosides, macrolides, b-lactams, vancomycin, fluoroquinolone, phenols, etc., are the overall utilized drugs that are used against pathogenic microorganisms in living organisms. Moreover, these are also the major contributors to environmental pollution. These antibiotics cause ecosystem damage and disturbance in the function of algae communities.[5] Hence, we can say that the monitoring of the environment is the major concern for today's world.

Several metal particles discarded into the environment stay undecayed or exist forever as sophisticated compounds. This has grown exponentially with contemporary civilization. It has a negative influence on both human health and the environment. As a result, developing a high-speed, sensitive, and fundamental demonstration approach for detecting and monitoring metal particles' toxicity in soil and water is critical. Nowadays, ultrasensitive 2D nanocomposites adapted terminals are used mostly to identify metals such as mercury, cadmium, lead, copper, iron, and zinc with incredible accuracy.[6,7]

10.2 UNIQUE FEATURES OF TWO-DIMENSIONAL MATERIALS IN SENSING

Evolving 2D materials have sparked a revolution in electronics, photonics, optical, thermal, and sensors. These are inherited with some unique properties, which makes them apt material for sensing application. These properties incorporate simplicity of functionalization, enormous surface area, high chemical stability, high electrical conductivity, and high mechanical strength due to associated structural resemblance with graphene.[8–10] The 2D material possesses a higher surface-to-volume ratio and ultrathin planer surface than 0D, 1D, and 3D materials, making it suitable for enhanced performance in the sensor. Due to huge surface area and vigorous surface

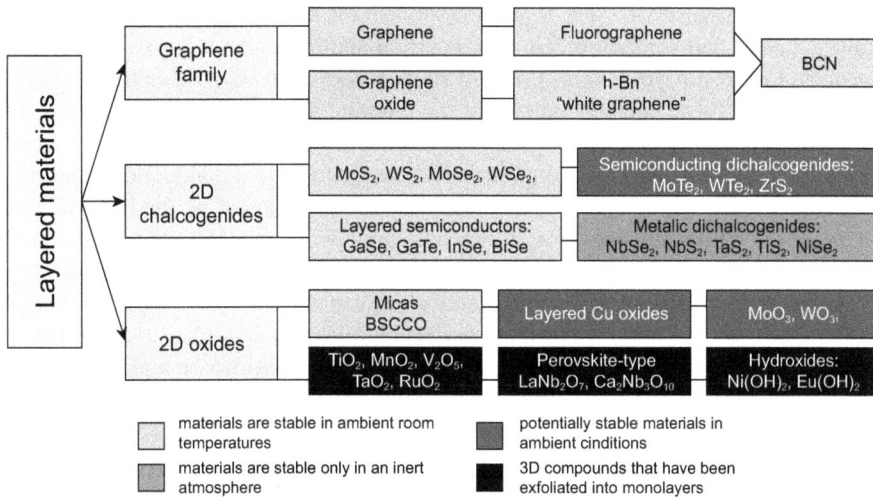

FIGURE 10.1   The classification of layered materials based stability.[12]

activities, graphene and other 2D materials demonstrate extraordinary applications in gas monitoring. The material used in sensing equipment requires large tensile strength, and metal oxide nanomaterials possess some characteristics that make them suitable for use in sensing applications. Their high surface-to-volume ratio makes more surface area available for analytes binding, allowing for more rapid and sensitive detection. Apart from these collective properties, some 2D materials also have exceptional properties, such as superior thermal transport of BN nanosheets, tuned energy bandgap, and high reactivity of molybdenum disulfide (MoS_2) toward hydrogen evolute.[11] Hence, owing to these properties, 2D materials are used as environmental sensors. Based on these 2D materials, different sensing techniques are evolved for environment pollutant, sensing such as fluorescence, chemiluminescence, plasma mass spectrometry, electrochemical, and surface-enhanced Raman spectroscopy. The classification of layered materials based stability is shown in Figure 10.1.[12]

10.3 CLASSIFICATION OF TWO-DIMENSIONAL MATERIALS BASED ENVIRONMENTAL SENSORS

10.3.1 GRAPHENE

Graphene has been regarded as a marvel material since its discovery. This exceptional material is distinct because of its incredible optical, electrical, and magnetic properties. It was accidentally created in small amounts, hundreds of years before actual adventure, using pencil and other graphite-based materials on electron microscopes in 1962. The material was subsequently found and described in 2004 by Andre Geim and Konstantin Novoselov, and they were granted the Nobel Prize in Physics in 2010 for the same.[13] Isolating high-quality graphene, which had previously seemed unachievable, was no longer so tough.

Graphene consists of a thin layer of carbon atoms that are densely arranged by a covalent bond and sp^2 hybridized. It is a carbon allotrope with a honeycomb lattice made up of a monolayer of atoms. The name *graphene* is an association of "graphite" and the suffix "-ene" demonstrates that the graphite allotrope of carbon comprises stacked graphene layers. Each atom in a graphene sheet is connected by a σ-bond to its three close neighbors and contributes one electron to a conduction band that extends over the whole sheet.[14] As a result, it can be regarded as the integral part of other allotropes; for instance, fullerenes, graphite, and CNTs can be pictured as wrapped-up sheets of graphene, while graphite can be visualized as stacked-up graphite sheets with relatively low distances between them around 3.37 Å. Although these allotropes are not created from graphene, in reality it is plausible. Graphene is stable and, at room temperature, exhibits magnificent electronic and mechanical properties like thermal conductivity of 5,000 W m^{-1} K^{-1}, the electrical conductivity of up to 6,000 S cm^{-1}, charge-carrier mobility of 250,000 cm^2 V^{-1} s^{-1}, and a large theoretical specific surface area of 2,630 m^2 g^{-1}. Moreover, it is highly transparent, with a retention of <2.3% toward visible light. All these properties make graphene and its derivative an excellent candidate to be used in various applications, specifically as an environment sensor. Various research groups used pristine graphene for detecting gas molecules like NO_2, CO_2, NH_3, etc.[15–17]

10.3.2 TRANSITION METAL DICHALCOGENIDES

Transition metal dichalcogenides (TMDs) materials are formulated as AX_2, where A represents transition metals and X belongs to chalcogens. Some of its properties are distinct, which are better than other 2D materials, for example, graphene with zero bandgap and lesser value of on/off ratio, which restricts its applications in detecting, and another material is black phosphorus (BP), having insufficient chemical stability and durability in inert conditions.[12] MoS_2 has a layered-like structure with high carrier mobility (60 cm^2/Vs at 250 K) with an energy gap that lies in the range of 1.2 to 1.8 eV, an elevated on/off ratio (approximately 108), and good stability in environmental conditions.[18] In FET sensor, two-dimensional MoS_2 is good channeling material and has applications in detecting different analytes such as nitrogen dioxide,[19,20] ammonia,[21] metal ion,[22] small molecule,[23] biomaterials for example microorganism,[24] protein,[25,26] etc. MoS_2, WS_2 nanosheets, etc. have a few fascinating optical, electric, magnetic, and chemical properties. Because of these properties, these materials have gained interest among researchers. The FET sensors based on TMDC also show a few unique characteristics. Gas sensors based on direct physisorption of gases on TMDC surface also depend on charge transfer, dipole-dipole interactions, and these operate in clean and chemically inert environments.[27] Sensing of water contaminants ordinarily works in aqueous media whose monitoring mechanism is complicated because of solution chemistry.

10.3.3 BLACK PHOSPHORUS (BP)

Initially, Bridgman put his emphasis on preparing black phosphorus (BP) in the year 1914. It was acquired by warming white phosphorus under high pressure (1.2 GPa).

After that, BP was rediscovered as 2D layered material. The BP has colossal properties in sensing devices given its interesting direct bandgap and outstanding electron versatility. Nanostructure catalysts based on semiconductors like TMDs and graphene exhibited superior catalytic properties. But usage of low-energy photons and minimizing the recombination of charge carriers is more challenging. Phosphorus-based layered 2D materials have gained more attention because of their large absorption range covering the full spectrum of light. The BP is unstable, and its properties are affected by changes in humid environments, which helps give to humid sensor applications.

Monolayered BP (phosphorene) consists of puckered honeycomb structure. That exemplifies the benefits of a basic 2D material while having an inherent bandgap near that of silicon.[28] The bandgap of BP for a single layer can be tuned from 1.51 eV to 0.59 eV for a five-layer system.[29] Because of its unique direct bandgap and amazing electron versatility, BP performs exceptionally well in sensors and is appropriate for creating gas sensors to recognize gas. The gases that are dangerous to human health, like NO_2 and NO, have high absorption energy with BP compared to other 2D materials based sensors. At a concentration of 200 ppb, the response of sensors to NO_2 can be expanded by a wide-range excess of 23%. The BP has neglected influence on cellular metabolism that is honored with acceptable similarity for biological cells, which makes it suitable for biological application. MicroRNA (miRNA) is a marker that can be utilized to anticipate and analyze disease and other modern medical problems. All the same, BP can be used to detect microRNA quickly and precisely.[30] The sensitivity of the detector is given by the formula $S = \Delta\theta\Delta n$, where $\Delta\theta$ is change in angle and Δn is change in the refractive index.[31]

10.3.4   2D METAL OXIDES

Metal oxides show an array in structural properties. It is classified into two classes: 2D layered MO (MoO_3, WO_3) and 2D non-layered MO (SnO_2 and ZnO nanosheets), which rely on the presence and absence of van der Waals layer structure in mass.[32] The 2D metal oxides exhibit large surface area due, so more atoms interact with the atmosphere. This makes it more suitable for gas sensors as compared to other sensors. The surface characteristics of MO are decided by the ionic character of the metal-oxygen bond and the number of oxygen particles. The conductivity of MO increases under ambient conditions, raising the temperature to T>250°C and keeping it below their melting point, then MO behaves as semiconductive material. This property permits MO to be broadly carried out in the gas sensor advances. Customary gas sensors made of metal oxides, for example, generally work at 100–400°C, prompting high power consumption and decreased sensor lifetime and stability due to instigated development of grains in the metal oxide. The manufacturing cost of MO is low in comparison to other sensing devices.

10.3.5   2D METAL-ORGANIC FRAMEWORK

Metal-organic frameworks are fashioned via self-assembly of molecules of inorganic metallic nodes with organic ligands. First, the concept of MOF was reported by

Yashi and his group.[33] In 1995, 60% of porosity and 2,900 m^2g^{-1} specific surface area was reported by them[34] on account of their properties and structure as high density of active sites, high surface area, high catalytic activity. Tailorable pore size MOF based sensor has been used for environmental contaminant detector together with gases, organic compounds, anions, and heavy metal ions. Electrochemical, optical, and field-effect transistor sensors are MOF-based environment sensors. The MOFs can have a high Langmuir surface area (>10,000 m^2g^{-1}), which makes it highly adsorbent toward analytes.[35] In water environments, heavy metals are one of the biodegradable pollutants. Even at trace levels, several heavy metal ions such as chromium, copper, cadmium, mercury, etc. are exceptionally harmful and risky to human health. Subsequently, a lot of MOF sensors for substantial metal particle identification in water have been reported.

10.3.6 MXenes Based 2D Materials

MXenes is another group of 2D materials that belong to the IV group. More than 80 MXnes have been theoretically predicted, and about 40 different MXene have been synthesized.[36] MXenes consist of ternary nitrides and carbides having formula $M_{n-1}AX_n$, where M denotes TM as Sc, V, Ta, Cr, etc. A belongs to group IIIA and group IVA elements, and X can either be C or N, where n = 1, 2, and 3 to form MXenes. In 2011, the primary MXenes material Ti_3C_2 was synthesized using layered hexagonal ternary carbide, by specifically etching the Al atoms in the layer, to form Ti_3AlC_2 with hydrofluoric acid at room temperature. MXenes have a layered structure that drives from MAX phases due to fast intercalation of various cations. The metallic conductivity of MXenes is up to 24,000 Scm^{-1} and has a hydrophilic surface, which make it a good sensing material. The CNTs integrated between MXenes nanosheets prevent self-restacking and a fast ion diffusion process. In addition to their surface terminations, electronic features of MXenes rely upon the nature of M, X. Hypothetical investigations demonstrated that, for specific composition with energy gaps about 0.25–2.0 eV, MXenes might behave metallic to semiconductor, except some heavy transition metal (e.g., Cr, Mo) containing MXenes, which may behave as an insulator. The preparation method for MXenes influences their electronic properties because of the termination and different extent of imperfections. MXenes possess intense absorption properties in UV-visible to NIR regions. MXene has excellent charge carrier mobility, making it an excellent medium for potential electrical interaction between the protein and the MXene based electrodes.

10.4 MONITORING OF ENVIRONMENTAL CONTAMINANTS

Various types of analytical tests are done to characterize the prepared sensor and for the knowledge of its performance. Some common tests are scanning electron microscopy (SEM), scanning transmission electron microscopy (STEM), cyclic voltammetry (CV), square wave voltammetry (SWV), electrochemical impedance spectroscopy (EIS), square-wave anodic stripping voltammetry (SWASV), X-ray photoelectron spectroscopy (XPS), atomic force microscopy (AFM), the transmission electron microscopy (TEM), thermogravimetric analysis (TG), UV-Vis spectroscopy, and differential pulse voltammetry (DPV). Different contaminants are described in the next section.

10.4.1 HEAVY METALS

Heavy metals like Pb, Ag, Au, Th, Co, Cr, Hg, Zn, etc. are used in construction of alloys, batteries, coating, electroplating, pigments, and so on. But pouring, processing, burning of fossil fuel, mining, inappropriate disposal, casting have made nonbiodegradable and extremely virulent heavy metals a threat to humans and environment.[37] Cd^{2+} may be carcinogenic and causes proteinuria, lung cancer, osteomalacia; whereas increased amount of Pb^{2+} may cause kidney injury, neurobehavioral development, high blood pressure, and anemia.[38] Excess consumption of Hg causes minamata, kidney and respiratory failure, attention deficiency in children, pulmonary edema,[39] cardio vascular disease, nephrotic syndrome, loss of IQ, cyanosis syndrome, etc.; whereas surplus amount of As^{3+} in water may lead to black foot and palm, cancer of bladder and mutations.[40] Cu^{2+}, Zn^{2+} are also toxic enough. Normal ways of detecting heavy metal ions are X-ray fluorescence, inductively coupled plasma mass spectroscopy, atomic absorption spectroscopy (AAS), inductively coupled plasma atomic emission spectroscopy. However these are lab-based techniques and cannot be used for on-site analysis. MoS_2 film, which is decorated with DNA functionalized Au Nps, can potentially serve as an FET sensor. The formation of which is shown in Figure 10.2.[22] Hg detection can be done using it with LOD lower than 0.1 nM and

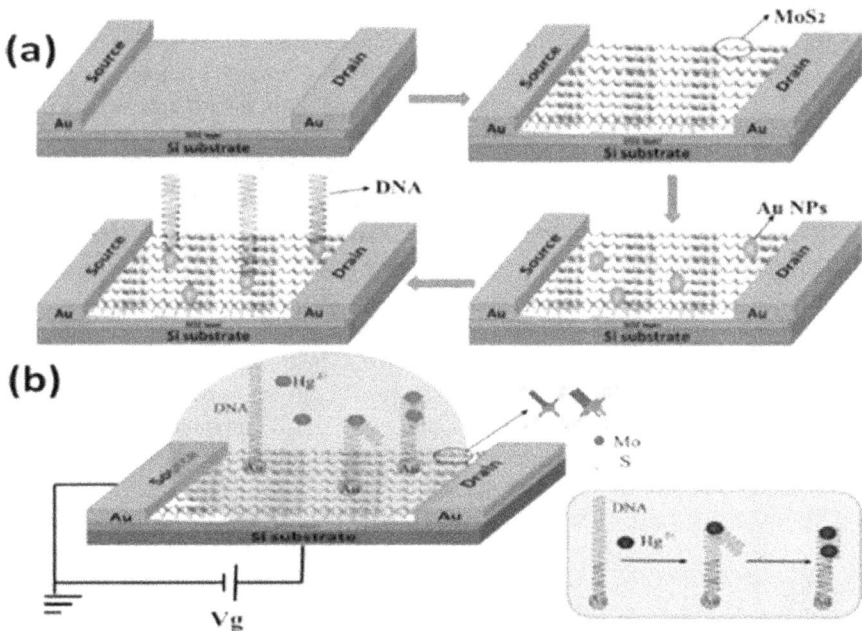

FIGURE10.2   (a) Schematic diagram for the fabrication process of the MoS_2/DNA-Au NPs hybrid structure. (b) FET sensor platform based on the hybrid structure.[22]

response time as low as 1–2 s. The detector even provides benefits like low price and high selectivity against other metal ions and acceptable real-time analysis.[22] Another FET device can be obtained by using polydimethylsiloxane coupled with MoS_2, which offers detection limit of 30 pM for Hg^{2+}.[41] Organic Hg is one of the most threatening segments of them.

Due to its lipophilic nature, CH_3Hg causes several damages to the human body. Aldehyde fabricating industry is one of the reasons of formation of CH_3Hg. 3D element doped haeckelite BN based sensor was reported for CH_3Hg sensor. Milks is a staple food all over the world, but it may contain various amounts of toxic heavy metals. A sensing electrode is reported to detect these kinds of heavy metals in milk. The electrochemical transducer used a novel bismuth film electrode based on electrochemically reduced graphene oxide (ERGNO) modified with SPE via electrodeposition method. It was used to detect Cd^{2+} and Pb^{2+} and shows a linear relationship in concentration of 1–60 µg L^{-1} with peak current. LOD for Cd^{2+} was 0.5 µg L^{-1} and for Pb^{2+} was 0.8 µg L^{-1}. Deposition potential and deposition time for both the metal were −1.2 V and 150 s, respectively.[42]

10.4.2 ORGANIC COMPOUNDS

Phenolic compounds are extensively used in various sectors, like pharmaceuticals, pesticides productions, wood preservation, petrochemical manufacturing, etc.[43] Due to its toxic nature, phenolic compounds have become a threat to humans and environment. A Pd-graphene/1-Butyl-3-methylimidazolium tetrafluoroborate/nafion composite electrode can detect 2-chlorophenol(2-Cp). Mixture of IL and sonochemically prepared Pd-graphene was dropped on polished GCE. Then the composite was dried, and again nafion was dropped. Detection range was 4–800 µmol L^{-1}, and detection limit was 1.5 µmol L^{-1}. Electrode shows excellent reproducibility, but 4-Cp interferes the peak current.[44] A CuO/nafion/GCE can be prepared by dropping mixture of nafion and CuO synthesized through aqueous chemical growth, on polished GCE. Figure 10.3 shows Ipa response increases linearly with increasing concentration of 2, 4, 6 Trichlorophenol LOD, and LOQ was found to be 0.046 and 0.154 µM, respectively.[45]

4-nitrophenol (4-Np) is one of the pollutants found in the environment and is extremely hazardous. A GCE modified by polyvinylpyrrolidone (PVP) assisted ultrasonically by exfoliated BP can potentially detect trace amount of 4-Np. The electrode shows DPV oxidation response at 1.09V for 20 µM. The oxidation peak current shows linearity with concentration in a range 0.10–5.0 µM, giving LOD 0.028µM. But the peak current decreases slightly in presence of similar compounds.[46] Presence of explosive or explosive-like substances is a big question to the homeland security and environmental safety. Thus, in this concern, J. Wu et al. manufactured TNT-specific peptide functionalized AuNPs/MoS_2 sensor. AuNPs/MoS_2 was prepared through one step conjugation and functionalized with TNT-specific peptide designed with sulfydryl group. TNT and peptide molecule manifested charge transfer interaction and thus optical spectra. Addition of TNT shifts the fundamental absorption peak from 541 nm. Upon increasing concentration of TNT from 2×10^{-7} to 1×10^{-14}, M absorption peak value decreases linearly. The sensor is highly selective to TNT, but DNT shows slight response.[47]

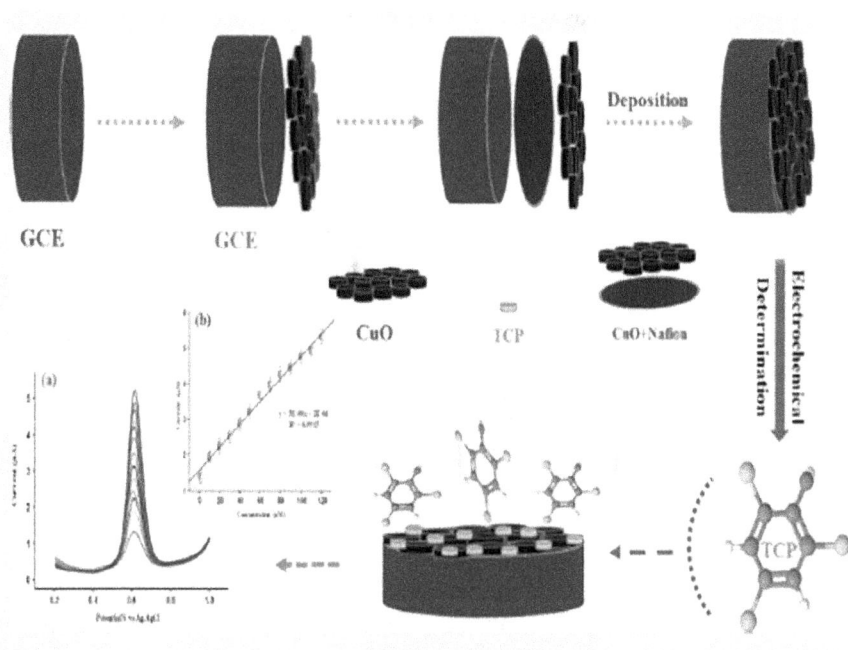

FIGURE 10.3 Schematic diagram of formation of CuO/nafion/GCE electrode and Ipa response curve.[45]

TABLE 10.1
Other Examples of Organic Molecule Detection Using 2D Materials

Organic Molecule	Sensor	Detection Range	LOD	References
Hydroquinone	Ti$_3$C$_2$-MWCNTs/GCE	2 µM to 150 µM	6.6 nM	48
Catacol			3.9 nM	
Nitromethane	Hb-CS/rGO-CS	5 µM to1.46 mM	1.5 µM	49
Phenol	m-GMN/c-PGEs	1 µg/mL to 7 µg/mL	0.35 µg/mL	50
Nitrobanzene	Cd MOF	0.010 mM to 0.060 mM	0.1 µM	51
Tartrazine	g-C3N4/PGE	$1.0 \times 10{-}7$ to $1.0 \times 10{-}5$ mol L^{-1}	0.21 µmol L^{-1}	52
Sunset yellow	GCE/ERGO	50–1,000 nM	19.2 nM	53

10.4.3 PESTICIDES

Pesticides are the second largest man-made chemicals after fertilizers.[54] Pesticides are chemicals that have agricultural applications in controlling or eliminating pests and unwanted weeds. Five major classifications of pesticides according to their composition are carbamate, organophosphate, pyrethroids, pyrethrin, and organo-chlorine. Abuse, inappropriate handling, and accidental release of these may cause serious venture to humans as well as environment.[55] Gas Chromatography (GC), gas chromatography/mass spectrometry (GC/MS), fluorometry, liquid chromatography/ mass spectrometry (LC-MS), high-performance liquid chromatography (HPLC), mass spectrometry (MS) are the traditional ways of determining pesticides.[56] But extreme complicated instrumentation, pre- and post-treatment of sample, require- ment of trained staff, high analytical cost, large time requirement restricts the real-time application, especially in emergency. In agricultural production, organo- phosphorus pesticides (OPs) are extensively used. The most poisonous OP is called paraoxon ethyl (PXL), which may cause poor visualization, diarrhea, hunger edema, tumor, neurotoxic effect, acute eye, even death.[57] PXL can be detected by using Pd NPs modified boron nitride (BN) heterojunction. First Pd NPs/BN HJ was fabri- cated by sonochemical method and further drop casted on glassy carbon electrode. Pd NPs/BN HJ provides working range from 0.09 to 210 μM and LOD 0.05 μM.[58] Other Ops, methyl parathion (MP), diazinon (DIA), and chlorpyrifos (CHL), in water sample can be determined by using BN quantum dot decorated GO obtained through ultrasonication and further drop casting of them on GCE. Linearity range was found to be 1.0×10^{-12}– 1.0×10^{-8} M, while LOD of MP, DIA, CHL are 3.1×10^{-13}, 6.7×10^{-14}, and 3.3×10^{-14} M, respectively. Good reproducibility, sensitivity, and stability were obtained for the sensor in water and apple juice as well.[59] A nanocom- posite of Prussian blue (PB) functionalized on $Ti_3C_2T_x$ was prepared by reduction of Fe(III) and $Fe(CN)_6^{3-}$ mixture with $Ti_3C_2T_x$. Then acetylcholinesterase (AchE) was immobilized on a PB@ $Ti_3C_2T_x$ modified GCE. It can potentially detect malathion in the range 0.0001–1.0 nM and LOD 1.3×10^{-16} M.[60] A nonenzymatic sensor was prepared for the detection of nitroaromatic OPs. First, silicon passivation was done on a silver screen printed electrode through roll-to-roll slot die coating method. Then GNPs were air brushed on the electrode in a roll-to-roll process. Last, ZrO_2 were electrosprayed in roll-to-roll process. The fabricated electrode showed rapid response time (10 min), high stability of 89.4% after 30 days, extreme sensitivity of 0.2 ppm with LOD 1 μM for methyl parathion.[61]

10.4.4 Antibiotic Compounds

Due to the nondegradable characteristics of antibiotics in drinking water, food products, and environment, it may cause many hazardous effects like disturbance in ecological equilibrium, allergic reactions, antibiotic-resistant bacterial strain, etc. Traditional analytical methods to detect antibiotics are capillary electrophoresis, sur- face enhanced Raman scattering (SERS), HPLC coupled with UV, LC-MS, super- critical fluid chromatography/mass spectrometry (SFC-MS), spectrophotometry, flow injection chemiluminescence (FI-CL), capillary zone electrophoresis (CZE), thin- layer chromatography, etc.[62,63] But these techniques are inconvenient as it requires expert staff for handling, long time of detection, complicated process. Some antibiot- ics that are reported to be detected simply using 2D materials are chloramphenicol, linezolid, kanamycin, ornidazole, nitrofuran, sulfamethoxazole.[64–66] Penicillin is a β-lactam based multipurpose antibiotic that is used in preventing and controlling various bacterial diseases in humans and animals. But relatively 5–10% population suffers from hypersensitivity of penicillin, causing allergic reaction for ultralow concentration as 1 ppb.[67] Y. Wu et al. constructed an ultrasensitive penicillin sensor using layer-by-layer method. First, single graphene nanosheet was prepared reducing GO (synthesized via Hummer's method) with $NaBH_4$. Further dispersed mixture of GO and hematein was drop casted on polished GCE. Next trihexyltetradecylphosphonium bis (trifluoromethylsulfonyl) imide (IL) and penicillinase were immobilized on the modified GCE one by one. Detection range was 1.25×10^{-13}– 7.5×10^{-3} M with detection limit 10^{-13} M (0.04 ppt). Good storability and selectivity were offered by the electrode with successful real-time sample detection.[68] Metronidazole (MNZ) is a widely used antibiotic that used in treatment of nosema apis,[69] infections, and acne diseases. But accumulation of MNZ is proven to be mutagenic and carcinogenic to humans. For the detection of MNZ, a molecular imprinted polyimide (MIP)/2D Sn_3O_4 was prepared. Sn_3O_4 was developed via hydrothermal method, above which polypyrrole particles were formed by chemical oxidative polymerization. The DPV peak current was linear with concentration of MNZ in the range 0.025–2.5 µM with LOD as low as 0.0032 µM. The sensor shows good selectivity, stability (92% after three weeks in room temperature) and practicability.[70] A popular drug used against gram positive bacteria is midecamycin (MD). A GO drop casted on GCE can potentially detect MD with detection range 3×10^{-7}– 2×10^{-4} mol L^{-1} with LOD 1×10^{-7} mol L^{-1}.[71] A lanthanum vanadate/functionalized boron nitride (LaV/BN) nanocomposite were prepared by sonochemical method for the detection of furazolidone (FZD). Figure 10.4 shows procedure of formation of LaV/BN/GCE electrode and CV signal of LaV/BN/GCE in different concentrations of FZD.[72] The composite was added on GCE by dropping. Peak current increases with increasing concentration of FZD from 50 to 250 µM, where the detection limit is 0.003 µM. Acceptable stability and real-time sample analysis is offered by this electrode.[72]

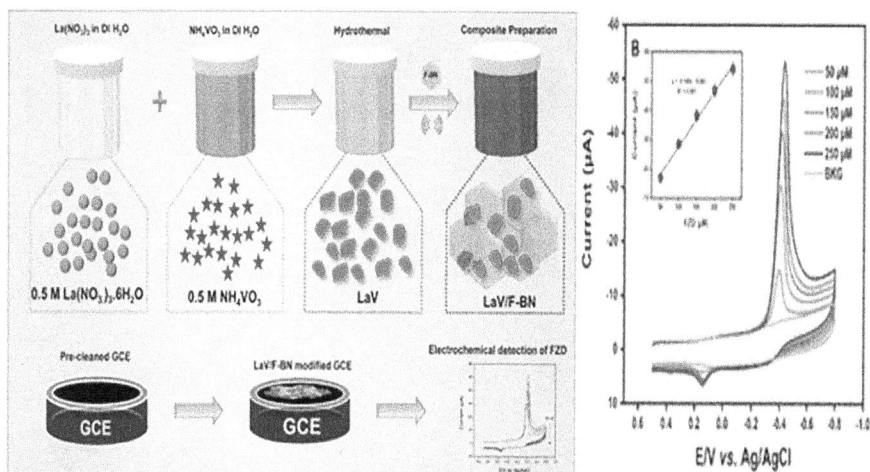

FIGURE 10.4   Procedure of formation of LaV/BN/GCE electrode and CV signal of LaV/BN/GCE in different concentrations of FZD.[72]

10.4.5 BACTERIA

Few foodborne bacteria are *Salmonella typhimurium, Escherichia coli* O157:H7, *Staphylococcus aureus, Listeria monocytogenes*, etc.[73] Detection of bacteria is normally done by processes that use culture techniques. But bacteria culture techniques are very much time engrossing and have very low positive rate (50–60%). New different ways are proposed like enzyme-linked immunosorbent assays (ELISA), polymerase chain reaction (PCR), nucleic acid amplification tests (NAATs).[74–76] But to perform ELISA, one requires enzyme-supporting antibodies while PCR is time-consuming, whereas for NAATs, highly equipped laboratory and skilled personnel are required. 2D materials are mainly functionalized with aptamers or antibodies so that they can bind with the bacteria.[77] A new idea to identify *E. coli* UT189 was proposed by using amine group functionalized PEI (polyethyleneimine) modified rGO electrode, which was further modified by anti-fimbrial antibody *E. coli* antibodies along with PEG (polyethylene glycol) modified pyrene units. LOD was 10 cfu mL^{-1}, and only 2.8% peak current was lost upon storage in PBS buffer at 4°C for two weeks[78] Figure 10.5 shows a holey reduced GO (hRGO) FET sensor that was functionalized with an antimicrobial peptide, magainin I that can potentially sense gram-negative bacteria.[79] Through reduction in the conductance, the device gives signal of the presence of bacteria. LOD was found to be 803 cfu mL^{-1} but the device is not reusable.[79]

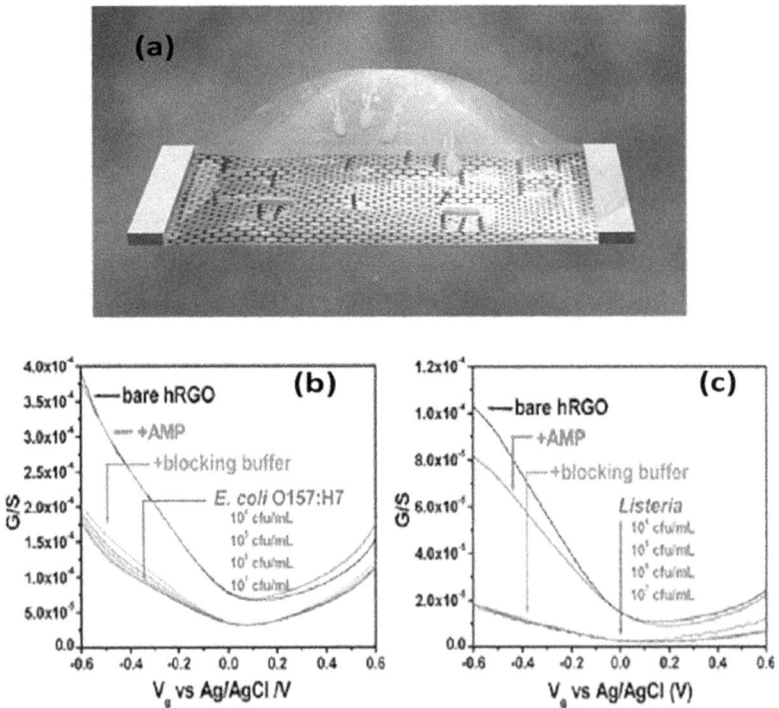

FIGURE 10.5   (a) Schematic of bacteria binding on a 2D sensor, (b) and (c) conductance (G) versus gate voltage (Vg) of bare hRGO FET devices in presence of *E. coli* and *Listeria*.[79]

E. coli causes most of the diseases like diarrhea, pneumonia, urinary tract infections, etc. An aminoterepthalic acid functioned GO based sensor was derived for *E. coli* detection. Aminoterepthalic acid adds amine functional group on the GO. The produced sensor shows good reproducibility and stability (up to 60 days) with LOD 2 cfu mL⁻¹ and four-minute response time.[80] *Mycobacterium tuberculosis* (MT) is one the common bacteria that causes tuberculosis. But recently they are reported to be drug resistant and often mutate their genes.

10.4.6 GASES

Gas detection is important in sectors like industrial productions, indoor air quality, greenhouse gas detection, and automobile exhaust measurements. Current commercial gas sensors are majorly based on metal oxide semiconductors, which works in elevated temperature. But to generate high temperature, the device needs high power and application of them in portable device. Nitrogen dioxide (NO_2) is one of the most toxic and death-dealing gas, which is harmful to humans over 1 ppm concentration.[81] NO_2 is one of the gases that majorly is produced in various industries and vehicles. It is a perilous and death-dealing gas that may cause numerous health problems. It may cause respiratory diseases like emphysema, bronchitis, and cardiovascular diseases. It has other harmful effects, like acid rain, primary cause of photochemical smog. It is formed as a byproduct of internal combustion engines, food processing, metal and petrol refinery, and thermal power stations. NO_2 is also used as biomarkers in diagnostic processes like infection of lungs tissue and gastrointestinal disorder symptoms.[82] A composite was prepared by using SnS_2 decorated rGO, which shows increased gas sensing ability in relatively lower temperature. The composite was tested for various gases (NO_2, NH_3, CH_4, C_2H_5OH, H_2O, and acetone). But the film shows no response for CH_4, C_2H_5OH, and acetone. NH_3 shows 15.3% response for 99 ppm only above 80°C. At 25°C and 50°C 11.9 ppm, NO_2 shows an irreversible ~85% and ~78% response, respectively. At 80°C, a 56.8% reversible response is obtained upon 5 minute exposure to 11.9 ppm NO_2, and minimum ~2.8 minute response time is required.[83] Due to the strong interaction of NO_2 and sp^2 carbons, graphene and CNTs show unsatisfactory high response time despite having great response. But 2D chalcogenides may be a better option. It was proved that suspended BP can give much better reproducibility and gas response than supported BP (23% increased at 200 ppm).[84] Multilayer BP was utilized in preparation of a field-effect transistor (FET) for NO_2 gas detection. It owns magnificent selectivity of 5 ppb and response time of ~280–350 s for different concentrations.[85]

H. Cui et al. prepared a tellurene gas sensor, which can detect NO_2 in the range from 25–150 ppb in dry air at room temperature.[86] This sensor provides LOD as low as 0.214 ppb and even stable till 100°C. Response and recovery time was found to be 83 s and 485 s for 25 ppb and 26 s and 240 s for 100 ppb NO_2.[86] Humidity sensing is important in food industry, health care, weather forecasting, and chemical industry.[87] For the first time $TiSi_2$ was tested as humidity sensing materials. Integrated electrode was formed by depositing 2D nanoflakes of $TiSi_2$. New sensor shows impedance sensitivity of 63 KΩ% RH at 1 KH$_z$. It shows good reproducibility and detection range from 0–100% relative humidity with stability up to 360 minutes.[88] An ammonia sensor electrode was prepared with scattered ZnO on silver paste, which was layered on an electrode. With concentration of NH_3 less than 2 ppm, the sensor provides reversible signal, which decreases slightly with increasing temperature from 250°C to 300°C. It can detect in the range of 600 ppb to 3 ppm NH_3 at 250°C. Rapid maximum response of 80% was obtained with good recovery. But baseline drift was observed, which was manipulated by baseline manipulation method.[89]

10.4.7 VOLATILE ORGANIC COMPOUNDS

Volatile organic compounds (VOCs) are organic compounds with low boiling point and high vapor pressure. These are common air pollutant that forms many carcinogenic compounds, thus harmful to human health.[90] Xylene may cause pneumonia, poisoning, dermatitis, whereas formaldehyde causes sick building syndrome.[91] VOCs can be detected via many sophisticated analytical techniques like fluorescence spectroscopy, flow-tube mass spectrometry, optical spectrometry, surface acoustic wave sensing, GC-MS, and quartz crystal microbalance techniques.[92] But these are inconvenient and time-consuming. An alternative is use of 2D materials in sensing devices, which can give fast results and in relatively low-cost and time. When Au decorated chemically exfoliated MoS_2 was exposed to various VOCs, Figure 10.6

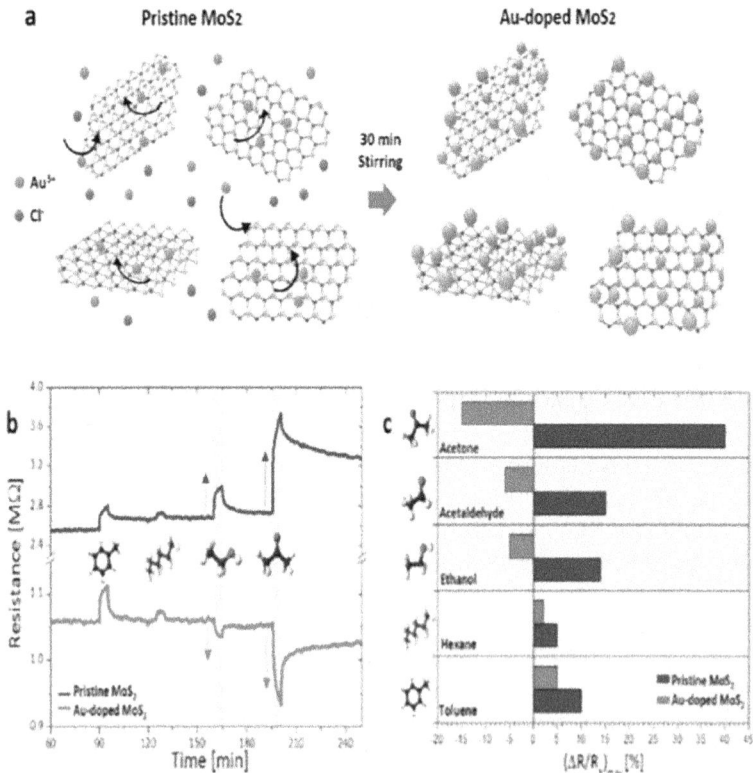

FIGURE 10.6 (a) Preparation of the Au-doped MoS_2 gas sensor, (b) real-time resistance, and (c) normalized response ($\Delta R/Rb$) of the pristine MoS_2 and Au-doped MoS_2 sensors exposed to various VOCs.[93]

TABLE 10.2
Other Examples of VOC Detections

VOCs	2D Materials	Detection Range	LOD	References
xylene	$NiCo_2O_4$ hollow microtubules	100 ppm	1 ppm	95
toluene	$CuO/Ti_3C_2T_x$MXene	50 ppm	0.32 ppm	96
acetone	$Ti_3C_2T_x$	50–1,000 ppb	0.011 ppb	97
ethanol		100–1,000 ppb	0.13 ppb	
isopropanol	$WO_3{\cdot}0.33H_2O$	100 ppm	1 ppm	98
glucose	NiO/GNS (graphene nanosheet)	5 μM–4.2 mM	5.0 μM	99
formaldehyde	Au/In_2O_3	50 ppm	1.42 ppb	100

shows how toluene, hexane gives positive response, whereas ethanol, acetone gives negative response.[93] But in the case of aldehyde, response depends on the doping ratio of Au and MoS_2.[93] An inkjet-printed titanium carbide $Ti_3C_2T_x$Mxene/WSe_2 (Tungsten diselenide) composite was designed by $Ti_3C_2T_x$ nanosheet and WSe_2 nanoflakes formed by liquid phase exfoliation. EtOH response was found to decrease with increasing amount of WSe_2 and increasing film thickness. It takes only 9.7 s to respond toward EtOH and 6.6 s for recovery at room temperature. The sensor is quite stable in low humidity that it offers 92% of its initial response after ten days in dry condition. It shows maximum response for EtOH and then MeOH.[94]

10.5 FUTURE PERSPECTIVE AND CONCLUSIONS

2D materials show exclusive interesting properties like enhanced surface area, absorption property. But they also have several problems. Practically, graphene is inert in nature and thus needs other treatment to activate it, which may lead to degradation of some of its properties. BP is also not really stable in environment. MoS_2 being a typical sulfide, its stability in aqueous media can be questioned, and it may oxidize easily in the presence of some oxidizing agent leading to formation of harmful ion like SO_4^{2-}.[101] These challenges need to be overcome. Though MoS_2 and WSe_2 are explored extensively, other examples of 2D chalcogenides are very rare. Environment stability of BP can be enhanced by surface functionalization or polymer coating.[102] But these ways need to be explored. Hybridization of 2D materials with other compounds imparts a great success, and there are plenty of examples of sensors for detection of various compounds. But very few of them are practically possible. Preparation of these sensor in commercial scale is a big challenge. Except all these improvement in LOD, selectivity, sensitivity, long-term usability, cost is essential. Majorly, LOD can be archived in ppb range with 2D materials. Though there are few examples of ppt range too. Some sensors require high temperature, specifically gas sensors. It required further demand to room temperature gas sensors. Some vital sensors are also not reusable due to the strong interaction with the analyte. Biofouling—that is, unwanted growth of microorganisms, algae, plants, etc. on sensor surface, which affects its performance, in presence of moisture—is also a challenge that requires attention. By polymer coating, it can be controlled to some extent. Last, understanding of analyte sensor interaction and sensing mechanism is very much important and needs to be fully known as only this can be the key to open the benefits of 2D materials completely.

REFERENCES

1. Bi, H.; Han, X., 10—Chemical sensors for environmental pollutant determination. In *Chemical, Gas, and Biosensors for Internet of Things and Related Applications*, Mitsubayashi, K.; Niwa, O.; Ueno, Y., Eds. Elsevier: 2019; pp. 147–160.
2. Rai, V.; Hapuarachchi, H. C.; Ng, L. C.; Soh, S. H.; Leo, Y. S.; Toh, C.-S., Ultrasensitive cDNA detection of dengue virus RNA using electrochemical nanoporous membrane-based biosensor. *PLOS ONE* **2012,**7 (8), e42346.
3. Zamora-Ledezma, C.; Negrete-Bolagay, D.; Figueroa, F.; Zamora-Ledezma, E.; Ni, M.; Alexis, F.; Guerrero, V. H., Heavy metal water pollution: A fresh look about hazards, novel and conventional remediation methods. *Environmental Technology & Innovation* **2021,**22, 101504.
4. Mohamed, A.-M. O.; Maraqa, M. A.; Howari, F. M.; Paleologos, E. K., Chapter 9—Outdoor air pollutants: sources, characteristics, and impact on human health and the environment. In *Pollution Assessment for Sustainable Practices in Applied Sciences and Engineering*, Mohamed, A.-M. O.; Paleologos, E. K.; Howari, F. M., Eds. Butterworth-Heinemann: 2021; pp. 491–554.
5. Godoy, M.; Sánchez, J., Chapter 12—Antibiotics as emerging pollutants in water and its treatment. In *Antibiotic Materials in Healthcare*, Kokkarachedu, V.; Kanikireddy, V.; Sadiku, R., Eds. Academic Press: 2020; pp. 221–230.
6. Wang, J.; Chen, X.; Wu, K.; Zhang, M.; Huang, W., Highly-sensitive electrochemical sensor for Cd2+ and Pb2+ based on the synergistic enhancement of exfoliated graphene nanosheets and bismuth. *Electroanalysis* **2016,**28 (1), 63–68.
7. Singh, R. S.; Li, D.; Xiong, Q.; Santoso, I.; Yu, X.; Chen, W.; Rusydi, A.; Wee, A. T. S., Anomalous photoresponse in the deep-ultraviolet due to resonant excitonic effects in oxygen plasma treated few-layer graphene. *Carbon* **2016,**106, 330–335.
8. Chhowalla, M.; Shin, H. S.; Eda, G.; Li, L. J.; Loh, K. P.; Zhang, H., The chemistry of two-dimensional layered transition metal dichalcogenide nanosheets. *Nature Chemistry* **2013,**5 (4), 263–75.
9. Singh, R. S.; Gautam, A.; Rai, V., Graphene-based bipolar plates for polymer electrolyte membrane fuel cells. *Frontiers of Materials Science* **2019,**13 (3), 217–241.
10. Singh, R. S.; Jansen, M.; Ganguly, D.; Kulkarni, G. U.; Ramaprabhu, S.; Choudhary, S. K.; Pramanik, C., Shellac derived graphene films on solid, flexible, and porous substrates for high performance bipolar plates and supercapacitor electrodes. *Renewable Energy* **2022,**181, 1008–1022.
11. He, K.; Poole, C.; Mak, K. F.; Shan, J., Experimental demonstration of continuous electronic structure tuning via strain in atomically thin MoS$_2$. *Nano Letters* **2013,**13 (6), 2931–2936.
12. Tyagi, D.; Wang, H.; Huang, W.; Hu, L.; Tang, Y.; Guo, Z.; Ouyang, Z.; Zhang, H., Recent advances in two-dimensional-material-based sensing technology toward health and environmental monitoring applications. *Nanoscale* **2020,**12 (6), 3535–3559.
13. Varghese, S. S.; Varghese, S. H.; Swaminathan, S.; Singh, K. K.; Mittal, V., Two-dimensional materials for sensing: graphene and beyond. *Electronics* **2015,**4 (3), 651–687.

14. Tan, C.; Cao, X.; Wu, X.-J.; He, Q.; Yang, J.; Zhang, X.; Chen, J.; Zhao, W.; Han, S.; Nam, G.-H.; Sindoro, M.; Zhang, H., Recent advances in ultrathin two-dimensional nanomaterials. *Chemical Reviews* **2017,***117* (9), 6225–6331.

15. Ko, G.; Kim, H. Y.; Ahn, J.; Park, Y. M.; Lee, K. Y.; Kim, J., Graphene-based nitrogen dioxide gas sensors. *Current Applied Physics* **2010,***10* (4), 1002–1004.

16. Yoon, H. J.; Jun, D. H.; Yang, J. H.; Zhou, Z.; Yang, S. S.; Cheng, M. M.-C., Carbon dioxide gas sensor using a graphene sheet. *Sensors and Actuators B: Chemical* **2011,***157* (1), 310–313.

17. Romero, H. E.; Joshi, P.; Gupta, A. K.; Gutierrez, H. R.; Cole, M. W.; Tadigadapa, S. A.; Eklund, P. C., Adsorption of ammonia on graphene. *Nanotechnology* **2009,***20* (24), 245501.

18. Chen, X.; Liu, C.; Mao, S., Environmental analysis with 2D transition-metal dichalcogenide-based field-effect transistors. *Nano-Micro Letters* **2020,***12* (1), 95.

19. Radisavljevic, B.; Radenovic, A.; Brivio, J.; Giacometti, V.; Kis, A., Single-layer MoS₂ transistors. *Nature Nanotechnology* **2011,***6* (3), 147–150.

20. Han, Y.; Huang, D.; Ma, Y.; He, G.; Hu, J.; Zhang, J.; Hu, N.; Su, Y.; Zhou, Z.; Zhang, Y.; Yang, Z., Design of hetero-nanostructures on MoS₂ nanosheets to boost NO2 Room-temperature sensing. *ACS Applied Materials & Interfaces* **2018,***10* (26), 22640–22649.

21. Late, D. J.; Doneux, T.; Bougouma, M., Single-layer MoSe2 based NH3 gas sensor. *Applied Physics Letters* **2014,***105* (23), 233103.

22. Zhou, G.; Chang, J.; Pu, H.; Shi, K.; Mao, S.; Sui, X.; Ren, R.; Cui, S.; Chen, J., Ultrasensitive mercury ion detection using DNA-functionalized molybdenum disulfide nanosheet/gold nanoparticle hybrid field-effect transistor device. *ACS Sensors* **2016,***1* (3), 295–302.

23. Zheng, C.; Jin, X.; Li, Y.; Mei, J.; Sun, Y.; Xiao, M.; Zhang, H.; Zhang, Z.; Zhang, G.-J., Sensitive molybdenum disulfide based field effect transistor sensor for real-time monitoring of hydrogen peroxide. *Scientific Reports* **2019,***9* (1), 759.

24. Lee, H. W.; Kang, D.-H.; Cho, J. H.; Lee, S.; Jun, D.-H.; Park, J.-H., Highly sensitive and reusable membraneless field-effect transistor (FET)-type tungsten diselenide (WSe2) biosensors. *ACS Applied Materials & Interfaces* **2018,***10* (21), 17639–17645.

25. Singhal, C.; Khanuja, M.; Chaudhary, N.; Pundir, C. S.; Narang, J., Detection of chikungunya virus DNA using two-dimensional MoS₂ nanosheets based disposable biosensor. *Scientific Reports* **2018,***8* (1), 7734.

26. Yoo, G.; Park, H.; Kim, M.; Song, W. G.; Jeong, S.; Kim, M. H.; Lee, H.; Lee, S. W.; Hong, Y. K.; Lee, M. G.; Lee, S.; Kim, S., Real-time electrical detection of epidermal skin MoS₂ biosensor for point-of-care diagnostics. *Nano Research* **2017,***10* (3), 767–775.

27. Cho, B.; Hahm, M. G.; Choi, M.; Yoon, J.; Kim, A. R.; Lee, Y.-J.; Park, S.-G.; Kwon, J.-D.; Kim, C. S.; Song, M.; Jeong, Y.; Nam, K.-S.; Lee, S.; Yoo, T. J.; Kang, C. G.; Lee, B. H.; Ko, H. C.; Ajayan, P. M.; Kim, D.-H., Charge-transfer-based Gas Sensing Using Atomic-layer MoS₂. *Scientific Reports* **2015,***5* (1), 8052.

28. Tran, V.; Soklaski, R.; Liang, Y.; Yang, L., Layer-controlled band gap and anisotropic excitons in few-layer black phosphorus. *Physical Review B* **2014,***89* (23), 235319.

29. Shirotani, I., Growth of large single crystals of black phosphorus at high pressures and temperatures, and its electrical properties. *Molecular Crystals and Liquid Crystals* **1982,***86* (1), 203–211.

30. Walia, S.; Sabri, Y.; Ahmed, T.; Field, M. R.; Ramanathan, R.; Arash, A.; Bhargava, S. K.; Sriram, S.; Bhaskaran, M.; Bansal, V.; Balendhran, S., Defining the role of humidity in the ambient degradation of few-layer black phosphorus. *2D Materials* **2016,***4* (1), 015025.

31. Zhang, Y.; Jiang, Q.; Lang, P.; Yuan, N.; Tang, J., Fabrication and applications of 2D black phosphorus in catalyst, sensing and electrochemical energy storage. *Journal of Alloys and Compounds* **2021,***850*, 156580.

32. Suman, P. H.; Felix, A. A.; Tuller, H. L.; Varela, J. A.; Orlandi, M. O., Comparative gas sensor response of SnO2, SnO and Sn3O4 nanobelts to NO2 and potential interferents. *Sensors and Actuators B: Chemical* **2015,***208*, 122–127.

33. Yaghi, O. M.; Li, H., Hydrothermal synthesis of a metal-organic framework containing large rectangular channels. *Journal of the American Chemical Society* **1995,***117* (41), 10401–10402.

34. Yaghi, O. M.; Li, G.; Li, H., Selective binding and removal of guests in a microporous metal-organic framework. *Nature* **1995,***378* (6558), 703–706.

35. Li, H.; Eddaoudi, M.; O'Keeffe, M.; Yaghi, O. M., Design and synthesis of an exceptionally stable and highly porous metal-organic framework. *Nature* **1999,***402* (6759), 276–279.

36. Li, Q.; Li, Y.; Zeng, W., Preparation and application of 2D MXene-based gas sensors: a review. *Chemosensors* **2021,***9* (8), 225.

37. Wu, Y.; Pang, H.; Liu, Y.; Wang, X.; Yu, S.; Fu, D.; Chen, J.; Wang, X., Environmental remediation of heavy metal ions by novel-nanomaterials: a review. *Environmental Pollution* **2019,***246*, 608–620.

38. Shen, L.-L.; Zhang, G.-R.; Li, W.; Biesalski, M.; Etzold, B. J. M., Modifier-free microfluidic electrochemical sensor for heavy-metal detection. *ACS Omega* **2017,***2* (8), 4593–4603.

39. Zahir, F.; Rizwi, S. J.; Haq, S. K.; Khan, R. H., Low dose mercury toxicity and human health. *Environmental Toxicology and Pharmacology* **2005,***20* (2), 351–360.

40. Mandal, B. K.; Suzuki, K. T., Arsenic round the world: a review. *Talanta* **2002,***58* (1), 201–235.

41. Jiang, S.; Cheng, R.; Ng, R.; Huang, Y.; Duan, X., Highly sensitive detection of mercury(II) ions with few-layer molybdenum disulfide. *Nano Research* **2015,***8* (1), 257–262.

42. Ping, J.; Wang, Y.; Wu, J.; Ying, Y., Development of an electrochemically reduced graphene oxide modified disposable bismuth film electrode and its application for stripping analysis of heavy metals in milk. *Food Chemistry* **2014,***151*, 65–71.

43. de Morais, P.; Stoichev, T.; Basto, M. C.; Vasconcelos, M. T., Extraction and preconcentration techniques for chromatographic determination of chlorophenols in environmental and food samples. *Talanta* **2012,***89*, 1–11.

44. Shi, J.-J.; Zhu, J.-J., Sonoelectrochemical fabrication of Pd-graphene nanocomposite and its application in the determination of chlorophenols. *Electrochimica Acta—ELECTROCHIM ACTA* **2011,***56*, 6008–6013.

45. Buledi, J. A.; Solangi, A. R.; Memon, S. Q.; Haider, S. I.; Ameen, S.; Khand, N. H.; Bhatti, A.; Qambrani, N., Nonenzymatic electrochemical detection of 2,4,6-trichlorophenol using CuO/Nafion/GCE: A practical sensor for environmental toxicants. *Langmuir* **2021,***37* (10), 3214–3222.

46. Jian, S.; Liu, L.; Huang, W.; Wu, K., Polyvinylpyrrolidone-assisted solvent exfoliation of black phosphorus nanosheets and electrochemical sensing of p-nitrophenol. *Analytica Chimica Acta* **2021,***1167*, 338594.

47. Wu, J.; Lu, Y.; Wu, Z.; Li, S.; Zhang, Q.; Chen, Z.; Jiang, J.; Lin, S.; Zhu, L.; Li, C.; Liu, Q., Two-dimensional molybdenum disulfide (MoS$_2$) with gold nanoparticles for biosensing of explosives by optical spectroscopy. *Sensors and Actuators, B: Chemical* **2018,***261*, 279–287.

48. Huang, R.; Chen, S.; Yu, J.; Jiang, X., Self-assembled Ti(3)C(2) /MWCNTs nanocomposites modified glassy carbon electrode for electrochemical simultaneous detection of hydroquinone and catechol. *Ecotoxicology and Environmental Safety* **2019,***184*, 109619.

49. Wen, Y.; Wen, W.; Zhang, X.; Wang, S., Highly sensitive amperometric biosensor based on electrochemically-reduced graphene oxide-chitosan/hemoglobin nanocomposite for nitromethane determination. *Biosensors and Bioelectronics* **2016,***79*, 894–900.

50. Congur, G., Development of a novel methyl germanane modified disposable sensor and its application for voltammetric phenol detection. *Surfaces and Interfaces* **2021,***25*, 101268.

51. Yang, L.; Liu, Y.-L.; Liu, C.-G.; Ye, F.; Fu, Y., A luminescent sensor based on a new Cd-MOF for nitro explosives and organophosphorus pesticides detection. *Inorganic Chemistry Communications* **2020,***122*, 108272.

52. Karimi, M. A.; Aghaei, V. H.; Nezhadali, A.; Ajami, N., Graphitic carbon nitride as a new sensitive material for electrochemical determination of trace amounts of tartrazine in food samples. *Food Analytical Methods* **2018,***11* (10), 2907–2915.

53. Tran, Q. T.; Phung, T. T.; Nguyen, Q. T.; Le, T. G.; Lagrost, C., Highly sensitive and rapid determination of sunset yellow in drinks using a low-cost carbon material-based electrochemical sensor. *Analytical and Bioanalytical Chemistry* **2019,***411* (28), 7539–7549.

54. Gould, F.; Brown, Z. S.; Kuzma, J., Wicked evolution: Can we address the sociobiological dilemma of pesticide resistance? *Science* **2018,***360* (6390), 728–732.

55. Dincer, C.; Bruch, R.; Costa-Rama, E.; Fernández-Abedul, M. T.; Merkoçi, A.; Manz, A.; Urban, G. A.; Güder, F., Disposable sensors in diagnostics, food, and environmental monitoring. *Advanced Materials* **2019,***31* (30), 1806739.

56. Lara-Ortega, F. J.; Robles-Molina, J.; Brandt, S.; Schütz, A.; Gilbert-López, B.; Molina-Díaz, A.; García-Reyes, J. F.; Franzke, J., Use of dielectric barrier discharge ionization to minimize matrix effects and expand coverage in pesticide residue analysis by liquid chromatography-mass spectrometry. *Analytica Chimica Acta* **2018,***1020*, 76–85.

57. Worek, F.; Aurbek, N.; Wetherell, J.; Pearce, P.; Mann, T.; Thiermann, H., Inhibition, reactivation and aging kinetics of highly toxic organophosphorus compounds: pig versus minipig acetylcholinesterase. *Toxicology* **2008,***244* (1), 35–41.

58. Vengudusamy, R.; Ramachandran, B.; Chen, S.-M.; Kokulnathan, T., Coherent design of palladium nanostructures adorned on the boron nitride heterojunctions for the unparalleled electrochemical determination of fatal organophosphorus pesticides. *Sensors and Actuators B: Chemical* **2019,***307*, 127586.

59. Yola, M. L., Electrochemical activity enhancement of monodisperse boron nitride quantum dots on graphene oxide: Its application for simultaneous detection of organophosphate pesticides in real samples. *Journal of Molecular Liquids* **2019,***277*, 50–57.

60. He, Y.; Zhou, X.; Zhou, L.; Zhang, X.; Ma, L.; Jiang, Y.; Gao, J., Self-reducing prussian blue on Ti3C2Tx MXene nanosheets as a dual-functional nanohybrid for hydrogen peroxide and pesticide sensing. *Industrial & Engineering Chemistry Research* **2020,***59* (35), 15556–15564.

61. Ulloa, A. M.; Glassmaker, N.; Oduncu, M. R.; Xu, P.; Wei, A.; Cakmak, M.; Stanciu, L., Roll-to-roll manufactured sensors for nitroaromatic organophosphorus pesticides detection. *ACS Applied Materials & Interfaces* **2021,***13* (30), 35961–35971.

62. Santos, S. M.; Henriques, M.; Duarte, A. C.; Esteves, V. I., Development and application of a capillary electrophoresis based method for the simultaneous screening of six antibiotics in spiked milk samples. *Talanta* **2007,***71* (2), 731–7.

63. Wang, Y.; Zhang, P.; Jiang, N.; Gong, X.; Meng, L.; Wang, D.; Ou, N.; Zhang, H., Simultaneous quantification of metronidazole, tinidazole, ornidazole and morinidazole in human saliva. *Journal of Chromatography B: Analytical Technologies in the Biomedical and Life Sciences* **2012,***899*, 27–30.

64. Cardoso, A. R.; Marques, A. C.; Santos, L.; Carvalho, A. F.; Costa, F. M.; Martins, R.; Sales, M. G. F.; Fortunato, E., Molecularly-imprinted chloramphenicol sensor with laser-induced graphene electrodes. *Biosensors and Bioelectronics* **2019,***124–125*, 167–175.

65. Chen, X.; Hao, S.; Zong, B.; Liu, C.; Mao, S., Ultraselective antibiotic sensing with complementary strand DNA assisted aptamer/MoS$_2$ field-effect transistors. *Biosensors and Bioelectronics* **2019,***145*, 111711.

66. Hu, S.; Fei, Q.; Li, Y.; Wang, B.; Yu, Y., Br-terminated 2D Bi2WO6 nanosheets as a sensitive light-regenerated electrochemical sensor for detecting sulfamethoxazole antibiotic. *Surfaces and Interfaces* **2021,***25*, 101302.

67. Castells, M.; Khan, D. A.; Phillips, E. J., Penicillin allergy. *The New England Journal of Medicine* **2019,***381* (24), 2338–2351.

68. Wu, Y.; Tang, L.; Huang, L.; Han, Z.; Wang, J.; Pan, H., A low detection limit penicillin biosensor based on single graphene nanosheets preadsorbed with hematein/ionic liquids/penicillinase. *Materials Science and Engineering: C* **2014,***39*, 92–99.

69. Gisder, S.; Genersch, E., Identification of candidate agents active against N. ceranae infection in honey bees: establishment of a medium throughput screening assay based on N. ceranae infected cultured cells. *PLOS ONE* **2015,***10* (2), e0117200.

70. Wang, J.; Du, W.; Huang, X.; Hu, J.; Xia, W.; Jin, D.; Shu, Y.; Xu, Q.; Hu, X., A novel metronidazole electrochemical sensor based on surface imprinted vertically cross-linked two-dimensional Sn3O4 nanoplates. *Analytical Methods* **2018,***10* (41), 4985–4994.

71. Xi, X.; Ming, L., A voltammetric sensor based on electrochemically reduced graphene modified electrode for sensitive determination of midecamycin. *Analytical Methods* **2012,***4* (9), 3013–3018.

72. Kokulnathan, T.; Almutairi, G.; Chen, S.-M.; Chen, T.-W.; Ahmed, F.; Arshi, N.; AlOtaibi, B., Construction of lanthanum vanadate/functionalized boron nitride nanocomposite: the electrochemical sensor for monitoring of furazolidone. *ACS Sustainable Chemistry & Engineering* **2021,***9* (7), 2784–2794.

73. Pashazadeh, P.; Mokhtarzadeh, A.; Hasanzadeh, M.; Hejazi, M.; Hashemi, M.; de la Guardia, M., Nano-materials for use in sensing of salmonella infections: recent advances. *Biosensors and Bioelectronics* **2017,***87*, 1050–1064.

74. Wang, H.; Chen, H.-W.; Hupert, M. L.; Chen, P.-C.; Datta, P.; Pittman, T. L.; Goettert, J.; Murphy, M. C.; Williams, D.; Barany, F.; Soper, S. A., Fully integrated thermoplastic genosensor for the highly sensitive detection and identification of multi-drug-resistant tuberculosis. *Angewandte Chemie International Edition* **2012,***51* (18), 4349–4353.

75. Rai, V.; Nyine, Y. T.; Hapuarachchi, H. C.; Yap, H. M.; Ng, L. C.; Toh, C.-S., Electrochemically amplified molecular beacon biosensor for ultrasensitive DNA sequence-specific detection of Legionella sp. *Biosensors and Bioelectronics* **2012,***32* (1), 133–140.

76. Rai, V.; Toh, C.-S., Electrochemical amplification strategies in DNA nanosensors. *Nanoscience and Nanotechnology Letters* **2013,***5* (6), 613–623.

77. So, H.-M.; Park, D.-W.; Jeon, E.-K.; Kim, Y.-H.; Kim, B. S.; Lee, C.-K.; Choi, S. Y.; Kim, S. C.; Chang, H.; Lee, J.-O., Detection and titer estimation of escherichia coli using aptamer-functionalized single-walled carbon-nanotube field-effect transistors. *Small* **2008,***4* (2), 197–201.

78. Jijie, R.; Kahlouche, K.; Barras, A.; Yamakawa, N.; Bouckaert, J.; Gharbi, T.; Szunerits, S.; Boukherroub, R., Reduced graphene oxide/polyethylenimine based immunosensor for the selective and sensitive electrochemical detection of uropathogenic Escherichia coli. *Sensors and Actuators B: Chemical* **2018,***260*, 255–263.

79. Chen, Y.; Michael, Z. P.; Kotchey, G. P.; Zhao, Y.; Star, A., Electronic detection of bacteria using holey reduced graphene oxide. *ACS Applied Materials & Interfaces* **2014,***6* (6), 3805–3810.

80. Gupta, A.; Bhardwaj, S. K.; Sharma, A. L.; Deep, A., A graphene electrode functionalized with aminoterephthalic acid for impedimetric immunosensing of Escherichia coli. *Microchimica Acta* **2019,***186* (12), 800.

81. Novikov, S.; Lebedeva, N.; Satrapinski, A.; Walden, J.; Davydov, V.; Lebedev, A., Graphene based sensor for environmental monitoring of NO2. *Sensors and Actuators B: Chemical* **2016,***236*, 1054–1060.

82. Ou, J. Z.; Yao, C. K.; Rotbart, A.; Muir, J. G.; Gibson, P. R.; Kalantar-zadeh, K., Human intestinal gas measurement systems: in vitro fermentation and gas capsules. *Trends in Biotechnology* **2015,***33* (4), 208–213.

83. Shafiei, M.; Bradford, J.; Khan, H.; Piloto, C.; Wlodarski, W.; Li, Y.; Motta, N., Low-operating temperature NO2 gas sensors based on hybrid two-dimensional SnS2-reduced graphene oxide. *Applied Surface Science* **2018,***462*, 330–336.

84. Lee, G.; Kim, S.; Jung, S.; Jang, S.; Kim, J. H., Suspended black phosphorus nanosheet gas sensors. *Sensors and Actuators B-Chemical* **2017,***250*, 569–573.

85. Abbas, A. N.; Liu, B.; Chen, L.; Ma, Y.; Cong, S.; Aroonyadet, N.; Köpf, M.; Nilges, T.; Zhou, C., Black phosphorus gas sensors. *ACS Nano* **2015,***9* (5), 5618–5624.

86. Cui, H.; Zheng, K.; Xie, Z.; Yu, J.; Zhu, X.; Ren, H.; Wang, Z.; Zhang, F.; Li, X.; Tao, L.-Q.; Zhang, H.; Chen, X., Tellurene nanoflake-based NO2 sensors with superior sensitivity and a sub-parts-per-billion detection limit. *ACS Applied Materials & Interfaces* **2020,***12* (42), 47704–47713.

87. He, J.; Xiao, P.; Shi, J.; Liang, Y.; Lu, W.; Chen, Y.; Wang, W.; Théato, P.; Kuo, S.-W.; Chen, T., High performance humidity fluctuation sensor for wearable devices via a bioinspired atomic-precise tunable graphene-polymer heterogeneous sensing junction. *Chemistry of Materials* **2018,***30* (13), 4343–4354.

88. Shaukat, R. A.; Khan, M. U.; Saqib, Q. M.; Chougale, M. Y.; Kim, J.; Bae, J., All range highly linear and sensitive humidity sensor based on 2D material TiSi2 for real-time monitoring. *Sensors and Actuators B: Chemical* **2021,***345*, 130371.

89. Kanaparthi, S.; Govind Singh, S., Highly sensitive and ultra-fast responsive ammonia gas sensor based on 2D ZnO nanoflakes. *Materials Science for Energy Technologies* **2020,***3*, 91–96.

90. Kampa, M.; Castanas, E., Human health effects of air pollution. *Environmental Pollution* **2008,***151* (2), 362–7.

91. Akiyama, T.; Ishikawa, Y.; Hara, K., Xylene sensor using double-layered thin film and Ni-deposited porous alumina. *Sensors and Actuators B: Chemical* **2013,***181*, 348–352.

92. Buszewski, B.; Ulanowska, A.; Ligor, T.; Denderz, N.; Amann, A., Analysis of exhaled breath from smokers, passive smokers and non-smokers by solid-phase microextraction gas chromatography/mass spectrometry. *Biomedical chromatography : BMC* **2009,***23*, 551–6.

93. Cho, S.-Y.; Koh, H.-J.; Yoo, H.-W.; Kim, J.-S.; Jung, H.-T., Tunable volatile-organic-compound sensor by using Au nanoparticle incorporation on MoS$_2$. *ACS Sensors* **2017,***2* (1), 183–189.

94. Chen, W. Y.; Jiang, X.; Lai, S.-N.; Peroulis, D.; Stanciu, L., Nanohybrids of a MXene and transition metal dichalcogenide for selective detection of volatile organic compounds. *Nature Communications* **2020,***11* (1), 1302.

95. Du, L.; Song, X.; Liang, X.; Liu, Y.; Zhang, M., Formation of NiCo2O4 hierarchical tubular nanostructures for enhanced xylene sensing properties. *Applied Surface Science* **2020,***526*, 146706.

96. Hermawan, A.; Zhang, B.; Taufik, A.; Asakura, Y.; Hasegawa, T.; Zhu, J.; Shi, P.; Yin, S., CuO nanoparticles/Ti3C2Tx MXene hybrid nanocomposites for detection of toluene gas. *ACS Applied Nano Materials* **2020,***3* (5), 4755–4766.

97. Kim, S. J.; Koh, H.-J.; Ren, C. E.; Kwon, O.; Maleski, K.; Cho, S.-Y.; Anasori, B.; Kim, C.-K.; Choi, Y.-K.; Kim, J.; Gogotsi, Y.; Jung, H.-T., Metallic Ti3C2Tx MXene gas sensors with ultrahigh signal-to-noise ratio. *ACS Nano* **2018,***12* (2), 986–993.

98. Perfecto, T. M.; Zito, C. A.; Mazon, T.; Volanti, D. P., Flexible room-temperature volatile organic compound sensors based on reduced graphene oxide—WO3·0.33H2O nano-needles. *Journal of Materials Chemistry C* **2018,***6* (11), 2822–2829.

99. Zeng, G.; Li, W.; Ci, S.; Jia, J.; Wen, Z., Highly dispersed NiO nanoparticles decorating graphene nanosheets for non-enzymatic glucose sensor and biofuel cell. *Scientific Reports* **2016,***6* (1), 36454.

100. Gu, F.; Di, M.; Han, D.; Hong, S.; Wang, Z., Atomically dispersed Au on In2O3 nanosheets for highly sensitive and selective detection of formaldehyde. *ACS Sensors* **2020,***5* (8), 2611–2619.

101. Sabaraya, I. V.; Shin, H.; Li, X.; Hoq, R.; Incorvia, J. A. C.; Kirisits, M. J.; Saleh, N. B., Role of electrostatics in the heterogeneous interaction of two-dimensional engineered MoS(2) Nanosheets and natural clay colloids: influence of ph and natural organic matter. *Environmental Science & Technology* **2021,***55* (2), 919–929.

102. Xu, Y.; Wang, W.; Ge, Y.; Guo, H.; Zhang, X.; Chen, S.; Deng, Y.; Lu, Z.; Zhang, H., Quantum dots: stabilization of black phosphorous quantum dots in PMMA nanofiber film and broadband nonlinear optics and ultrafast photonics application (Adv. Funct. Mater. 32/2017). *Advanced Functional Materials* **2017,***27* (32).

11 2D Nanomaterials in Diagnostics and Therapy of Cardiovascular Diseases

Pooja Yadav, Samir K. Beura, Abhishek R. Panigrahi, Abhinaba Chatterjee, Jyoti Yadav, and Sunil K. Singh

CONTENTS

11.1 INTRODUCTION

With the era of advancement in nanotechnology, 2D nanomaterials have attracted ample interest for next-generation materials. The focus on 2D material starts with the discovery of graphene sheet[1] and advances with the mechanical cleavage of graphene from graphite.[2] Since then, other 2D nanomaterials such as transition metal dichalcogenides (TMDs), black phosphorus nanosheets, boron nitride nanosheet (BNNS),

DOI: 10.1201/9781003247890-11

metal-organic framework (MOF) nanosheets, and others have been investigated to be applied in various fields.[3] The synthesis of 2D nanomaterials can be carried out by two major approaches, including top-down or bottom-up approaches.[4] 2D nanomaterials display a wide range of promising applications including electronics,[5] energy storage (e.g., wearable batteries, supercapacitors),[6] catalysis,[7] sensors as well as bio-medicine.[8] Unique properties of 2D materials such as ultralow weight, high strength, high reactivity,[9] optical properties (like polarized photoluminescence),[10] plasmonic behavior,[11] unique physicochemical responsiveness, layer-dependent band structure,[12] ability to fabricate[13] make them ideal agents to be employed for diagnostics as well as therapeutic purpose in the biomedical field. Several reports are available to embark the 2D nanomaterials in health care applications, such as tissue engineering, regen-erative nanomedicines,[14] bone therapy,[15] cancer therapy,[3] neuronal regeneration,[16] diabetic detection,[17] and wound healing.[18] In this chapter, we review the theranostic potential of 2D nanomaterials in cardiovascular diseases (Figure 11.1).

Cardiovascular diseases (CVDs) impose serious health concerns among the global community as they are the widespread reason for death in developing as well as developed countries.[19] Cardiomyopathy refers to the complex group of heart dis-eases, where heart muscle becomes damaged, and further replaced with noncontrac-tile scar tissue, thus making it unable to pump enough blood throughout the body to maintain a normal functioning of a system. This could lead to the heart failure (HF) and ultimately death of a person.[20] The survival of patient in cardiomyopathy requires early diagnosis and appropriate treatments. The major limitation associated with the diagnosis process is that most of the cardiac-related complications exhibit similar symptoms. It introduces the need for cardiac biomarker detection in this field, which exhibits its specificity to a particular disease among the cocktail of cardiac disorders. High specificity, increased sensitivity, quick release, and quantitative assay mark the ideal characteristics for a cardiac biomarker.[21] Increased cardiac biomarker concentration in serum exhibits a strong correlation with recurrent cardiovascu-lar complications; hence the detection of biomarkers with the appropriate device also imparts a critical role in successful treatments of the patient. In recent years,

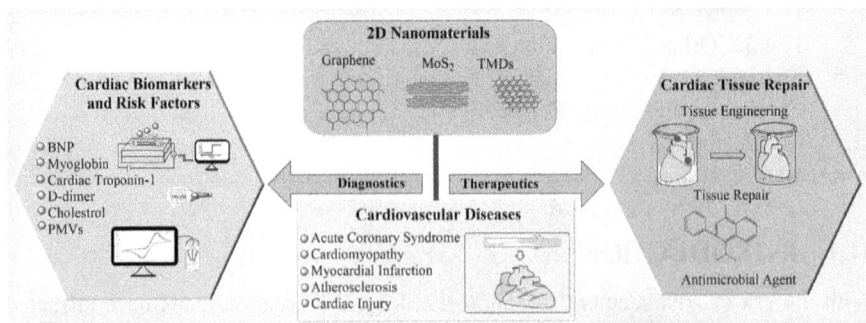

FIGURE 11.1   2D nanomaterials in diagnostics and therapeutics of cardiovascular disease.

* BNP: brain natrieuretic hormone; PMV: platelet microvesicle; MoS_2: molybdenum disulfide; TMD: transition metal dichalcogenide

biosensing emerges as the most promising platform in the early detection of cardiac marker.[22,23,24] In addition to diagnosis, treatment strategies also imbibe its crucial role in the survival of patients, which includes tissue repair and regeneration program. Incorporation of nanomaterials in creating a biocompatible scaffold to replace the heart's scar tissue has gained significant interest in current era.[25]

In this chapter, we explore the cardiac disease association with 2D nanomaterials in terms of diagnostics as well as therapeutics. In summary, we explain general 2D nanomaterials like graphene, MoS_2, and others with their significant contribution in biosensing of cardiac biomarkers as well as their therapeutic potential implied in tissue repair, which occurs during cardiac diseases (Figure 11.1).

11.2   2D NANOMATERIALS

2D nanomaterials are ultrathin materials that are single or few atomic layers thick in one of their dimensions, while the rest two dimensions are outside the 100 nm range. In sheetlike 2D nanomaterials, their transverse dimensions are larger than 100 nm, whereas their thickness is usually measured to be less than 5 nm. The unique shape of 2D material provides broad surface area and anisotropic physical/chemical properties to it.[26] Based on their synthetic strategies, 2D materials are categorized under top-down and bottom-up approaches. In top-down approach, the nanoscale materials are created by removal of materials from its bulk size in a controlled manner, whereas bottom-up approach uses atomic or molecular precursors and permits them to self-assemble for the development of more complex structures.[4]

11.2.1   GRAPHENE

Graphene is an allotrope of carbon arranged in two-dimensional (2D) honeycomb sheetlike structure. Graphene is also considered to be the thinnest material having higher mechanical strength than steel.[27,28] Generally, graphene nanosheets are generated by two conventional approaches, that is, top-down (by carbon nanotube (CNT) splitting and/or chemical/mechanical exfoliation of graphite) and bottom-up (by total organic synthesis and/or chemical vapor deposition from precursor molecules).[28,29,30] Furthermore, graphene has some unique physical and mechanical properties, including excellent electrical conductivity because of its sp^2 hybridized orbitals (10^8 S/m) and high thermal conductivity (5,000 W/mK).[31,32] Apart from these, it also possesses an enlarged specific area, strong mechanical strength, and an easily modifiable (chemically) surface (e.g., carboxylation, polyethylene glycation, etc.). Optically, graphene can produce gate-dependent transition and can potentially display photoluminescence.[33,34] Hence, considering all these characteristic properties, graphene is an obvious choice for being a widely used nanomaterial in diagnostic and theranostic applications.

11.2.2   MoS$_2$

Molybdenum disulfide (MoS_2) comes under a large family of compounds known as transition metal dichalcogenides (TMD). It is the second most widely studied 2D nanomaterial after graphene for its plausible applications, such as drug delivery and biosensors designing. In general, chemical vapor deposition[31] or atomic layer deposition[35] are used for the synthesis of MoS$_2$. The crystalline structure of MoS$_2$ is arranged in such a way that a layer of Mo atom is sandwiched between two layers of S atoms. With carrier mobility of 200 cm^2/V/S.[36] MoS$_2$ has an advantage over graphene by forming a direct bandgap, which makes it a preferable choice for field-effect transistors (FET). Like graphene, MoS$_2$ is also considered a potential candidate for biosensor designing due to its salient properties, including photoluminescence, tunable energy bandgap, intercalated morphology, and liquid media stability.[37,38]

11.2.3 Other 2D Nanomaterials

11.2.3.1 TMD

Transition metal dichalcogenides (TMDs) provide a promising alternative to graphene in the field of 2D nanomaterials. TMDs are made up of one atom of transition metal (e.g., molybdenum (Mo), tungsten (W), etc.), which combines with two atoms of chalcogen (e.g., sulfur (S), selenium (Se), the metalloid tellurium (Te), etc.) making a generalized molecular formula MX$_2$. Atomic-scale thickness, direct bandgap, favorable electronic and mechanical properties make researchers enthusiastic about their possible application in the biomedical field.[39] Synthesis of TMD involves chemical vapor deposition, mechanical exfoliation, heterostructure restructuring, and topochemical intercalation.[40] In a study, Wang et al. were the first to develop a horseradish peroxidase HRP/MoS$_2$ based glassy carbon electrode (GCE) to detect an electrocatalytic reduction of H$_2$O$_2$ when ubiquinone was used as a mediator.[41] Tungsten disulfide (WS$_2$) based fluorescent biosensor for detection of T4 polynucleotide kinase (PNK) activity was also prepared due to excellent optical property exhibited by TMDs.[42] In another report, Liu et al. developed a composite carrier (MoS$_2$-PEG) by functionalizing MoS$_2$ nanosheets with lipoic acid-modified polyethylene glycol (PEG) to be utilized as a multifunctional drug carrier for combined cancer therapy.[43] Thus, when it comes to the application of graphene-like nanomaterials, attempts to exploit the properties of TMD obviously draw the attention of researchers.

11.2.3.2 BN-Nanosheet

Boron nitride (BN) nanosheets are formed by the alternative arrangement of boron (B) and nitrogen (N) atoms in a honeycomb lattice structures, which are isoelectric to carbon. These hexagonal BN nanosheets possess stronger mechanical strength with Young's modulus about 0.865 TPa as compared to graphene. Furthermore, the mechanical strength of BN nanosheets decreases less dramatically when the thickness is increased.[44] Other than mechanical characteristics, better surface adsorption, wide bandgap, thermal stability,[45] better oxidation resistance,[46] and excellent dielectric properties trigger interest among scientists. BN nanosheets are also excellent proton conductors.[26] BN nanosheets can be synthesized by chemical vapor deposition,[47] mechanical cleavage, solvent exfoliation,[47] solid-state reaction,[4] and sonication.[48] Apart from these, multifunctional BN nanosheet-AuNP (gold nanoparticle) nanocomposite based immunosensors have been developed to detect

IL-6.[49] Moreover, hexagonal boron nitride (h-BN) graphene quantum dot exhibits strong fluorescence property, and for being very low cytotoxic, this can also be used in cell imaging.[50]

11.2.3.3 g-C$_3$N$_4$ Nanosheet

Graphitic carbon nitride (g-C$_3$N$_4$) is a nanopolymer composed of tris-triazine-based patterns with the carbon (C) and nitrogen (N) along with a small amount of H. g-C$_3$N$_4$ has electron-rich properties apart from H-bonding motifs, which makes it a potential candidate in the application of biomaterial science.[51] g-C$_3$N$_4$ nanosheet shows an excellent thermal conductivity, better chemical stability with a bandgap of ca. 2.7 eV, and inherent photoluminescence (PL) for its unusual structure.[52] Synthesis of g-C$_3$N$_4$ is a challenging task because of its low thermodynamic stability.[53] Recently, ultra-rapid synthesis of the polymer has been reported using direct microwave heating for field emission.[54] Zhang et al. developed graphene carbon nitride and manganese oxide (C$_3$N$_4$-MnO$_2$) sandwich nanocomposite for the detection of glutathione,[55] thus indicating possible biosensing application of this material in near future.

11.2.3.4 Black Phosphorus

Black phosphorus, otherwise known as phosphorene, exhibits extraordinary electrical conductivity and electron transfer capability. In phosphorene, each phosphorus atom is covalently bound with three other phosphorus atoms forming a bilayer structure.[56] Layer-dependent tunable bandgap, high carrier mobility,[57] high current on/off ratio makes phosphorene an eligible candidate for FET-sensing applications. Mechanical cleavage and liquid exfoliation is the most commonly used method for the synthesis of phosphorene.[58,59,60] Aptamer functionalized poly-L-lysine and black phosphorus (PLL-BP) nanosheet has been successfully designed to detect myoglobin.[61] In addition, poly(3,4-ethylenedioxythiophene)-poly(styrenesulfonate) (BP-PEDOT:PSS) BP nanocomposite has also been designed to detect hemoglobin.[62] By the side of diagnostic, 2D black phosphorus also finds application in the therapeutic field. For example, it can be noted that BP nanosheets are widely used in photothermal therapy (PTT) and photodynamic therapy (PDT).[63] In a study, Yang et al. developed BP-Au-Fe$_3$O$_4$ nanocomposite for combined PTT/PDT cancer therapy.[64] Therefore, from the previous discussion, one can conclude that not only graphene or MoS$_2$ but also other graphene-like nanomaterials possess certain properties that are well suited for biosensing as well as therapeutic applications.

11.3 CARDIOVASCULAR DISEASES (CVDs)

Cardiovascular disease is an umbrella term for a wide array of disease complications, commonly described as cerebrovascular disease, coronary heart disease, venous thromboembolism, peripheral artery disease, and rheumatoid and congenital heart diseases. About 31% of global deaths are caused by CVDs, making it one of the highly fatal complications among all classes of people worldwide.[65] Apart from these, there are specific cardiac diseases, which specifically affect the heart muscles resulting in an inefficient pumping of blood by the heart to other body parts, collectively called as cardiomyopathy. There are several types of cardiomyopathy, including dilated cardiomyopathy (DCM), hypertrophic cardio-myopathy (HCM), restrictive cardiomyopathy (RCM), arrhythmogenic cardiomy-opathy (ACM), etc.[20] For example, the occurrence of myocardial infarction (MI) among all these heart diseases has been significantly increasing every year. MI is the death of cardiomyocytes due to decreased blood flow in the heart followed by depletion of oxygen supply or ischemia. The predominant cause behind prolonged ischemia-induced cardiac tissue death is considered to be atherosclerosis or nar-rowing of blood vessel due to atheromatous plaques formation.[66] The pathophysi-ology of MI starts with the accumulation of low-density lipoprotein (LDL) in the tunica intima layer of arteries, which results in the formation of oxidized-LDL (ox-LDL). This ox-LDL activates a cascade of inflammatory cytokine and che-mokine production and leads to the infiltration of lymphocytes and monocytes. These events eventually form atherosclerotic plaque, and its subsequent rupture results in thrombosis, thus bringing the onset of MI.[67] The major risk factors for atherosclerosis includes diabetes, hypertension, hyperlipidemia, smoking, in addi-tion to gender and age.[66]

11.3.1 CARDIOVASCULAR BIOMARKERS AS DIAGNOSTIC PARAMETERS

There is a considerable number of cardiovascular biomarkers, which are assumed to complement the clinical tools meant for cardiovascular disease (CVD) symptom assessment, including electrocardiogram (ECG) and imaging techniques. These bio-markers aid in the evaluation of risk stratification, early prognosis, clinical diagno-sis, as well as disease management of patients having cardiovascular complications, including acute myocardial infarction (AMI) and heart failure (HF)[68] (Figure 11.1). An ideal biomarker needed for the early detection of myocardial injury requires to be predominantly expressed within the cardiac tissue exclusively at relatively higher levels, having a rapid release phenomenon after myocardial injury, having signifi-cantly higher clinical sensitivity, as well as specificity in blood early after the onset of disease symptoms[69] (Table 11.1).

The most reliable cardiac biomarker, which is reported to be used for the early diagnosis of cardiac injury in CVDs especially in AMI is cardiac troponin (cTn). The cTn complex plays a crucial role in cardiomyocyte contraction by mediating the interaction between actin and myosin filaments. This complex is composed of three isoforms, including TnC (binds with calcium), TnI (blocks actin-myosin interaction), and TnT (binds with tropomyosin).[70,69] A dynamic change equal to or higher than 20% in plasma concentration of cTn over 6–9 hours can be interpreted as an affirma-tion for AMI. Therefore, the elevated concentration of serum cardiac troponin (cTn) is still considered the gold standard for the early diagnosis of cardiac injury.[71] Among all isoforms, cTnI is exclusively found in cardiomyocytes, hence the detection of cTnI is extremely specific for diagnosing MI. Interestingly, cTn is secreted into blood after 6–8 hours of myocardial injury, gains its highest circulatory concentration after 12–24 hours, and remains there for seven to ten days. The only disadvantage is the difficultly of predicting recurrent MI as cTn remains elevated in blood for a long time, thus prolonging its clearance.[66]

TABLE 11.1

2D Nanomaterials Based Diagnostic Sensors for Cardiovascular Diseases

S.N.	2D Nanomaterial(s)	Cardiac Biomarker/Risk Factor	Modifications	Detection Range/Lower Detection Limit	Application Used	References
1.	Graphene	Cardiac troponin I	Porous graphene oxide (PrGO) is partially reduced and immobilized with the anticardiac troponin I (anti-cTnI) antibody on glassy carbon electrode (GC) (anti-cTnI/PrGO/GC)	Linearity range: 0.1–10 ng/mL Detection limit: 0.07 ng/mL	EIS	22
		Cardiac troponin I	Nitrogen-doped reduced graphene oxide (N-prGO) modified with 1-pyrenecarboxylic acid (py-COOH) and poly(ethylene glycol) modified pyrene (py-PEG) and integrated with Tro4 aptamer (Tro4 aptamer/ py-COOH/py-PEG /N-prGO/GC)	Detection limit:1 pg/mL	DPV	98
		Cardiac troponin I	Amine functionalized graphene (f-GN) modified with monoclonal anti-cTnI antibodies (Anti-cTnI/ f-GN)	Linear range: 0.01–1 ng/mL Detection limit: 0.01 ng/mL	CV, LSV, EIS	83
		Cardiac troponin I	6-carboxyfluorescein (6-FAM) based anti-cTnI aptamers immobilized on the surface of graphene oxide (GO) sheet	Linear range: 0.10–6.0 ng/mL Detection limit: 0.07 ng/mL	FRET	99
		Cardiac troponin I	Transition metal chalcogenide (nMo$_3$Se$_4$) embedded upon reduced graphene oxide (rGO) and functionalized by 3-aminopropyltriethoxy silane (APTES) and this composite further immobilized with monoclonal anti-cardiac troponin I (anti-cTnI) anti-cTnI /APTES/nMo$_3$Se$_4$-rGO/ITO	Linear range: 1 fg–100 ng/mL Detection limit: 1 fg/mL	CV	100
		BNP-32	Screen-printed electrodes (SPE) modified by a polyethyleneimine (PEI)/reduced graphene oxide (rGO) nanocomposite and immobilized with A10 aptamer (A10 aptamer /SPE-rGO/PEI)	Linear range:1 pg–1 μg/ mL Detection limit: 0.9 pg/mL	DPV	23
		BNP	Reduced graphene oxide (rGO) fabricated on FET chip and subsequently decorated by platinum nanoparticles (PtNPs)	Detection limit: 100 fM/mL	FET	82
		D-dimer	Graphene oxide (GO) was conjugated on surface plasmon resonance (SPR) chip surface and immobilized with anti-D-dimer. (anti-D-dimer/GO/SPR chip)	Linear range: 10–150 ng/mL Detection limit: 5.08 ng/mL	SPR	84

(Continued)

TABLE 11.1
(Continued)

S.N.	2D Nanomaterial(s)	Cardiac Biomarker/Risk Factor	Modifications	Detection Range/Lower Detection Limit	Application Used	References
		Fibrinogen	Anti-fibrinogen antibody attached with graphene (12HC-G) and deposited on screen printed electrodes (SPE) (12HC-G/SPE)	Linear range: 938–44542 mg/dL Detection limit: ~ 500 mg/dL	EIS	85
		Thrombin	Thrombin binding aptamer (TBA) used graphene oxide as modifier and deposited on pencil graphite electrode (TBA/GO/PGE)	Linear range: 0.1–10 nM Detection limit of 0.07 nM	DPV	101
		Thrombin	Graphene oxide (GO) sheet enriched with silver nanoparticles (AgNP) and modified with thrombin binding aptamer (TBA2-thrombin-TBA1-AgNP-GO)	Linear range: 0.05–5 nM/L Detection limit: 0.03 nmol/L	CV	86
		CRP	Anti-CRP antibody linked with reduced graphene oxide (rGO) and deposited on indium tin oxide (ITO) electrode (Anti-CRP/rGO/ITO)	Linear range: 1–10,000 ng/mL Detection limit: 0.08 ng/mL	EIS	87
		Cholesterol	Protonation of the graphene carbon nitride sheet (g-C$_3$N$_4$H$^+$) and doping with cylindrical spongy polypyrrole (CSPPy) and immobilized with cholesterol oxidase (ChOx) (ChOx-CSPPy-g-C$_3$N$_4$H$^+$/GCE)	Linear range: 0.02–5.0 mM detection limit: 8.0 μM	EIS, CV	102
		PMP	Graphene oxide combined with anti-$\alpha_{IIb}\beta_3$ antibody and deposited on glassy carbon electrode (Anti- $\alpha_{IIb}\beta_3$/GO/GCE)	Linear range: 100 to 7000/μL)	EIS	103
2.	MoS$_2$	Cardiac Troponin T	Nanocouple formulated using anti cTnT-labelled carbon dot (CD) and molybdenum disulfide (MoS$_2$) (anti cTnT-CD/MoS$_2$)	Linear range: 0.1–50 ng/mL Detection limit: 0.12 ng/mL	FRET	89
		Cardiac troponin I	MoS$_2$ nanosheets linked with cTnI based aptamer	Linear range: 10 pM–1.0 μM Detection limit: 0.95 pM	EIS	90
		Cardiac troponin I	Copper nanowires/molybdenum disulfide/reduced graphene oxide composite linked with cTnI based aptamer (cTnI-aptamer/CuNWs/MoS$_2$/rGO)	Linear range: 5.0 × 10^{-13} to 1.0 × 10^{-10} g/mL Detection limit: 1.0 × 10^{-13} g/mL	EIS	91
		Myoglobin	Copper sulfide–molybdenum disulfide (CuS–MoS$_2$) hybrid nanostructures immobilized with anti-myoglobin detection antibody	Linear range: 0.005–20 ng/mL Detection limit: 1.2 pg/mL	CV	24

No.	Nanomaterial	Analyte	Description	Linear range / Detection limit	Method	Ref.
		microRNA-499, microRNA-133a, and microRNA-1.	Duplex specific nuclease and molybdenum disulfide (DSN-MoS₂) conjugated with aptamer	Detection limit: ~100 fM for all three microRNAs	Fluorescence assay	75
		Thrombin	MoS₂ nanosheets conjugated with ssDNA aptamer	Linear range: 53–854 pM; Detection limit: 53 pM	EIS	92
		MMP9	MoS₂ field-effect transistor (FET) functionalized with Amyloid β_{1-42}	Linear range: 1 pM–10 nM	FET	93
		Glucose	Glassy carbon electrode modified with molybdenum disulfide (MoS₂) nanosheets decorated with gold nanoparticles (AuNPs) and immobilized with glucose oxidase (GOxAuNPs@MoS₂/GCE)	Linear range: 10–300 mM; Detection limit: 2.8 µM	EIS, CV	94
		CRP	MoS₂ and gold nanoparticles fabricated with black gated field-effect transistor (Au-NPs/MoS₂ BG-FET)	Detection limit: 8.38 fg/mL	(FET)	104
3.	**Others**					
3.1	Tungsten trioxide	Troponin I	Tungsten trioxide nanosheets (WO₃ NS) were functionalized with 3-aminopropyletriethoxysilane (APTES) and immobilized with cardiac troponin I antibodies (Anti-cTnI/APTES/WO₃ NS/ITO)	Linear range: 0.1–100 ng/mL	EIS	97
3.2	Metal organic framework (MOF)	Cardiac troponin I	MOF functionalized with carboxyl-rich tris(4,4'-dicarboxylic acid-2,2'-bipyridyl) ruthenium(II) (Ru(dcbpy)₃²⁺)	Linear range: 1 fg–10 ng/mL; Detection limit: 0.48 fg/mL	ECL immunosensing	105
3.3	2D nickel MOF nanosheets	h-FABP	2D nickel MOF nanosheets was fabricated PEI and luminol	Linear range:100 fg–100 pg/mL; Detection limit: 44.5 fg/mL	ECL immunosensing	106
3.4	Boron nitride	Myoglobin	Deposition of AuNPs on boron nitride sheets and immobilization of thiol-functionalized DNA aptamer and deposited on fluorine-doped tin oxide (FTO) (Apt/AuNPs/BNNSs/FTO)	Linear range: 0.1–100 µg/mL; Detection limit: 34.6 ng/mL	CV, EIS, DPV	96
3.5	Black phosphorus	Myoglobin	Poly-l-lysine (PLL) deposited on black phosphorus nanosheets and facilitate the binding of anti-Mb DNA aptamers	Linear range: 1 pg/mL–16 µg/mL; Detection limit: ~0.524 pg/mL	CV, EIS	61
		Thrombin	Fluorescent-dye-labeled single-strand DNA aptamer adsorbed on calcium-cation–doped polydopamine-modified 2D black phosphorus nanosheets	Linear range: 10–25 nM; Detection limit: 0.02 nM	Fluorescence assay	95
3.6	2D polyaniline layer	BNP	None	Detection limit: 50 pg/mL	FET	107

Apart from cTn, creatine kinase (CK), a cellular energy metabolism regulator as well as its isoform, i.e., CK-myocardial band (CK-MB), is also considered as a gold standard for the detection of cardiac injury. It has been reported that CK-MB constitutes about 30% of CK in the myocardium, thus making this enzyme a crucial biomarker of myocardial injury. Moreover, a rise beyond 5% of total CK activity in blood suggests possible damage to the heart muscle. In addition, CK-MB appears in circulation 4–6 hours after the onset of myocardial injury, gains its highest circulatory concentration after 10–12 hours, and remains there for about two days.[72,73] This biomarker has about 91% sensitivity as well as specificity for the disease diagnosis for which it has become widely used in clinical emergency departments. The expression of CK as well as CK-MB is not restricted to the heart only, thus reducing their diagnostic specificity for AMI.[71]

Myoglobin (Mb) is an iron-binding protein located primarily in the cardiac and skeletal muscle tissues, where it carries oxygen to muscle cells. It is a very sensitive biomarker for the prediction of muscle injury, especially in myocardial injury. Mb appears in circulation 1 hour after myocardial damage, gains its highest circulatory concentration after 8–10 hours, and remains there for about 24 hours. The major disadvantage of this biomarker is its poor specificity despite being useful in evaluating reperfusion as well as infarct size in AMI.[73,72]

Recent research has supplemented a few novel sets of biomarkers for the diagnosis of myocardial necrosis, including sarcomeric cardiac myosin-binding protein C (cMyC), ischemia-modified albumin (IMA), and heart-type fatty acid–binding protein (hFABP).[71] The h-FABP is found predominantly in cardiomyocytes, where it helps in fatty acid transport and metabolism. Furthermore, recent studies have confirmed that h-FABP is rapidly secreted into the cytoplasm upon MI, thus making this protein an emerging potential biomarker for MI.[74,70] Similarly, cMyC is a myosin-associated protein, which is exclusively expressed by cardiomyocytes, thus making it a specific biomarker for myocardial injury.[71] IMA is assumed to be N-terminal-modified albumin having reduced cobalt binding affinity, where its level is attributed upon myocardial injury, thus making this protein a potential biomarker for MI.[73] Recent studies have also confirmed that there are certain miRNAs that can predict the potential future onset of CVD. For instance, miR-21, miR-27b, miR-126, miR-130a, miR-221, miR-222, and let-7 are involved in the patho-progression of angiogenesis, atherosclerosis, congenital heart defects, and coronary artery disease.[68] In addition, miRNA-499, miRNA-133a, and miRNA-1 are also considered to be the reliable biomarkers for MI.[75]

Besides these, several other potential biomarkers are often detected during CVDs, especially MI and atherosclerosis, including N-terminal pro-B-type natriuretic peptide (NT-proBNP), mid-regional pro-adrenomedullin (MR-proADM), mid-regional pro-atrial natriuretic peptide (MR-proANP), C-terminal pro-endothelin 1 (CT-proET-1), high sensitivity C-reactive protein (hs-CRP), low-density lipoprotein (LDL), pregnancy-associated plasma protein-A (PAPP-A), myeloperoxidase (MPO), and copeptin.[66,69,70,71]

In addition to cardiac biomarkers, cardiac risk factors also play significant roles in the assessment of cardiovascular disease. Cardiac risk factors are described as a person's risk for the development of cardiac disease in the near future. High level of measurable cardiac risk factors such as glucose,[76] thrombin,[77] fibrinogen,[78] choles-terol,[79] etc. help to predict the prognosis of diseases.

11.4 2D NANOMATERIALS IN DIAGNOSTICS

As it has been noted previously, increased concentration of cardiac biomarkers and risk factors played a significant role in the early diagnosis of cardiovascu-lar complications.[66,69,70] In recent years, biosensing emerges as the most promis-ing area in diagnostic aspects of cardiac markers.[23] Electrochemical biosensing is most widely used in this area, where an immobilized recognition element selectively binds with an analyte, and this biding is transformed to an electri-cal signal, which is seen as a digital imprint on the connected computer screen. Based on their transducer, biosensors are usually classified into potentiometric, amperometric, and impedimetric. The main methodologies implied in these bio-sensors are electrochemical impedance spectroscopy (EIS), cyclic voltammetry (CV), linear sweep voltammetry (LSV), and field-effect transistors (FET) based biosensing.[80,80] The translation of these devices into clinical practices allows the 2D nanomaterials based sensors to be employed as a diagnostic tool for cardiac biomarkers. Therefore, this section deals with the diagnostic aspects of 2D nano-materials in cardiac diseases (Table 11.1).

11.4.1 GRAPHENE

Graphene has been preferred as a 2D nanomaterial to be used in a biosensor due to its robust mechanical strength, high surface area, and better electrical conduc-tivity capacity. The graphene and graphene-modified biosensors detect several cardiac biomarkers, including myoglobin, cTn, and BNP, which are supposed to be released upon cardiac injury.[25] In a recent experiment, Singh et al. designed a screen-printed electrode (SPE) coated with a composite consisting of gold nanoparticles and reduced graphene oxide (AuNPs-rGO) to be used as an immu-nosensor for the detection of cardiac myocardial myoglobin. The immunosensor displayed a dynamic range of detection of myocardial myoglobin, i.e., 1–1400 ng/mL with a significantly better detection limit of ~0.67 ng/mL in contrast to the detection limit of ELISA (~4 ng/mL).[81] In a similar study, Kazemi et al. designed a nano-immunosensor (cTnI/PrGO/GC) by coating a glassy carbon (GC) elec-trode with porous GO (prGO) and further functionalized with cTnI antibodies for the serum detection of human cTnI. This nano-immunosensor had displayed an outstanding sensitivity as well as selectivity for the detection of myocardial cTnI in the dynamic range. i.e., 0.1–10 ng/mL detection limit of ~0.07 ng/mL.[22] In another study, Lei et al. demonstrated that reduced graphene oxide (rGO) field-effect transistor (FET) chip decorated with platinum nanoparticles (PtNPs) and further immobilized with BNP antibody can be an effective biosensor for the detection of BNP in human blood. This designed biosensor can detect blood BNP at as low as 100 fM, thus making it an effective one with high sensitivity and specificity for the prediction of HF and CVDs.[82] Tuteja et al. demonstrated amine functionalized graphene (f-GN) for the detection of cardiac troponin I with the linearity range of 0.01–1 ng/mL.[83] The graphene-based biosensor also implied to detect other biomarkers or risk factors such as BNP,[23,82] D-dimer,[84] fibrinogen,[85] thrombin,[86] CRP,[87] etc. (Table 11.1).

11.4.2 MoS$_2$

MoS$_2$ is the second-most studied 2D nanomaterial after graphene. Its unique optical, physiochemical, and biological properties presented its wide potential in diagnostic tools. Typically, MoS$_2$ is used in industries such as dry lubricants and negative electrode material in lithium-ion batteries, but its 2D form implies its potential in the biomedical field such as sensing, therapy, as well as imaging.[88] Accordingly, MoS$_2$ is implied in the diagnostics tools for cardiac biomarker and risk factor detection. Recently, Gogoi et al. demonstrated a Förster resonance energy transfer (FRET) technique to detect the cardiac troponin T (cTnT), which is based on the MoS$_2$ nanosheet combined with anti-cTnT labelled carbon dots (CDs). The FRET-based sensor displayed a linearity range of 0.1–50 ng/mL and implied a lower detection limit of 0.12 ng/mL.[89] Similarly, Qiao et al. used electrochemical impedance spectroscopy (EIS) and displayed aptamer-based cardiac troponin I biosensor with the usage of MoS$_2$ nanosheets as matrix material. It exhibits a linear detection range of 10 pM to 1.0 µM/mL with a lower detection limit of 0.95 pM/mL.[90] Another cardiac troponin I detection biomarker used aptamer/CuNWs/MoS$_2$/rGO composite and its linearity displayed in 5.0×10^{-13} to 1.0×10^{-10} g/mL, having a detection limit of 1.0×10^{-13} g/mL.[91] Zhang et al. used hybrid nanostructures (CuS-MoS$_2$) along with monoclonal mouse antihuman myoglobin capture antibody for myoglobin detection. This immune complex had a relative response with myoglobin concentration in the range of 0.005–20 ng/mL, with a detection limit of 1.2 pg/mL.[24] MoS$_2$ has also been used in the multiplexing cardiac biomarker biosensor such as Zhu et al. detected microRNA-499, microRNA-133a, and microRNA-1 via fluorescence assay having lower detection limits around 100 fM for all three microRNAs.[75] Additional cardiac marker and risk factors including thrombin,[92] MMP9,[93] glucose,[94] CRP[94] have also been detected via MoS$_2$ based biosensing techniques.

11.4.3 Other 2D Nanomaterials

Apart from graphene and MoS$_2$, other 2D nanomaterials also exhibit their diagnostic applications in cardiovascular diseases. Kumar et al. used black phosphorus nanosheets and functionalized them with poly-l-lysine (PLL) for facilitative binding of anti-Mb DNA aptamers for the detection of myoglobin. The highly sensitive developed biosensor displayed a dynamic response range from 1 pg/mL to 16 µg/mL with the detection limit of ~0.524 pg/mL.[61] Recently, black phosphorus has also been modified with calcium-cation-doped polydopamine (PDA) for the detection of thrombin.[95] In another finding, Adeel et al. developed an electrochemical aptasensor using boron nitride nanosheets decorated with gold nanoparticles for the detection of myoglobin.[96] Another 2D nanomaterial, that is, tungsten trioxide nanosheets (WO$_3$NS) has also been used to develop biosensor for troponin I detection with linearity response of 0.1 to 100 ng/mL.[97]

11.5 THERAPEUTIC APPLICATION OF 2D NANOMATERIALS

Cardiac injury can lead to the death of functional cardiomyocytes, which cannot be regenerated by the heart itself, thus generating a permanent scar at the injury site. It can be detrimental to the healthy pumping function of a heart. If large enough, it can lead to heart failure also. Tissue engineering can provide a solution by blending cells and nanomaterials to construct biocompatible scaffolds and injectable hydrogels in order to replace and/or regenerate the lost or dead cardiomyocytes after cardiac injury, especially MI. Among all nanomaterials, graphene and its modified forms significantly supplement the biocompatible scaffolds for cardiac tissue engineering[25] (Table 11.2). Recently, Saravanan et al. demonstrated GO-Au nanosheets

TABLE 11.2
Role of 2D Nanomaterials in Cardiac Tissue Engineering

S.N.	2D Nano-material(s)	Modifications	Properties	Applications in Tissue Engineering	References
1	Graphene	rGO-GelMA (Gelatin Methacrylyol) Hydrogel	Enhanced electrical conductivity and mechanical properties	Promotes cell viability, proliferation, and maturation in cardiomyocytes. These cardiomyocytes exhibit stronger contractility and faster spontaneous beating rate on rGO-GelMA hydrogel.	116
		GO-Col (Collagen) Scaffold Matrix	Shows enhanced electrical signal propagation among cardiac cells	Promotes neonatal cardiomyocyte adhesion and upregulates the expression of the cardiac specific genes like *Cx43*, *Actin4*, and *Trpt-2*.	117
		rGO-Ag-PU (Polyurethane) Matrix	Shows enhanced electrical and mechanical properties	Promotes cell adhesion, increases cytocompatibility, upregulates the expression of cardiac specific genes, including GATA-4, TBX-18, cTnI, and α-MHC in hCPCs.	118
		GO-CS (Chitosan Scaffold) Matrix	Shows good swelling, porosity, and conductive properties	Promotes cell adhesion, intercellular networking, and cell viability in in H9C2 rat cardiomyoblasts.	109
		GO-Au-CS (Chitosan Scaffold) Matrix	Shows excellent porous architecture, improved swelling capacity, increased conductivity and biodegradability	Promotes cell adhesion in rat SMCs and mouse fibroblasts and potentiates growth with minimal cytotoxicity in a rat model of MI.	108

(Continued)

TABLE 11.2
(Continued)

S.N.	2D Nano-material(s)	Modifications	Properties	Applications in Tissue Engineering	References
		GO-OPF (oligo (poly (ethylene glycol) fumarate)) Hydrogel	Shows enhanced semiconductive properties that were lacking in pure OPF	Promotes cell adhesion of rat cardiac fibroblasts, generation of cytoskeletal structure, Ca^{2+} signal conduction, and intercalated disc assembly in cardiomyocytes of rat model of MI.	110
		GO-Alginate Microgel	Shows good swelling, porosity, and conductive properties	Promotes cell viability of hMSCs and enhances cardiac maturation in an oxidative stress condition in rat model of MI.	111
		GO- FA-PEG (Polyethyl-enimine- Folic Acid) Matrix	Shows better aqueous stability than individual GO-PEG	Mitigates ROS in macrophages and promotes cardiac tissue repair by reducing fibrosis in mouse model of MI.	113
2	Molybdenum Disulfide (MoS_2)	MoS_2 Nanoparticles	Shows good electrical conductivity, excellent biocompatibility, and high surface area/mass ratio	Promotes the proliferation of H9C2 cardiomyocytes, and this nanomaterial protects H9C2 cells from oxidative stress induced by H_2O_2.	115
		MoS_2-Nylon Nanofiber Scaffold	Shows enhanced mechanical properties and electrical conductivity	Promotes cellular attachment and cell proliferation along with upregulating cardiac functional genes, including GATA-4, c-TnT, Nkx 2.5 and α-MHC in mECCs.	114
		MoS_2-SF (Silk Fibroin) Nanofiber Scaffold	Shows enhanced mechanical properties, elastic properties, and electrical conductivity	Promotes growth and differentiation of TBX18-hiPSCs into cardiac cells. It also enhances cell attachment and viability by upregulation of cardiac functional genes, including GATA-4, c-TnT, and α-MHC.	119

in CS scaffold to be an excellent porous biomaterial having controlled degradation properties along with the desired swelling capacity for cardiac tissue engineering. In addition, it not only exhibited a better electrical conductivity (improved QRS interval) but also supported cell adhesion and growth with minimal cytotoxicity in a rat model of MI.[108] In a similar study, Jiang et al. fabricated a Chitosan Scaffold (CS) by blending with GO, which displayed an excellent biomaterial for promoting cell adhesion, intercellular networking, as well as cell viability by upregulating the expression of cardiac-specific proteins, like connexin 43 in H9C2 cells.[109] In another study, Zhou et al. developed a novel conductive hydrogel by introducing GO into an oligo (poly(ethylene glycol) fumarate) (OPF) hydrogel. This OPF/GO hydrogel proved to improve cell adhesion of rat cardiac fibroblasts in vitro accompanied by the generation of cytoskeletal structure, Ca^{2+} signal conduction, and intercalated disc assembly in cardiomyocytes of rat model of MI.[110] In a recent study, Choe et al. demonstrated that GO/alginate microgel could be a potential encapsulated carrier for stem cells to initiate cardiac tissue regeneration. The group co-cultured the cardiomyocytes (CMs) with human mesenchymal stem cell (hMSCs) encapsulated with GO/alginate microgel and showed the increased potential of cell viability and cardiac maturation in an oxidative stress condition in the rat model of MI. This finding could provide an effective platform for stem cell therapeutic regeneration in cardiac diseases, including MI.[111] In a similar study, Zhao et al. developed a gel matrix (GO-RTG) by modifying reverse thermal gel (RTG) with GOs, which could enhance the growth, proliferation, and viability of neonatal rat ventricular myocytes (NRMs), thus could potentially be a therapeutic candidate for tissue regeneration after cardiac injury.[112] In a novel study, Han et al. formulated a macrophage-targeting/polarizing GO complex (MGC) by modifying GO with polyethylenimine (PEI) and folic acid-PEG (FA-PEG). Furthermore, the MGC decreased ROS in macrophages and induced polarization of M1 macrophage (secrets pro-inflammatory cytokines) into M2 macrophage (secrets anti-inflammatory cytokines), thus reducing fibrosis and improving cardiac tissue repair in a mouse model of MI[113] (Table 11.2).

Molybdenum disulfide (MoS_2) is a transition metal dichalcogenide that functions as a new generation 2D nanosheet having excellent properties like good electrical conductivity, excellent biocompatibility, and high surface area/mass ratio. New studies are providing emerging evidences for MoS_2 to be used in cardiac tissue engineering.[114] For example, Ming et al. demonstrated that MoS_2 nanoparticles could promote the growth and proliferation of H9C2 cardiomyocytes apart from protecting these cells from H_2O_2-induced ROS mediated oxidative stress. This finding suggests that MoS_2 can be used as a therapeutic option for cardiomyocyte proliferation during the repair of cardiac injury.[115] In another study, Nazari et al. designed a nanofiber scaffold, MoS_2-Nylon6 by conjugating MoS_2 nanosheets into nylon6 electrospun nanofibers, which displayed enhanced mechanical properties and electrical conductivity. This MoS_2-Nylon6 scaffold promoted cellular attachment and cell proliferation along with upregulating cardiac functional genes, including GATA-4, c-TnT, Nkx 2.5, and α-MHC in mouse embryonic cardiac cells (mECCs).[114] In a similar study, Nazari et al. incorporated MoS_2 into silk fibroin (SF) nanofibers and demonstrated that MoS_2-SF potentially promoted growth and differentiation of TBX18-hiPSCs (human-induced pluripotent stem cells) into cardiac cells. It also enhances cell attachment and viability by upregulation of cardiac functional genes, including GATA-4, c-TnT, and α-MHC.[115] Hence both GO as well as MoS_2 are being implemented in cardiac tissue engineering to regenerate or repair the lost cells during cardiac injury (Table 11.2).

11.6   CHALLENGES AND FUTURE PERSPECTIVE

2D nanomaterials are an emerging group of biomaterials having unique physical, chemical, optical, thermal, as well as electronic properties, making them remarkable potential candidates for diagnostic applications (e.g., biosensing and bioimaging), therapeutic applications (e.g., drug delivery), and regenerative medicines (e.g., tissue engineering) (Figure 11.1). As in this chapter, we had earlier discussed in detail about the need of therapeutic alternative(s) in cardiac complications like MI. The 2D nanomaterials have exhibited high efficiency in biosensing the cardiac biomarkers in CVDs, especially in MI, signifying their propitious utilization in biomedical practices (Table 11.1). The application of 2D nanomaterials in biosensors has been a promising avenue for the health care industry due to their excellent electrical conductivity and planar nature. The major challenge in the fabrication of biosensors is a low-cost yet controllable synthesis route (management of surface charge density), which hinders in its implementation for mass production. Furthermore, the biocompatibility and environmental stability both in storage condition and testing medium remain challenging for real-life application of 2D nanomaterial-based biosensors.[8] Apart from the use in diagnostic biosensors, the advancement of 2D nanomaterials in the fabrication of biocompatible cardiac scaffolds has been commendable (Table 11.2). Although these 2D nanomaterial-based scaffolds are still in experimental phases owing to their cytotoxic effects,[120,121] which need to be optimized before being effectively applied in biomedical practices of cardiac tissue engineering. There are several challenges that need to be overcome, including the reproducibility of 2D nanomaterial behavior in animal models, along with their molecular interaction with other cellular and matrix biomolecules. Apart from these, the biodegradability of these 2D nanomaterials is a considerable threat in in vivo experiments, where they are prone to cause cytotoxic effects in a prolonged condition.[14] As we move from basic biotechnology toward translational theranostics, innovating strategies will emerge to develop reliable, safe, and highly efficient 2D nanomaterials for interdisciplinary research and biomedical applications.

11.7   CONCLUSIONS

2D nanomaterials are ultrathin nanomaterials that display extremely high degree of anisotropy in addition to good biochemical functionality. These 2D nanomaterials include carbon-based nanosheets like GO nanosheets, TMDs, MoS_2 nanosheets, boron nitride nanosheet, and metal-organic framework (MOF) nanosheets. Over the past few years, a variety of 2D nanomaterials have emerged as potential candidates for the restructuring, repair, and regeneration of damaged and/or injured tissues and organs (Figure 11.1). The use of 2D nanomaterials has dramatically revolutionized biomedical fields, especially for bone and skin repair, neural regeneration, wound healing, and cardiac tissue engineering. Among these, cardiac tissue injury has remained challenging due to the intrinsic inability of the damaged cardiac tissue to regenerate after an injury like stroke and MI. Hence, diagnostic tools are vital for determination of these complications. 2D nanomaterials have been successful in the construction of electrochemical biosensors for the detection of cardiac biomarkers after myocardial injury, especially in MI, owing to their high electrical conductivity and large surface area (Table 11.1). Apart from this, tissue engineering has also emerged as an effective tool for restructuring, repair, and regeneration of damaged and/or injured tissues and organs. In tissue engineering, biocompatible scaffolds are pivotal as they provide structural support for tissue repair along with serving as a template for tissue restructuring and regeneration. The construction of biocompatible scaffolds are possible due to their unique physiochemical properties like uniform shapes, good surface charge, high surface-to-volume ratio, and superior porosity, which enable these nanomaterials to be used in the field cardiac tissue engineering (Table 11.2). Although the potential application of 2D based nanomaterials in cardiac tissue engineering in still in its infancy, future research will elucidate these biomedical applications.

REFERENCES

1. Boehm, H. P., Setton, R. & Stumpp, E. International union of pure and applied chemistry inorganic chemistry division commission on high temperature and solid state chemistry* nomenclature and terminology of graphite intercalation compounds. *Pure Appl. Chem.* **66**, 1893–1901 (1994).
2. Novoselov, K. S. *et al.* Electric field effect in atomically thin carbon films. *Science* **306**, 666–669 (2004).
3. Wang, L., Xiong, Q., Xiao, F. & Duan, H. 2D nanomaterials based electrochemical biosensors for cancer diagnosis. *Biosens. Bioelectron.* **89**, 136–151 (2017).
4. Wang, J., Li, G. & Li, L. Synthesis Strategies about 2D Materials. *Two-dimensional Mater.—Synth. Charact. Potential Appl.* (2016). doi:10.5772/63918
5. Tao, L. *et al.* Silicene field-effect transistors operating at room temperature. *Nat. Nanotechnol.* **10**, 227–231 (2015).
6. Yi, F. *et al.* Wearable energy sources based on 2D materials. *Chem. Soc. Rev.* **47**, 3152–3188 (2018).
7. Zhu, Y. *et al.* Structural engineering of 2D nanomaterials for energy storage and catalysis. *Adv. Mater.* **30**, 1706347 (2018).
8. Bolotsky, A. *et al.* Two-dimensional materials in biosensing and healthcare: From in vitro diagnostics to optogenetics and beyond. *ACS Nano* **13**, 9781–9810 (2019).
9. Zhang, R. & Cheung, R. Mechanical properties and applications of two-dimensional materials. *Two-dimensional Mater.—Synth. Charact. Potential Appl.* (2016). doi:10.5772/64017
10. Wang, C., Zhang, G., Huang, S., Xie, Y. & Yan, H. The optical properties and plasmonics of anisotropic 2D materials. *Adv. Opt. Mater.* **8**, 1900996 (2020).
11. Sharma, A. K., Kaur, B. & Popescu, V. A. On the role of different 2D materials/heterostructures in fiber-optic SPR humidity sensor in visible spectral region. *Opt. Mater. (Amst).* **102**, 109824 (2020).
12. Cai, Y., Zhang, G. & Zhang, Y. W. Layer-dependent band alignment and work function of few-layer phosphorene. *Sci. Rep.* **4**, 1–6 (2014).

13. Pomerantseva, E. & Gogotsi, Y. Two-dimensional heterostructures for energy storage. *Nat. Energy.* **2**, 1–6 (2017).

14. Zheng, Y. *et al.* 2D Nanomaterials for tissue engineering and regenerative nanomedicines: Recent advances and future challenges. *Adv. Healthc. Mater.* **10**, 2001743 (2021).

15. Cheng, L. *et al.* Black phosphorus-based 2D materials for bone therapy. *Bioact. Mater.* **5**, 1026–1043 (2020).

16. Veeralingam, S. & Badhulika, S. X (metal: Al, Cu, Sn, Ti)-functionalized tunable 2D-MoS_2 nanostructure assembled biosensor arrays for qualitative and quantitative analysis of vital neurological drugs. *Nanoscale* **12**, 15336–15347 (2020).

17. Li, G. & Wen, D. Sensing nanomaterials of wearable glucose sensors. *Chinese Chem. Lett.* **32**, 221–228 (2021).

18. Chu, J. *et al.* PEGylated graphene oxide-mediated quercetin-modified collagen hybrid scaffold for enhancement of MSCs differentiation potential and diabetic wound healing. *Nanoscale* **10**, 9547–9560 (2018).

19. Pedrero, M., Campuzano, S. & Pingarrón, J. M. Electrochemical biosensors for the determination of cardiovascular markers: A review. *Electroanalysis* **26**, 1132–1153 (2014).

20. Braunwald, E. Cardiomyopathies: An overview. *Circ. Res.* **121**, 711–721 (2017).

21. Al-Hadi, H. A. & Fox, K. A. Cardiac markers in the early diagnosis and management of patients with acute coronary syndrome. *Sultan Qaboos Univ. Med. J.* **9**, 231 (2009).

22. Kazemi, S. H. K., Ghodsi, E., Abdollahi, S. & Nadri, S. Porous graphene oxide nanostructure as an excellent scaffold for label-free electrochemical biosensor: Detection of cardiac troponin I. *Mater. Sci. Eng. C.* **69**, 447–452 (2016).

23. Grabowska, I. *et al.* Electrochemical aptamer-based biosensors for the detection of cardiac biomarkers. *ACS Omega* **3**, 12010–12018 (2018).

24. Zhang, B. *et al.* Copper sulfide-functionalized molybdenum disulfide nanohybrids as nanoenzyme mimics for electrochemical immunoassay of myoglobin in cardiovascular disease. *RSC Adv.* **7**, 2486–2493 (2017).

25. Alagarsamy, K. N. *et al.* Carbon nanomaterials for cardiovascular theranostics: Promises and challenges. *Bioact. Mater.* **6**, 2261–2280 (2021).

26. Hu, S. *et al.* Proton transport through one-atom-thick crystals. *Nat.* **516**, 227–230 (2014).

27. Radadiya, T. M. A properties of Graphene Zaiko Warehouse management systems application platform view project tarun Radadiya a properties of Graphene. *Eur. J. Mater. Sci.* **2**, 6–18 (2015).

28. Whitener, K. E. & Sheehan, P. E. Graphene synthesis. *Diam. Relat. Mater.* **46**, 25–34 (2014).

29. Xu, Y., Cao, H., Xue, Y., Li, B. & Cai, W. Liquid-phase exfoliation of graphene: An overview on exfoliation media, techniques, and challenges. *Nanomater.* **8**, 942 (2018).

30. Kumar, N. *et al.* Top-down synthesis of graphene: A comprehensive review. *FlatChem* **27**, 100224 (2021).

31. Babaahmadi, V. & Montazer, M. Reduced graphene oxide/SnO2 nanocomposite on PET surface: Synthesis, characterization and application as an electro-conductive and ultraviolet blocking textile. *Colloids Surfaces A Physicochem. Eng. Asp.* **506**, 507–513 (2016).

32. Yang, G., Li, L., Lee, W. B., & Ng, M. C. (2018). Structure of graphene and its disorders: a review. *Science and technology of advanced materials*, **19**(1), 613–648. https://doi.org/10.1080/14686996.2018.1494493

33. Cao, L., Meziani, M. J., Sahu, S. & Sun, Y. P. Photoluminescence properties of graphene versus other carbon nanomaterials. *Acc. Chem. Res.* **46**, 171–182 (2012).

34. Wang, F. *et al.* Gate-variable optical transitions in graphene. *Science* **320**, 206–209 (2008).

35. Jurca, T. *et al.* Low-temperature atomic layer deposition of MoS 2 films. *Angew. Chemie* **129**, 5073–5077 (2017).

36. Choudhary, N., Patel, M. D., Park, J., Sirota, B. & Choi, W. Synthesis of large scale MoS_2 for electronics and energy applications. *J. Mater. Res.* **31**, 824–831 (2016).

37. Li, X. & Zhu, H. Two-dimensional MoS_2: Properties, preparation, and applications. *J. Mater.* **1**, 33–44 (2015).

38. Kalantar-Zadeh, K. & Ou, J. Z. Biosensors based on two-dimensional MoS_2. *ACS Sensors* **1**, 5–16 (2015).

39. López-Posadas, C. B. *et al.* Direct observation of the CVD growth of monolayer MoS_2 using in situ optical spectroscopy. *Beilstein J. Nanotechnol. 1057* **10**, 557–564 (2019).

40. Han, S. A., Bhatia, R. & Kim, S. W. Synthesis, properties and potential applications of two-dimensional transition metal dichalcogenides. *Nano Converg.* **2**, 1–14 (2015).

41. Wang, G. X., Bao, W. J., Wang, J., Lu, Q. Q. & Xia, X. H. Immobilization and catalytic activity of horseradish peroxidase on molybdenum disulfide nanosheets modified electrode. *Electrochem. Commun.* **35**, 146–148 (2013).

42. Liu, T. *et al.* Drug delivery with PEGylated MoS_2 Nano-sheets for combined photothermal and chemotherapy of cancer. *Adv. Mater.* **26**, 3433–3440 (2014).

43. Ge, J. *et al.* A WS2 nanosheet based sensing platform for highly sensitive detection of T4 polynucleotide kinase and its inhibitors. *Nanoscale* **6**, 6866–6872 (2014).

44. Falin, A. *et al.* Mechanical properties of atomically thin boron nitride and the role of interlayer interactions. *Nat. Commun. 2017 81* **8**, 1–9 (2017).

45. Zhu, H. *et al.* Highly thermally conductive papers with percolative layered boron nitride nanosheets. *ACS Nano* **8**, 3606–3613 (2014).

46. Li, L. H., Cervenka, J., Watanabe, K., Taniguchi, T. & Chen, Y. Strong oxidation resistance of atomically thin boron nitride nanosheets. *ACS Nano* **8**, 1457–1462 (2014).

47. Lin, Y. & Connell, J. W. Advances in 2D boron nitride nanostructures: Nanosheets, nanoribbons, nanomeshes, and hybrids with graphene. *Nanoscale* **4**, 6908–6939 (2012).

48. Chunyi Zhi, B. *et al.* Large-scale fabrication of boron nitride nanosheets and their utilization in polymeric composites with improved thermal and mechanical properties. *Adv. Mater.* **21**, 2889–2893 (2009).

49. Yang, G., Zhu, C., Du, D., Zhu, J. & Lin, Y. Graphene-like two-dimensional layered nanomaterials: Applications in biosensors and nanomedicine. *Nanoscale* **7**, 14217–14231 (2015).

50. Peng, J. *et al.* Fabrication of graphene quantum dots and hexagonal boron nitride nanocomposites for fluorescent cell imaging. *J. Biomed. Nanotechnol.* **9**, 1679–1685 (2013).

51. Patnaik, S., Sahoo, D. P. & Parida, K. An overview on Ag modified g-C3N4 based nanostructured materials for energy and environmental applications. *Renew. Sustain. Energy Rev.* **82**, 1297–1312 (2018).

52. Das, D., Shinde, S. L. & Nanda, K. K. Temperature-dependent photoluminescence of g-C3N4: Implication for temperature sensing. *ACS Appl. Mater. Interfaces* **8**, 2181–2186 (2016).

53. Maya, L., Cole, D. R. & Hagaman, E. W. Carbon-Nitrogen pyrolyzates: Attempted preparation of carbon nitride. *J. Am. Ceram. Soc.* **74**, 1686–1688 (1991).

54. Yu, Y., Zhou, Q. & Wang, J. The ultra-rapid synthesis of 2D graphitic carbon nitride nanosheets via direct microwave heating for field emission. *Chem. Commun.* **52**, 3396–3399 (2016).

55. Zhang, X.-L. *et al.* Turn-on fluorescence sensor for intracellular imaging of glutathione using g-C_3N_4 nanosheet-MnO_2 sandwich nanocomposite. *Anal. Chem.* **86**, 3426–3434 (2014).

56. Kou, L., Chen, C. & Smith, S. C. Phosphorene: Fabrication, properties, and applications. *J. Phys. Chem. Lett.* **6**, 2794–2805 (2015).

57. Liu, H. *et al.* Phosphorene: An unexplored 2D semiconductor with a high hole mobility. *ACS Nano* **8**, 4033–4041 (2014).

58. Batmunkh, M., Bat-Erdene, M. & Shapter, J. G. Phosphorene and phosphorene-based materials-prospects for future applications. *Adv. Mater.* **28**, 8586–8617 (2016).

59. Brent, J. R. *et al.* Production of few-layer phosphorene by liquid exfoliation of black phosphorus. *Chem. Commun. (Camb).* **50**, 13338–13341 (2014).

60. Guo, Z. *et al.* From black phosphorus to phosphorene: Basic solvent exfoliation, evolution of raman scattering, and applications to ultrafast photonics. *Adv. Funct. Mater.* **25**, 6996–7002 (2015).

61. Kumar, V. *et al.* Nanostructured aptamer-functionalized black phosphorus sensing platform for label-free detection of myoglobin, a cardiovascular disease biomarker. *ACS Appl. Mater. Interfaces* **8**, 22860–22868 (2016).

62. Li, X. *et al.* Black phosphorene and PEDOT: PSS-modified electrode for electrochemistry of hemoglobin. *Electrochem. Commun.* **86**, 68–71 (2018).

63. Qiu, M. *et al.* Omnipotent phosphorene: A next-generation, two-dimensional nanoplatform for multidisciplinary biomedical applications. *Chem. Soc. Rev.* **47**, 5588–5601 (2018).

64. Yang, D. *et al.* Assembly of Au plasmonic photothermal agent and iron oxide nanoparticles on ultrathin black phosphorus for targeted photothermal and photodynamic cancer therapy. *Adv. Funct. Mater.* **27**, 1700371 (2017).

65. Stewart, J., Manmathan, G. & Wilkinson, P. Primary prevention of cardiovascular disease: A review of contemporary guidance and literature. *JRSM Cardiovasc. Dis.* **6**, 204800401668721 (2017).

66. Mythili, S. & Malathi, N. Diagnostic markers of acute myocardial infarction. *Biomed. reports* **3**, 743–748 (2015).

67. Saleh, M. & Ambrose, J. A. Understanding myocardial infarction [version 1; referees: 2 approved]. *F1000Research* **7** (2018).

68. Ghantous, C. M. *et al.* Advances in cardiovascular biomarker discovery. *Biomedicines* **8**, 1–19 (2020).

69. Wang, X. Y., Zhang, F., Zhang, C., Zheng, L. R. & Yang, J. The biomarkers for acute myocardial infarction and heart failure. *Biomed Res. Int.* **2020** (2020).

70. Wu, Y. *et al.* Diagnostic and prognostic biomarkers for myocardial infarction. *Front. Cardiovasc. Med.* **7** (2021).

71. Tilea, I., Varga, A. & Serban, R. C. Past, present, and future of blood biomarkers for the diagnosis of acute myocardial infarction-promises and challenges. *Diagnostics (Basel, Switzerland)* **11** (2021).

72. Jacob, R. & Khan, M. Cardiac biomarkers: What is and what can be. *Indian J. Cardiovasc. Dis. Women—WINCARS* **03**, 240–244 (2019).

73. Aydin, S., Ugur, K., Aydin, S., Sahin, İ. & Yardim, M. Biomarkers in acute myocardial infarction: Current perspectives. *Vasc. Health Risk Manag.* **15**, 1–10 (2019).

74. Chan, D. & Ng, L. L. Biomarkers in acute myocardial infarction. *BMC Med.* **8** (2010).

75. Zhu, X. *et al.* Multiplexed fluorometric determination for three microRNAs in acute myocardial infarction by using duplex-specific nuclease and MoS_2 nanosheets. *Microchim. Acta* **187**, 1–9 (2020).

76. Jouven, X. *et al*. Diabetes, glucose level, and risk of sudden cardiac death. *Eur. Heart J.* **26**, 2142–2147 (2005).

77. Tracy, R. P. Thrombin, inflammation, and cardiovascular disease: An epidemiologic perspective. *Chest* **124**, 49S–57S (2003).

78. Lassé, M. *et al*. Fibrinogen and hemoglobin predict near future cardiovascular events in asymptomatic individuals. *Sci. Rep.* **11**, 1–10 (2021).

79. Zhang, Y. *et al*. Association between cumulative low-density lipoprotein cholesterol exposure during young adulthood and middle age and risk of cardiovascular events. *JAMA Cardiol.* (2021). doi:10.1001/JAMACARDIO.2021.3508

80. Bakirhan, N. K., Ozcelikay, G. & Ozkan, S. A. Recent progress on the sensitive detection of cardiovascular disease markers by electrochemical-based biosensors. *J. Pharm. Biomed. Anal.* **159**, 406–424 (2018).

81. Singh, S., Tuteja, S. K., Sillu, D., Deep, A. & Suri, C. R. Gold nanoparticles-reduced graphene oxide based electrochemical immunosensor for the cardiac biomarker myoglobin. doi:10.1007/s00604–016–1803-x

82. Lei, Y. M. *et al*. Detection of heart failure-related biomarker in whole blood with graphene field effect transistor biosensor. *Biosens. Bioelectron.* **91**, 1–7 (2017).

83. Tuteja, S. K., Kukkar, M., Suri, C. R., Paul, A. K. & Deep, A. One step in-situ synthesis of amine functionalized graphene for immunosensing of cardiac marker cTnI. *Biosens. Bioelectron.* **66**, 129–135 (2015).

84. Wang, J., Lu, Y., Zhang, Y., Ning, Y. & Zhang, G. J. Graphene oxide-assisted surface plasmon resonance biosensor for simple and rapid determination of D-dimer in plasma. *J. Nanosci. Nanotechnol.* **16**, 6878–6883 (2016).

85. Saleem, W. *et al*. Antibody functionalized graphene biosensor for label-free electrochemical immunosensing of fibrinogen, an indicator of trauma induced coagulopathy. *Biosens. Bioelectron.* **86**, 522–529 (2016).

86. Qin, B. & Yang, K. Voltammetric aptasensor for thrombin by using a gold microelectrode modified with graphene oxide decorated with silver nanoparticles. *Microchim. Acta* **185**, 1–9 (2018).

87. Yagati, A. K., Pyun, J. C., Min, J. & Cho, S. Label-free and direct detection of C-reactive protein using reduced graphene oxide-nanoparticle hybrid impedimetric sensor. *Bioelectrochemistry* **107**, 37–44 (2016).

88. Yadav, V., Roy, S., Singh, P., Khan, Z. & Jaiswal, A. 2D MoS_2-based nanomaterials for therapeutic, bioimaging, and biosensing applications. *Small* **15**, 1803706 (2019).

89. Gogoi, S. & Khan, R. Fluorescence immunosensor for cardiac troponin T based on Förster resonance energy transfer (FRET) between carbon dot and MoS_2 nano-couple. *Phys. Chem. Chem. Phys.* **20**, 16501–16509 (2018).

90. Qiao, X. *et al*. Novel electrochemical sensing platform for ultrasensitive detection of cardiac troponin I based on aptamer-MoS_2 nanoconjugates. *Biosens. Bioelectron.* **113**, 142–147 (2018).

91. Han, Y., Su, X., Fan, L., Liu, Z. & Guo, Y. Electrochemical aptasensor for sensitive detection of Cardiac troponin I based on CuNWs/MoS_2/rGO nanocomposite. *Microchem. J.* **169**, 106598 (2021).

92. Lin, K. C., Jagannath, B., Muthukumar, S. & Prasad, S. Sub-picomolar label-free detection of thrombin using electrochemical impedance spectroscopy of aptamer-functionalized MoS_2. *Analyst* **142**, 2770–2780 (2017).

93. Park, H. *et al*. MoS_2 field-effect transistor-amyloid-β1–42 hybrid device for signal amplified detection of MMP-9. *Anal. Chem.* **91**, 8252–8258 (2019).

94. Su, S. *et al.* Direct electrochemistry of glucose oxidase and a biosensor for glucose based on a glass carbon electrode modified with MoS{sub 2} nanosheets decorated with gold nanoparticles. *Microchim. Acta* **181**, 1497–1503 (2014).

95. Jiang, H., Xia, Q., Liu, D. & Ling, K. Calcium-cation-doped polydopamine-modified 2D black phosphorus nanosheets as a robust platform for sensitive and specific biomolecule sensing. *Anal. Chim. Acta* **1121**, 1–10 (2020).

96. Adeel, M., Rahman, M. M. & Lee, J.-J. Label-free aptasensor for the detection of cardiac biomarker myoglobin based on gold nanoparticles decorated boron nitride nanosheets. *Biosens. Bioelectron.* **126**, 143–150 (2019).

97. Deepika Sandil, Srivastava, S., Khatri, R., Sharma, K. & Puri, N. K. Synthesis and fabrication of 2D Tungsten trioxide nanosheets based platform for impedimetric sensing of cardiac biomarker. *Sens. Bio-Sensing Res.* **32**, 100423 (2021).

98. Chekin, F. *et al.* Sensitive electrochemical detection of cardiac troponin I in serum and saliva by nitrogen-doped porous reduced graphene oxide electrode. *Sensors Actuators B Chem.* **262**, 180–187 (2018).

99. Liu, D. *et al.* A novel fluorescent aptasensor for the highly sensitive and selective detection of cardiac troponin I based on a graphene oxide platform. *Anal. Bioanal. Chem.* **410**, 4285–4291 (2018).

100. Chauhan, D. *et al.* Nanostructured transition metal chalcogenide embedded on reduced graphene oxide based highly efficient biosensor for cardiovascular disease detection. *Microchem. J.* **155**, 104697 (2020).

101. Ahour, F. & Ahsani, M. K. An electrochemical label-free and sensitive thrombin aptasensor based on graphene oxide modified pencil graphite electrode. *Biosens. Bioelectron.* **86**, 764–769 (2016).

102. Shrestha, B. K., Ahmad, R., Shrestha, S., Park, C. H. & Kim, C. S. In situ synthesis of cylindrical spongy polypyrrole doped protonated graphitic carbon nitride for cholesterol sensing application. *Biosens. Bioelectron.* **94**, 686–693 (2017).

103. Kailashiya, J., Singh, N., Singh, S. K., Agrawal, V. & Dash, D. Graphene oxide-based biosensor for detection of platelet-derived microparticles: A potential tool for thrombus risk identification. *Biosens. Bioelectron.* **65**, 274–280 (2015).

104. Dalila, N. R., Arshad, M. K. M., Gopinath, S. C. B., Nuzaihan, M. N. M. & Fathil, M. F. M. Molybdenum disulfide–gold nanoparticle nanocomposite in field-effect transistor back-gate for enhanced C-reactive protein detection. *Microchim. Acta* **187**, 1–15 (2020).

105. Yan, M. *et al.* Ultrasensitive immunosensor for cardiac troponin i detection based on the electrochemiluminescence of 2D Ru-MOF Nanosheets. *Anal. Chem.* **91**, 10156–10163 (2019).

106. Gan, X. *et al.* A highly sensitive electrochemiluminescence immunosensor for h-FABP determination based on self-enhanced luminophore coupled with ultrathin 2D nickel metal-organic framework nanosheets. *Biosens. Bioelectron.* **171**, 112735 (2021).

107. Liu, P. *et al.* High yield two-dimensional (2-D) polyaniline layer and its application in detection of B-type natriuretic peptide in human serum. *Sensors Actuators B Chem.* **230**, 184–190 (2016).

108. Saravanan, S. *et al.* Graphene oxide-gold nanosheets containing chitosan scaffold improves ventricular contractility and function after implantation into infarcted heart. *Sci. Rep.* **8**, 1–13 (2018).

109. Jiang, L. *et al.* Preparation of an electrically conductive graphene oxide/chitosan scaffold for cardiac tissue engineering. *Appl. Biochem. Biotechnol.* **188**, 952–964 (2019).

110. Zhou, J. *et al.* Injectable OPF/graphene oxide hydrogels provide mechanical support and enhance cell electrical signaling after implantation into myocardial infarct. *Theranostics* **8**, 3317 (2018).

111. Choe, G. *et al.* Anti-oxidant activity reinforced reduced graphene oxide/alginate microgels: Mesenchymal stem cell encapsulation and regeneration of infarcted hearts. *Biomaterials* **225**, 119513 (2019).
112. Zhao, L. A novel graphene oxide polymer gel platform for cardiac tissue engineering application. *3 Biotech* **9**, 1–11 (2019).
113. Han, J. *et al.* Dual roles of graphene oxide to attenuate inflammation and elicit timely polarization of macrophage phenotypes for cardiac repair. *ACS Nano* **12**, 1959–1977 (2018).
114. Nazari, H., Heirani-Tabasi, A., Alavijeh, M. S., Jeshvaghani, Z. S., Esmaeili, E., Hosseinzadeh, S., ... & Soleimani, M. (2019). Nanofibrous composites reinforced by MoS_2 Nanosheets as a conductive scaffold for cardiac tissue engineering. *ChemistrySelect*, **4**(39), 11557–11563. https://doi.org/10.1002/slct.201901357
115. Ming, G., Kaidong, Z., Xiangyong, L. & Yuanzhe, J. Preparation of molybdenum disulfide nanoparticles and the cytoprotection on cardiac myocytes. *Chinese J. Appl. Chem.* **37**, 1010 (2020).
116. Shin, S. R. *et al.* Reduced graphene oxide-GelMA hybrid hydrogels as scaffolds for cardiac tissue engineering. *Small* **12**, 3677–3689 (2016).
117. Norahan, M. H., Amroon, M., Ghahremanzadeh, R., Mahmoodi, M. & Baheiraei, N. Electroactive graphene oxide-incorporated collagen assisting vascularization for cardiac tissue engineering. *J. Biomed. Mater. Res. A* **107**, 204–219 (2019).
118. Nazari, H. *et al.* Fabrication of graphene-silver/polyurethane nanofibrous scaffolds for cardiac tissue engineering. *Polym. Adv. Technol.* **30**, 2086–2099 (2019).
119. Nazari, H. *et al.* Incorporation of two-dimensional nanomaterials into silk fibroin nanofibers for cardiac tissue engineering. *Polym. Adv. Technol.* **31**, 248–259 (2020).
120. Singh, S. K. *et al.* Amine-modified graphene: Thrombo-protective safer alternative to graphene oxide for biomedical applications. *ACS Nano* **6**, 2731–2740 (2012).
121. Singh, S. K. *et al.* Thrombus inducing property of atomically thin graphene oxide sheets. *ACS Nano* **5**, 4987–4996 (2011).

Index

Taylor & Francis Group
an **informa** business

Taylor & Francis eBooks

www.taylorfrancis.com

A single destination for eBooks from Taylor & Francis
with increased functionality and an improved user
experience to meet the needs of our customers.

90,000+ eBooks of award-winning academic content in
Humanities, Social Science, Science, Technology, Engineering,
and Medical written by a global network of editors and authors.

TAYLOR & FRANCIS EBOOKS OFFERS:

A streamlined
experience for
our library
customers

A single point
of discovery
for all of our
eBook content

Improved
search and
discovery of
content at both
book and
chapter level

REQUEST A FREE TRIAL
support@taylorfrancis.com

Routledge
Taylor & Francis Group

CRC CRC Press
Taylor & Francis Group

For Product Safety Concerns and Information please contact our EU
representative GPSR@taylorandfrancis.com
Taylor & Francis Verlag GmbH, Kaufingerstraße 24, 80331 München, Germany

www.ingramcontent.com/pod-product-compliance
Lightning Source LLC
Chambersburg PA
CBHW060354220326
41598CB00023B/2918

9 7 8 1 0 3 2 1 6 2 8 8 1